World Soils Book Series

Series Editor

Prof. Alfred E. Hartemink
Department of Soil Science, FD Hole Soils Laboratory
University of Wisconsin–Madison
Madison
USA

For further volumes:
http://www.springer.com/series/8915

Aims and Scope

The *World Soils Book Series* brings together soil information and soil knowledge of a particular country in a concise and reader-friendly way. The books include sections on soil research history, geomorphology, major soil types, soil maps, soil properties, soil classification, soil fertility, land use and vegetation, soil management, and soils and humans.

International Union of Soil Sciences

Manuel Casanova • Osvaldo Salazar
Oscar Seguel • Walter Luzio

The Soils of Chile

Springer

Manuel Casanova
Osvaldo Salazar
Oscar Seguel
Walter Luzio
Department of Soil and Engineering
University of Chile
Santiago
Chile

ISSN 2211-1255 ISSN 2211-1263 (electronic)
ISBN 978-94-007-5948-0 ISBN 978-94-007-5949-7 (eBook)
DOI 10.1007/978-94-007-5949-7
Springer Dordrecht Heidelberg New York London

Library of Congress Control Number: 2012954627

Printed on acid-free paper

Springer is part of Springer Science+Business Media (www.springer.com)

Preface

Imagine the narrowest and longest country in the world, where high snow-covered mountains can be seen from the ocean and huge rivers sculpt the landscape, where frequent volcano eruptions cover the land with ash, where earthquakes shake the earth and tsunamis overwhelm the coastline, where enormous glaciers are retreating, and finally imagine a place where it never rains.... then you are seeing the majesty and magnificence of Chile.

It should be noted that within Chilean territory you can find almost all the soil types observed in the world, but unfortunately these represent a scarce and fragile natural heritage. Natural resources are one of more important economic assets in Chile, but to avoid over-exploitation of those considered nonrenewable, a transition toward sustainable development should be a priority.

The vision of local soil scientists about the problems that afflict Chilean soils has been extended to a broader concept than erosion, namely soil degradation. Such problems were unsuspected a few decades ago, but nowadays soils are studied in light of a wide range of complex and interconnected problems, which cast a long shadow over the future of fertile Chilean land and await the light of wisdom.

In response to increasing concerns about soil degradation and the sustainability of agricultural production potentials in almost all regions of Chile, many researchers and institutions have developed diverse and valuable initiatives. These efforts include resource inventories, the design and development of low-cost technological options, the development of ecologically sound cropping systems, and options designed to conserve and manage the agrobiodiversity and forest resources that exist in the country.

However, because the use and management of soils depends on many different actors, only limited progress is possible unless all are involved in planning and implementing programs to conserve this vital natural resource. In this regard, involvement takes on a very wide connotation, from having a deep knowledge of soil dynamics to planning management within an ethical context of this true *work of art by nature*.

Acknowledgments

A large number of people over a long period of time have greatly assisted in the development of soil science in Chile. A particular debt is owed to all those largely unknown stewards, the pioneer soil surveyors, who made it possible to understand the complex distribution of soils in Chile. We also thank all the soil scientists that have forged and are building daily soil knowledge in Chile, while apologizing to those who may have been inadvertently omitted in this book. Finally, the authors reserve special gratitude for the University of Chile, their current workplace and always revered alma mater, which allowed them to translate their passion for the Soils of Chile into this document.

Contents

1 General Chile Overview ... 1
1.1 Territory Formation: Geology and Geomorphology ... 1
1.2 Climate: From Desert to Glaciers ... 9
1.3 Vegetation ... 17
1.4 Land Use ... 20
References ... 23

2 Main Features of Chilean Soils ... 25
2.1 Soil Formation in Chile ... 25
2.2 Major Soil Zones ... 27
2.2.1 Soils of the Hyper-Arid to Semi-Arid Zone ... 27
2.2.2 Soils of the Mediterranean Zone ... 56
2.2.3 Soils of the Rainy and Patagonian Zone ... 71
2.2.4 Soils of the Insular (Easter–Juan Fernández) and Antarctic Zone ... 80
2.3 A Soil Map of Chile ... 93
References ... 93

3 Management of Soil Properties in Chile ... 99
3.1 Chemical Properties ... 99
3.1.1 Soil Reaction ... 99
3.1.2 Soil Salinity and Sodicity ... 102
3.1.3 Nutrient Availability ... 102
3.2 Physical Properties ... 106
3.2.1 Bulk Density ... 106
3.2.2 Particle Size Distribution and Water Retention ... 108
3.2.3 Structural Stability ... 109
3.2.4 Pore Functionality ... 111
3.3 Biological Properties ... 113
3.3.1 Soil Organic Carbon ... 113
3.3.2 Soil Biodiversity ... 115
References ... 116

4 Human-Induced Soil Degradation in Chile ... 121
4.1 Erosive Soil Degradation ... 121
4.1.1 Water Erosion ... 122
4.1.2 Wind Erosion ... 126
4.2 Non Erosive Soil Degradation (Physical, Chemical and Biological) ... 127
4.2.1 Soil Physical Degradation ... 127
4.2.2 Soil Chemical Degradation ... 136
4.2.3 Soil Biological Degradation ... 141

4.3 Desertification 144
4.3.1 Coquimbo Region and Patagonia, Two Emblematic Cases of Desertification in Extremes Zones of Chile 145
4.3.2 Easter Island, an Example of Collapse by Soil Degradation 149
4.4 Future of Soil Conservation in Chile 150
References 152

Appendix 159

Authors' Biographies 181

Index 183

General Chile Overview

1

Chile is a long (4,300 km) and narrow (180 km wide on average) country in the southwestern extreme of South America that presents varied and pristine landscapes truly unique in the world. Inhabited for around 10,000 year, its territory is bordered by Perú to the north through the Concordia line, Argentina and Bolivia to the east through the huge Andean altitude, the South Pole to the south and the Pacific Ocean along the western side.

Its continental length, between the northern and southern boundaries, is approximately 4,200 km. Including the Chilean Antarctic Territory, its longitude exceeds 8,000 km. A region of the Antarctic continent is also part of Chile, which forms a triangle ending in the South Pole. The continental and insular territory amounts to 756,915 km^2 and the Antarctic territory to 1,250,000 km^2 (Fig. 1.1).

Chilean territory is very asymmetrical in its length and width, 4,300 km and approximately 180 km on average, respectively. The maximum insular width is 468 km and is located at 52°S. The maximum continental width is found in Antofagasta (Region II), between the Mejillones and the Bolivian boundary (at 27°S; 380 km) and the minimum continental width at 31°37'S (90 km). As a nation, Chile became independent in 1818 and today is administratively divided into 15 Regions (Fig. 1.2), 50 Provinces and 341 Municipal Governments.

1.1 Territory Formation: Geology and Geomorphology

Western South America is one of the best known convergent margins on the Earth. The current cycle of ocean-continent convergence began in the Jurassic following the break-up of the Gondwana supercontinent and has been continuing ever since with varying degrees of obliquity. In the evolution of the Andean Orogen in Chile (c. 550 Ma geological history), it is possible to distinguish five separate main periods. The latest of these (Andean), occurring during late Early Jurassic to present, is characterised by continental break-up and represents the archetypal example of a subduction-related mountain belt.

Belts of active volcanoes, the most significant tectonic and geological events in the evolution of the Andes, have occurred since the Late Oligocene, after the break-up of the Farallon plate into the Cocos and Nazca plates at approximately 27 ± 2 Ma. This resulted in a change from oblique to more nearly orthogonal convergence between the Nazca and South American plates, as well as a greater than two-fold increase in convergence rates, which together produced a more than threefold increase in trench–normal convergence. This caused changes in subduction geometry which accelerated crustal shortening, thickening and uplift in the Northern Central Andes, but resulted initially in extension and crustal thinning in the Southern Central and Southern Andes. As a result of the increase in convergence rates, magmatic activity also increased along nearly all the Andean chain.

The Late Cenozoic tectonics of the coast of Northern Chile reflects processes related to the seismic coupling between the subducted Nazca Plate and the overriding South American Plate. Although these processes probably occur in all eroding convergent margins around the globe, only in Northern Chile is the record preserved due to the hyper-arid climate of the region (Allmendinger and González 2010).

The South American central volcanic zone (CVZ; 18–27°S) includes Chile and around 40 active volcanic centres (Fig. 1.3), as well as around 20 active minor centres and/or fields and at least 6 potentially active fields.

A zone where the passive Juan Fernández Ridge is subducting the continental margin is present between approx. 27 and 33°S, corresponding to a flat-slab subduction

M. Casanova et al., *The Soils of Chile*, World Soils Book Series, DOI: 10.1007/978-94-007-5949-7_1,

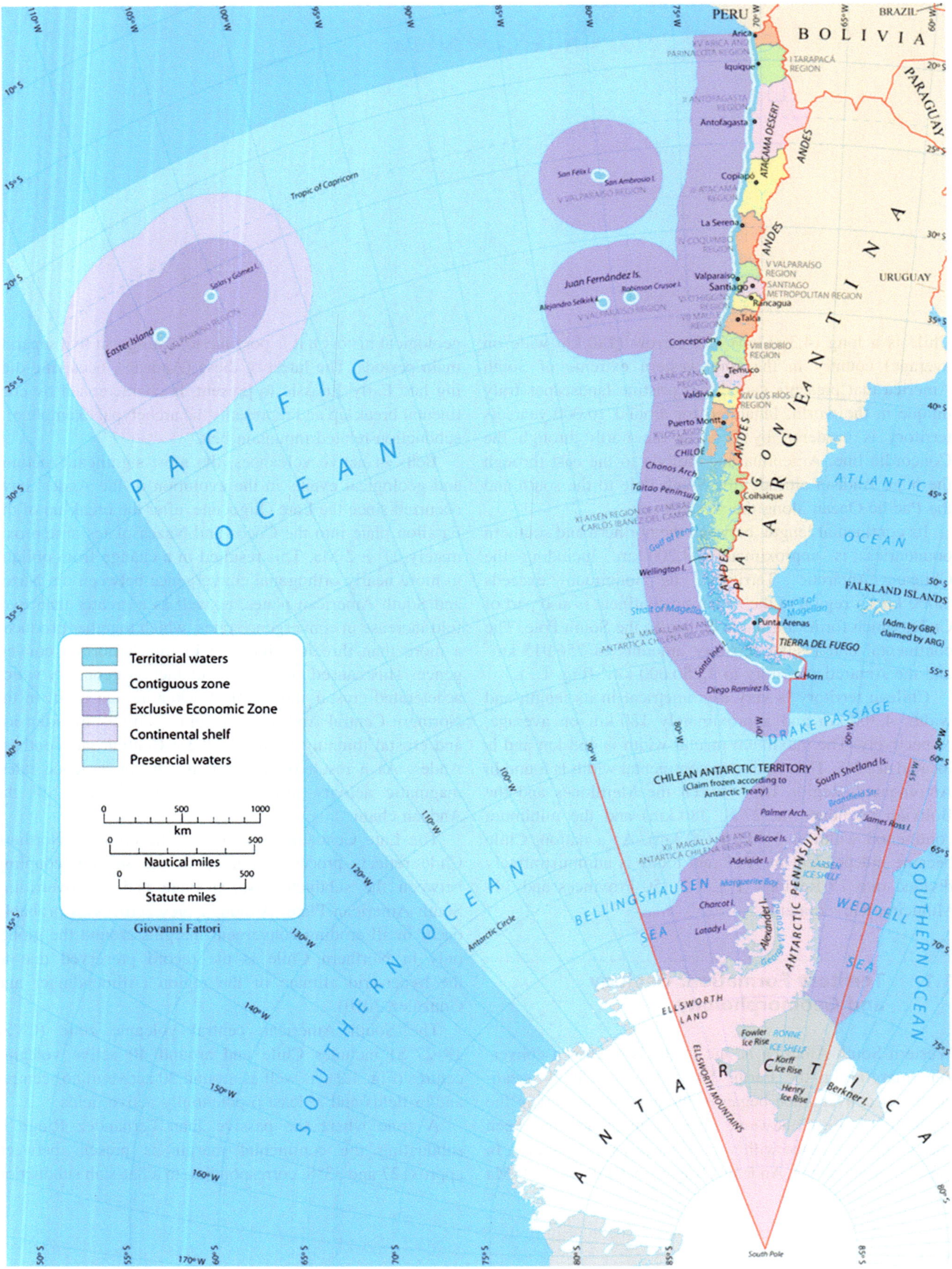

Fig. 1.1 Land and sea space of Chile (http://www.wikipedia.org, Accessed 30 November, 2012)

Fig. 1.2 Administrative division of Chile at present

Key	Region name	Capital
XV	Arica and Parinacota	Arica
I	Tarapacá	Iquique
II	Antofagasta	Antofagasta
III	Atacama	Copiapó
IV	Coquimbo	La Serena
V	Valparaíso	Valparaíso
VI	Libertador Bernardo O'Higgins	Rancagua
VII	Maule	Talca
VIII	Bío Bío	Concepción
IX	La Araucanía	Temuco
XIV	Los Ríos	Valdivia
X	Los Lagos	Puerto Montt
XI	Aysén del Gral. Carlos Ibañez del Campo	Coyhaique
XII	Magallanes and Antártica Chilena	Punta Arenas
MR	**Metropolitan**	**Santiago**

zone, whereas in the zones north and south of this aseismic segment the Wadati–Benioff zone is steeper.

The South American south volcanic zone (SVZ; 33–46°S) includes at least 60 historically and potentially active volcanic edifices in Chile and Argentina, as well as three giant silicic caldera systems and numerous minor eruptive centres. However, the continuity of the strike-parallel morphostructural units is interrupted in the regions where the Juan Fernández and the Chile ridges intersect the continental margin, causing segmentation of the orogen. The Chile Rise is an active spreading centre that marks the boundary between the Nazca Plate and the Antarctic Plate at the so-called Chile Triple Junction.

A gap in active volcanism occurs between 46 and 49°S to the south of the Chile Rise–Trench triple junction, where the south-east extension of the Chile Rise has been subducted during the last ~8 Ma without a Benioff zone of seismic activity below this volcanic gap.

Fig. 1.3 Volcanic zones in Chile. **a** Tacora volcano (Region XV), **b** Llaima volcano (Region IX) and **c** Hudson volcano (Region XI)

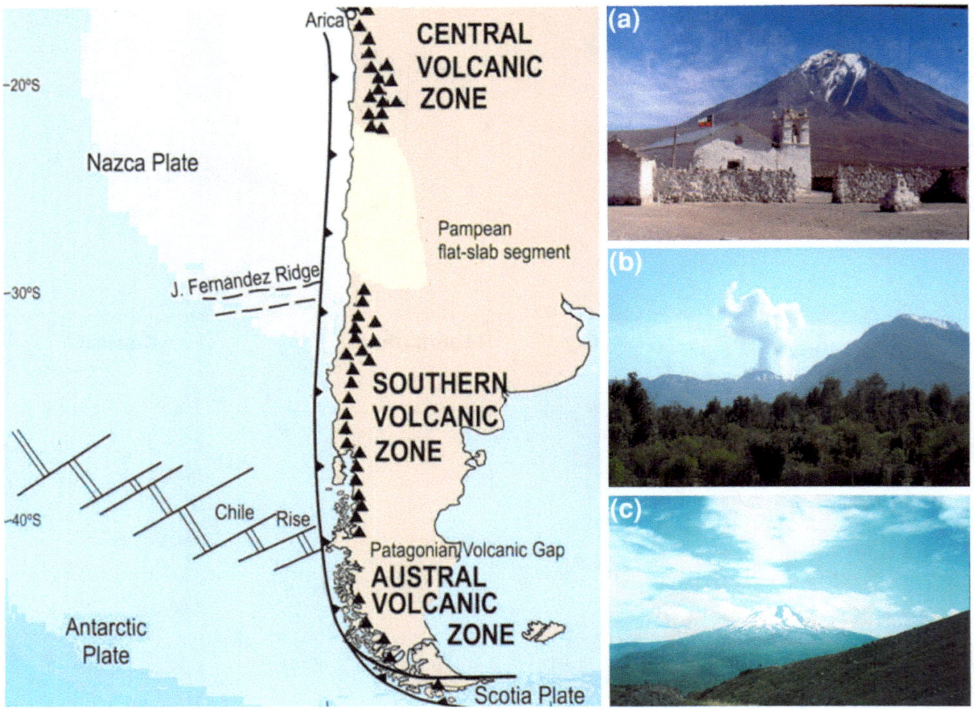

The Austral volcanic zone (AVZ; 49–55°S) consists of five stratovolcanoes and a small complex of Holocene domes and flows on Cook Island, the southernmost volcanic centre in the Andes south of the Magallanes Fault zone and therefore on the Scotia plate. Finally, the Antarctic Peninsula in the Chilean Antarctic Territory shares a lot of characteristics with the Andes and is sometimes considered to be an extension of those mountains.

The geological map of Chile shown in Fig. 1.4 (scale 1:2,500,000) gives a general view of the country. A digital geological map of Chile (scale 1:1,000,000) published by SERNAGEOMIN (2003) cannot be presented in this chapter because of the small, detailed units used. Cembrano and Lara (2009) include a map covering SVZ (Fig. 1.5), where regional-scale rock units are organised into several margin–parallel belts, ranging from Paleozoic plutonic and metamorphic rocks in the Coastal Range to Meso-cenozoic plutonic and volcano-sedimentary units in the Main Andes. The Central Valley, located in the middle, is characterised by Oligocene-Recent volcano-sedimentary rocks. Particular portions of Chilean territory have also been mapped on different scales by Muñoz et al. (2002) for the Atacama basin, Gioncada et al. (2010) for Easter Island, Calderón et al. (2007) for the Magallanes Fault zone, etc.

Considering different tectonic and morphostructural features of Chile, at least five macrozones along its territory are defined by Pankhurst and Hervé (2007):

(i) Coastal Range, a western coastal batholith (18 to ~42°S) of predominately Late Palaeozoic and Mesozoic igneous rocks, with paired belts of Palaeozoic metamorphic rocks cropping out south of ~34°S uplifted in an accretionary prism. It is the oldest and most western remnant of a magmatic arc formed at the birth of the modern Andes (195–130 Ma). With a moderate height, 1,000–2,000 m a.s.l., it disappears completely in Northern Chile near Arica.

(ii) Central Depression (125–90 Ma) or Longitudinal Central Valley, a tectonic downwarp with a Mesozoic to Quaternary sedimentary fill of volcanic, glacial, and fluvial origin, is only absent between 27 and 33°S (flat-slab subduction zone). However, according to Farías (2007), this zone is not a subsidence depression *sensu stricto* and valley denomination could be a more appropriate term due to a domain of erosive processes. According to Segerstrom (1964), it corresponds to an erosion surface, of Tertiary age, well denned in Central Chile by concordant summits, some of them are flat and extensive.

(iii) Main Andes, a chain of mountains that dates back to the Miocene, the emergence of which continues today. It is subdivided into three segments: Forearc Precordillera (78–37 Ma) and Western Cordillera (26 Ma to recent) from 18 to 27°S, High Andean Range at flat-slab subduction zone and Principal Cordillera (33°S to

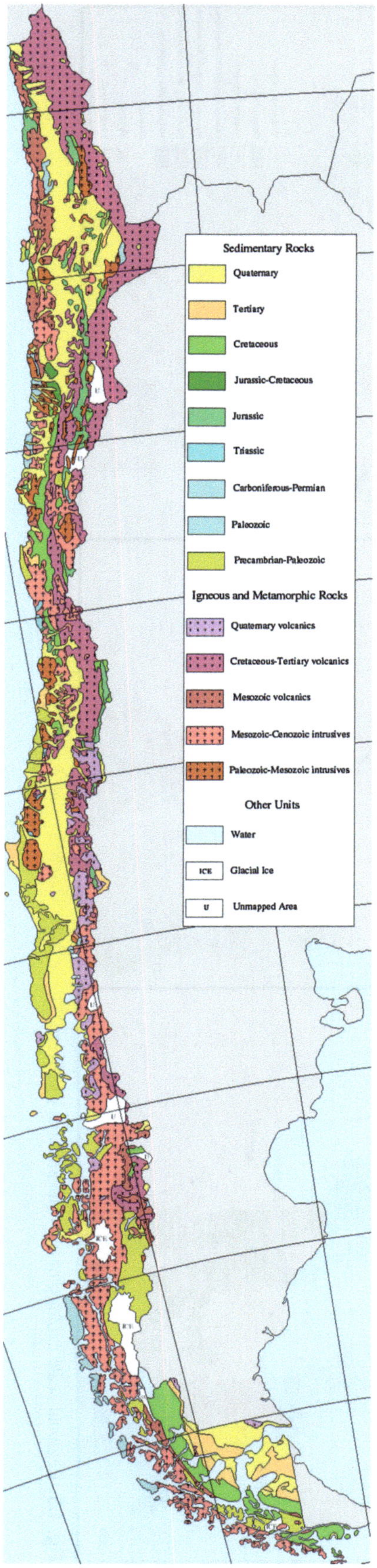

◂ **Fig. 1.4** General geologic map of Chile derived by Andrew Alden from US Geological Survey OFR 97-470D (http://geology.about.com/od/othernationgeomaps/ss/South-America-Geologic-Maps_4.htm. Accessed 8 April, 2010)

ca. 42°S). Petrologically distinct to the Coastal Range, the Andes are made up of Cenozoic age basaltic to andesitic volcanic rocks reaching to 1,500–4,000 m a.s.l. in Southern Chile and 4,000 m a.s.l. to nearly 7,000 m a.s.l. in Central and Northern Chile.

(iv) Patagonian Cordillera, the continuation of the Andes right down into *Tierra del Fuego* at the southern tip of Chile, with a continuous reduction in height. The origin of this low portion of the Andes has been related to an allochtonous Palaeozoic terrane. The southwestern margin of the land (42°S to the South) is formed by recent glaciations that carved the coastal areas into fjords and archipelagos comprising thousands of little islands and narrow channels. This estuarine system also resulted from the tectonic sinking of a longitudinal valley south of Puerto Montt (41°31′S) during the Quaternary.

(v) Andean foreland of the southern Patagonian Cordillera or Magallanes basin, consisting of Upper Jurassic to Early Cenozoic sedimentary deposits.

On the other hand, based on shortening rates and tectonic activity, the Central to Southern Andes of Chile is subdivided by Rehak et al. (2010) into only four tectonic provinces:

(1) The central part from 14 to 27°S comprises the Coastal Range, the Central Valley, the Precordillera and the Main Cordillera, with the internally drained intraorogenic plateaus of the Altiplano and Puna displaying mean elevations of 3,700 m. This sector exhibits pronounced crustal shortening in the Main Cordillera and the Subandean Ranges.

(2) A zone between 27 and 33°S with the above-mentioned related absence of late Neogene to Quaternary volcanism and a Central Valley. However, this zone comprises the broken foreland province of the *Sierras Pampeanas* that experiences active deformation and destructive earthquakes.

(3) South of 31°S, total shortening is significantly reduced and south of 37° shortening stopped at the end of the Miocene. South of 33°S the western onshore margin shows a pronounced morphotectonic segmentation integrating the forearc Coastal Cordillera, the Central Valley and the Main Cordillera.

(4) South of ~37°S the Main Cordillera is called Patagonian Cordillera. Here, deformation is partitioned into intra-arc strike-slip tectonics along a fault zone and thrust faulting.

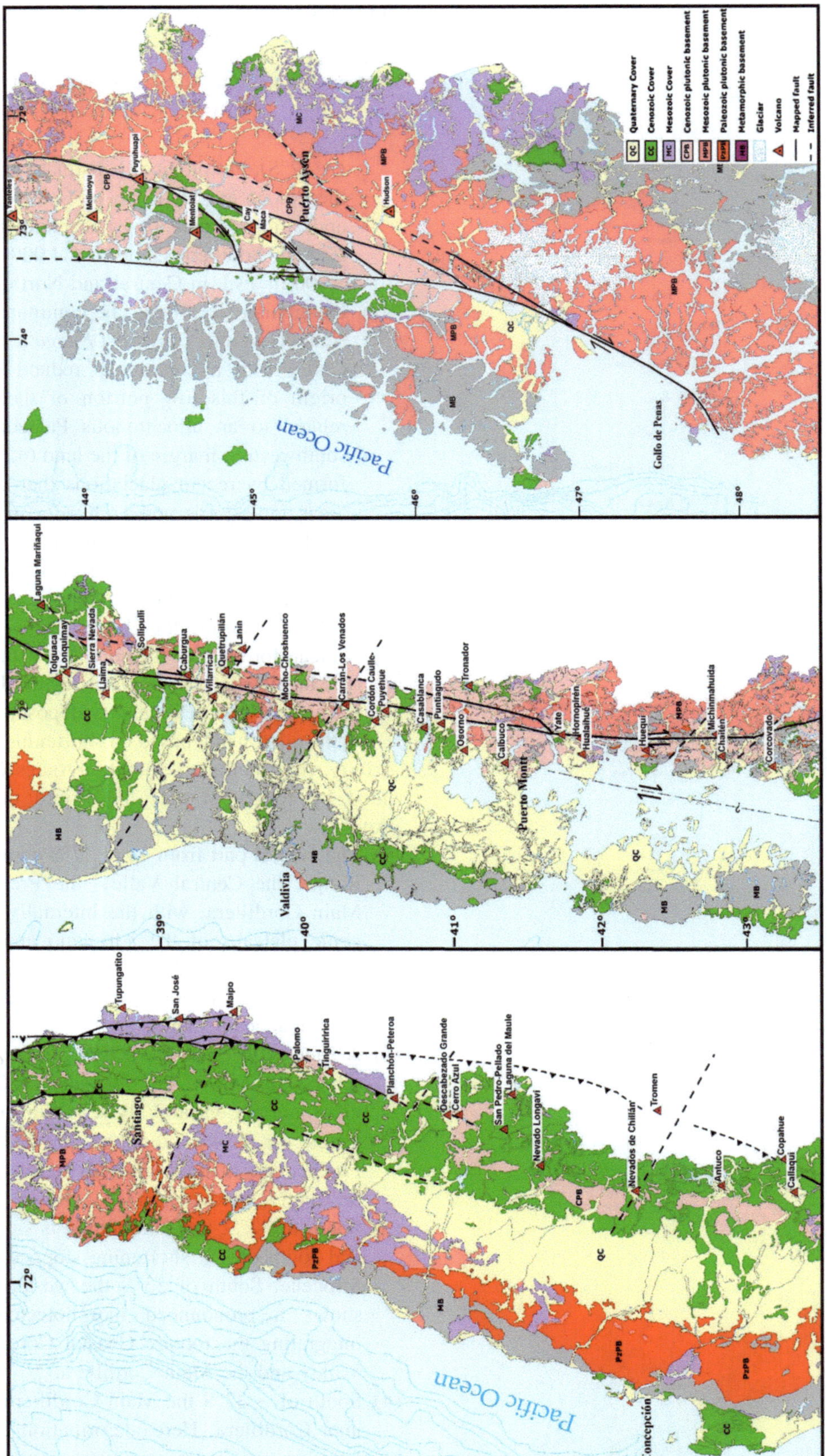

Fig. 1.5 Regional-scale geology of the Chilean Andes between 33 and 47°S (Cembrano and Lara 2009)

Emphasising the general theme of overall plate tectonic control of the geological evolution of the Chilean margin, there also appears to be a close relationship between subduction-related tectonic evolution, magmatic processes and climate along and across the Andean range. In addition, Rehak et al. (2010) suggest that the generation of local relief is a force balance between protracted fluvial erosion and glacial erosion acting as opposite agents; whereas glacial erosion appears to create local relief, persistent moderate rainfall above a critical threshold appears to smooth it. Figure 1.6 gives a general view of current geomorphological features of Chile, including principal transverse profiles east–west along continental territory.

In order to diagnose the imprint of different climate zones on the relief of mountain ranges, Rehak et al. (2010) correlated climate with relief parameters and investigated the impact of different geomorphological regimes on relief evolution. They concluded that the catchment-scale relief of the western flank of the Andean mountain chain distinctly reflects the dominant geomorphological process, which is determined by, and therefore represents, past and present regional climate regimes. However, they do not challenge the claim that the large-scale architecture of the Andes is maintained by tectonic processes and the reactivation of inherited basement structures. The geomorphological signature of the Western Andes between 28 and 35°S, however, expresses significant transient components, which may reflect erosion processes under different climate conditions during the geological past. Figure 1.7 describes basins along most Chilean territory, which according to their location can be grouped into: (a) Forearc, catchments drain only the Coastal Range and have not been affected by Quaternary glaciations; (b) Andean mountain front catchments also including the Central Valley and the Andean foothills (do not extend into the high Andes and have mostly not been glaciated during the Quaternary); (c) arc catchments extending from the Pacific to the principal Andean watershed and d) subcatchments that are parts of the arc catchments and constitute the uppermost headwater basins along the Andean main drainage divide.

Recent works published by Paskoff and Araya-Vergara (2010) and Araya-Vergara (2010) are included here to describe the western margin of Chilean territory (Fig. 1.8).

South of the Peruvian border along to Arica, the coast of the Atacama Desert is mainly cliffed. Lluta and Azapa are two alluvium valleys that are very important for agriculture in Region XV. The Lluta Valley is located only 10 km from the Peruvian border. This valley is formed by the 150 km long Lluta River that extends from the Tacora volcano to the sea. The 4,500 ha Camarones Valley is located 100 km south of Arica city and is crossed by the Camarones River, historically the most important water source in the region.

From Iquique, massive transgressive dunes have been deposited between the present beach and the local megacliff. Due to their position with respect to present sea level and the degree of weathering of their sands, they are thought to have formed during the Pleistocene, when sea level was lower. They are a striking feature, because dunes are scarce on the cliffy coast of the desert. North of Antofagasta (23°S), cliffs are cut into horizontally bedded Tertiary sandstones. In some parts, the coastline has embayments cut in sandstone between protruding headlands where harder rocks outcrop near mean sea level.

At 27°S, the Copiapó River (Region III), the first outlet, appears with permanent stream discharge to the Pacific Ocean, after more than 1,000 km of desert coast. There are Pleistocene marine terraces with deflated beaches on their surfaces, and the ergs of the Atacama Desert. A large part of the sand of the ergs has been supplied directly from the Copiapó River, but dune sands of the ergs located in basins inland have a marine origin. South of the Copiapó River, all the principal rivers (Elqui, Limarí and Choapa, for example, in Region IV) flow permanently to the sea through microtidal estuaries. Over more than 700 km of coastline, increasing supplies of fluvial sediment to the shore form beaches in zeta-form embayments, as at Tongoy (30°S) and Los Vilos (32°S).

Between the estuaries of the Petorca-Ligua and Maipo rivers (33°S), there is a southward transition from rias to estuarine deltas, as in the Aconcagua estuary (Region V). Between the Maipo river outlet and San Antonio harbour, the coastline has been straightened as the result of accretion along a breakwater.

Between Rapel (34°S) and Bío-Bío (37°S), river ebb and flow deltas are typical. The deltaic estuarine banks become more stable south to the Maule estuary (35°S), then decrease and become ephemeral down to the Bío-Bío estuary.

Zeta-form bays are common in this sector, and show increasing volumes of sand downdrift to the north, where there are three generations of dunes: early Holocene, middle Holocene and presently active. The Gulf of Arauco (37°S) is a zeta-form embayment backed by a cliff cut in soft sediment. The width of the beach and surf zones increases down drift (northward), indicating sand drift towards the Bío-Bío estuary. Similar features are seen in the embayment

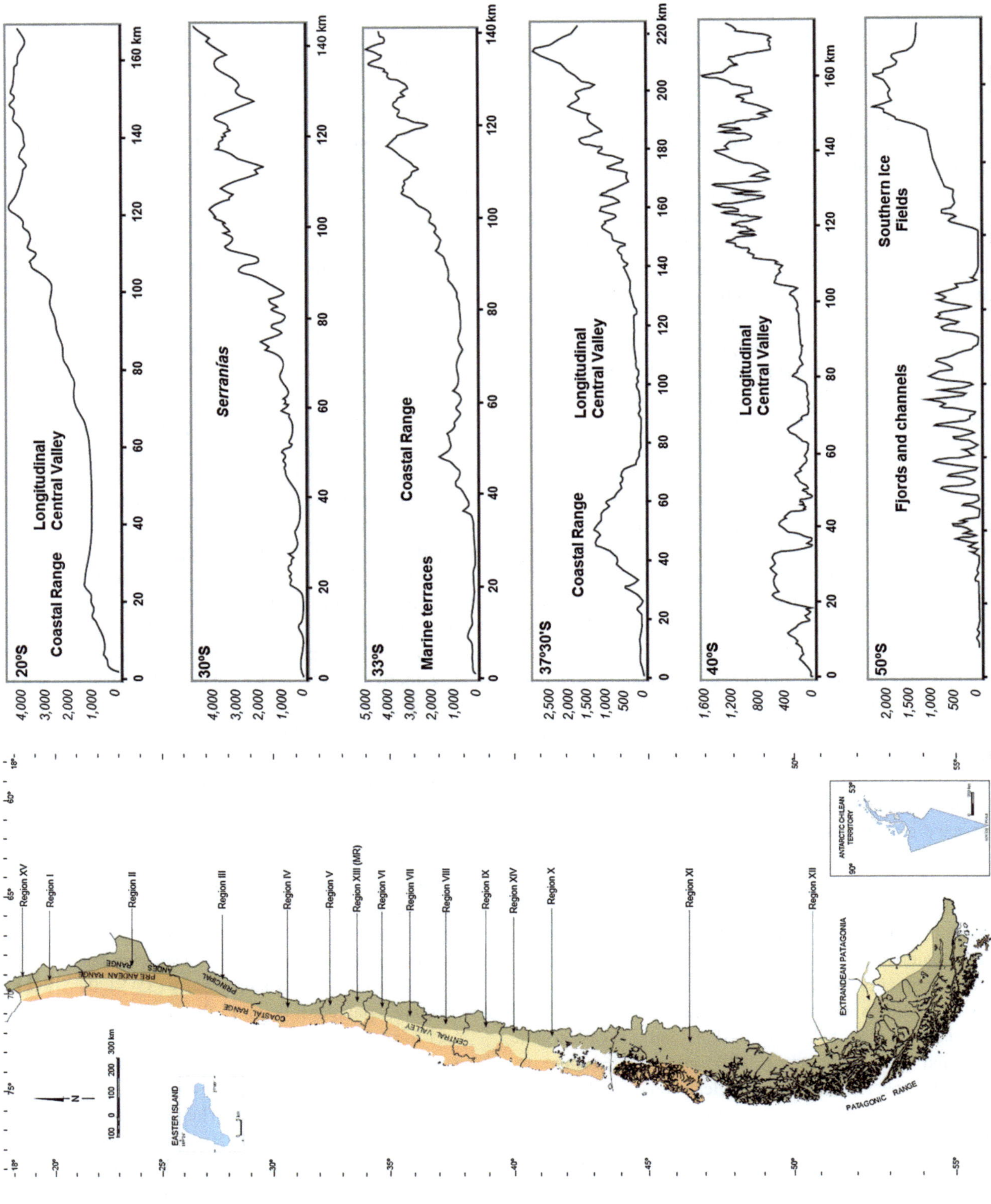

Fig. 1.6 Geomorphology of continental Chile and principal East–West topographical profiles

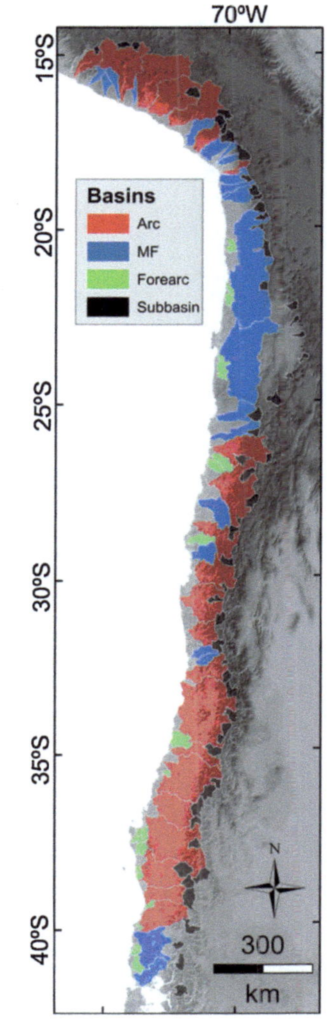

Fig. 1.7 Basins along the western Andean margin (Rehak et al. 2010)

between Lebu and Quidico. Associated with these are the Arauco dunes, the largest coastal erg in Chile, with small impersistent compound barchans because the climate is humid, and there are annual floods. The Imperial River (37°40′ to 38°50′S) meanders across a terrace which is inset in an ancient ria. To the south is a ria coast, formed by partial submergence of river valleys incised into the schistose coast ranges. The intervening sectors are cliffy.

Latitude 40°S is the northern limit of direct influence of the Last Glaciation, giving way south of the present outlet glacier to fjords, marked by a transition from rias to fjords, förde and submarine glacial piedmonts. The floors of the fjords have submerged terraces formed by deposition of fine sediment during Late Glacial and Holocene times, and there are steep bordering glaciated slopes. Deltas at the heads of fjords and mouths of side valleys have streams which discharge clouds of fine sediment that settle on the fjord floor.

To the south, periglaciated coastal slopes descend to the sea, undercut by marine cliffs on exposed sectors. Active periglaciation results in the formation of screes on steep slopes. In the inner coastline of Chiloé Island, the key landforms are förde, formed by marine submergence of former subglacial channels. There are many islands of soft glacial drift with coastlines smoothed by erosion, longshore drift and deposition. Tidal currents are strong because of the large tide range.

To the south, the coastline in front of the Patagonian ice fields (~48 to 51°S) is trenched by deep fjords floored by morainic banks, which were formed during the Last Glaciation and Late Pleistocene to Holocene terraces. Above the fjords are the Patagonian ice fields, with calving glaciers at fjord heads. Glacifluvial outwash produces turbidity plumes and deposits fine to coarse sediment in the fjords. The glaciers have been receding.

The Magellan region was shaped during the Last Glaciation, producing fjord and piedmont coastlines. The Strait of Magellan originated as a fjord in the western part, which is similar to the fjords in front to the Patagonian ice field, and a set of piedmont lobes in the eastern part with many cliffs cut in soft glacial, but the fjords are smaller. In the eastern part of the Strait of Magellan the tide range is large and intertidal zones, including well-developed shore platforms, are wide. Tidal currents are very strong in narrow straits, where they contribute to erosion of the coastline. Some valley mouths have small deltas formed in the Holocene.

Finally, Chile's oceanic islands are a group of offshore island territories located west of the Peru–Chile trench and the Nazca plate. All are of volcanic origin and represent the fraction under subaerial volcanic edifices that extend under sea level. Easter Island, Salas y Gómez, San Félix and San Ambrosio along the Juan Fernández archipelago are typical examples of tholeiitic intraplate alkaline volcanism associated with hot spots or hot lines. These islands or groups of islands are part of large chains of seamounts that in any event are independent volcanic complexes, with evolving and compositional differences. Most experts agree on the presence of an initial construction phase with evidence of underwater magma–water interaction and of global variations in sea level (Lara 2010). Figure 1.9 illustrate the simplified geology of this insular Chilean territory.

1.2 Climate: From Desert to Glaciers

In a broad perspective, a pronounced N–S and E–W asymmetry in the distribution of precipitation on both flanks of the Andes was observed by Strecker et al. (2007) as shown in Fig. 1.10.

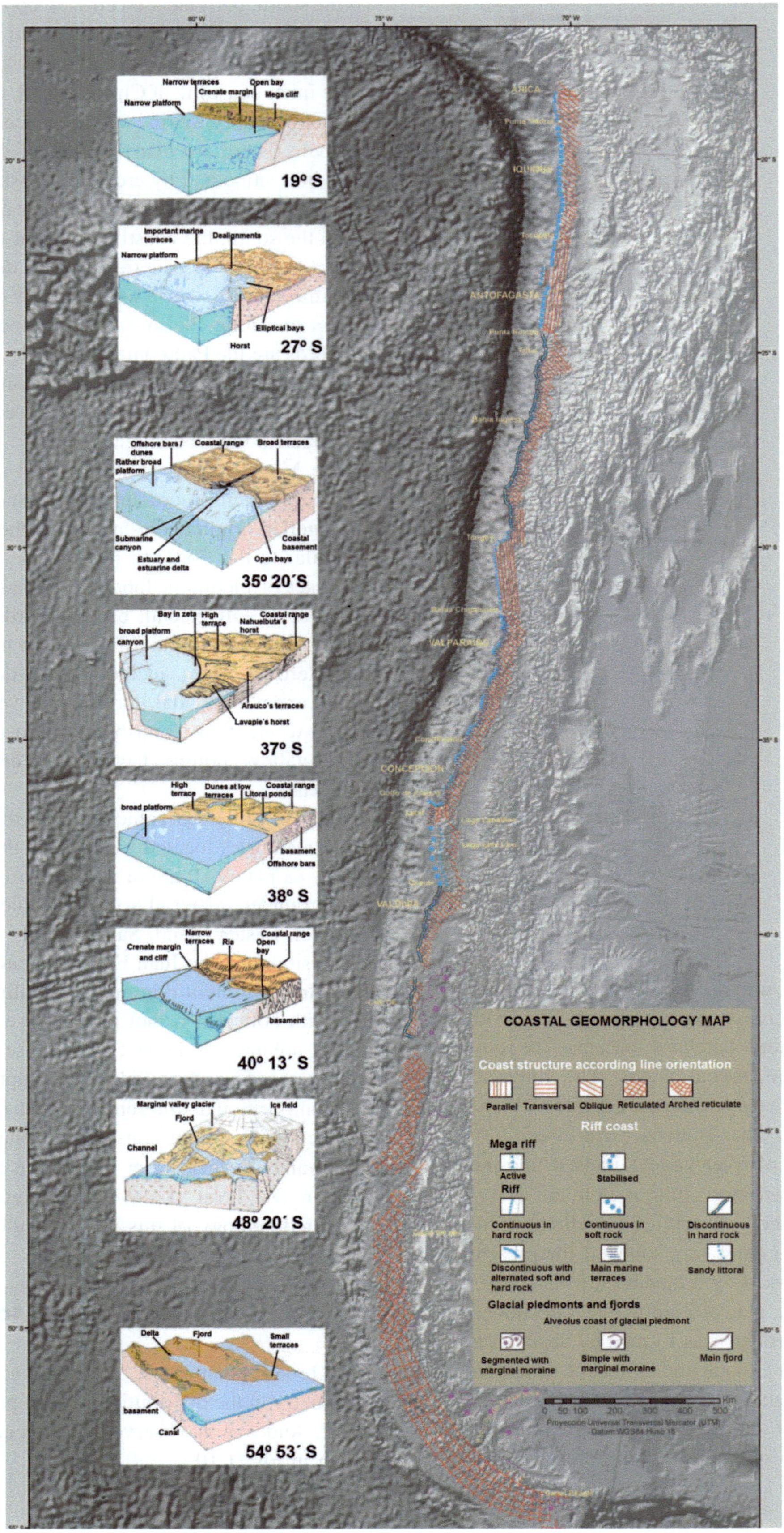

Fig. 1.8 Coastal geomorphology of Chile (Paskoff and Araya-Vergara 2010)

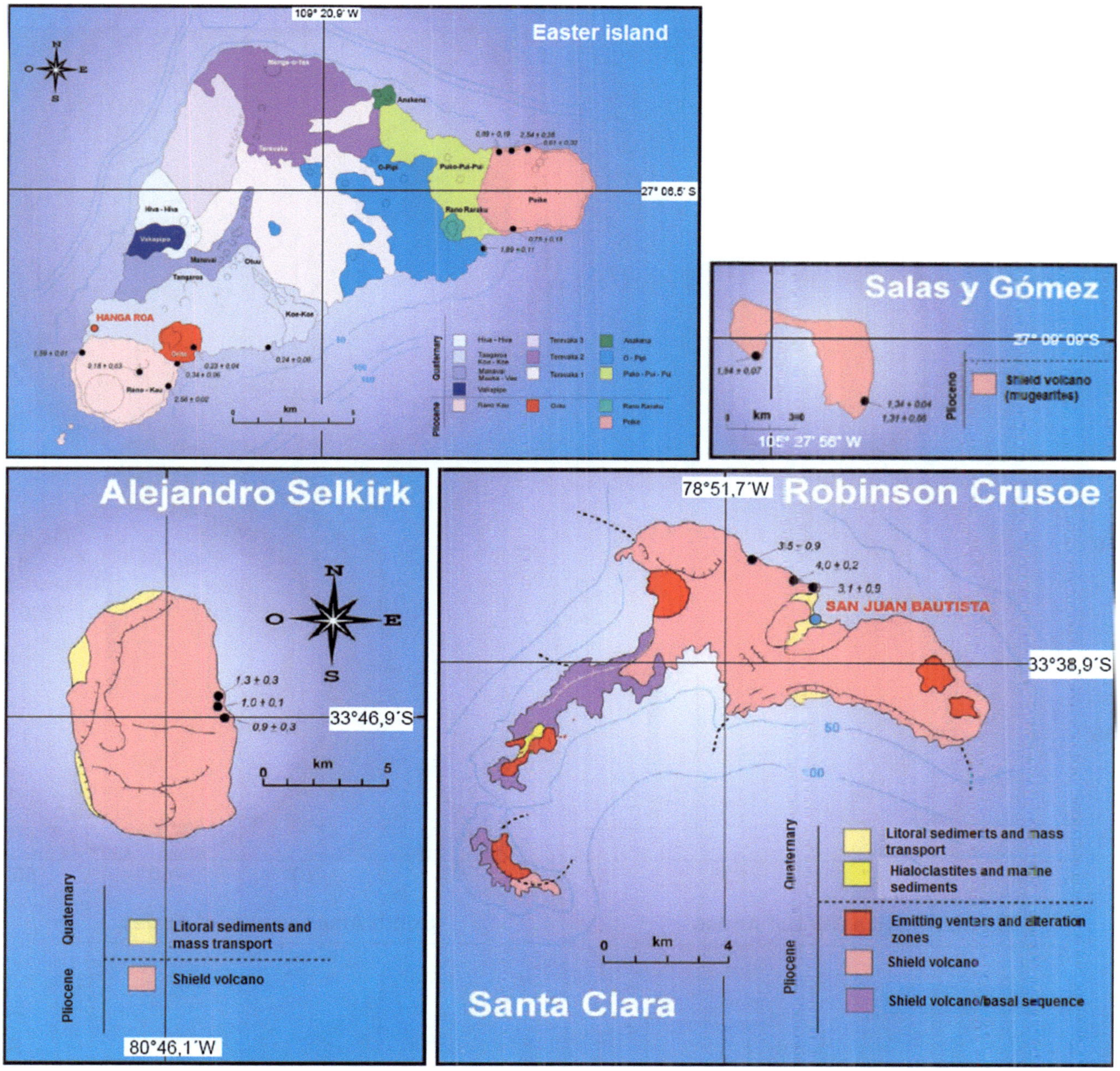

Fig. 1.9 Chilean oceanic islands geology (Lara 2010)

The main features of the weather, climate and climate variability over the Andes Range in South America have been described (Garreaud 2009, Garreaud et al. 2009), providing a basic physical understanding of the mean annual cycle of atmospheric circulation and precipitation over the mountains and adjacent lowlands (Fig. 1.11). In particular, the diversity of precipitation, temperature and wind patterns is interpreted in terms of the long meridianal extent of the Andes and the disruption of the large-scale circulation by this formidable topographical barrier. Among the many factors that determine the interannual climate variability in South America, ENSO (*El Niño* South Oscillation) plays a major role in many regions. ENSO is typically associated with anomalously wet conditions in the southeastern portion of the continent and Central Chile; opposite rainfall anomalies are typically observed during *La Niña* events

In this context, the Chilean climate in particular is extremely varied due to the abrupt relief, geographical position and elongated shape of the country. The geographical position determines atmospheric circulation,

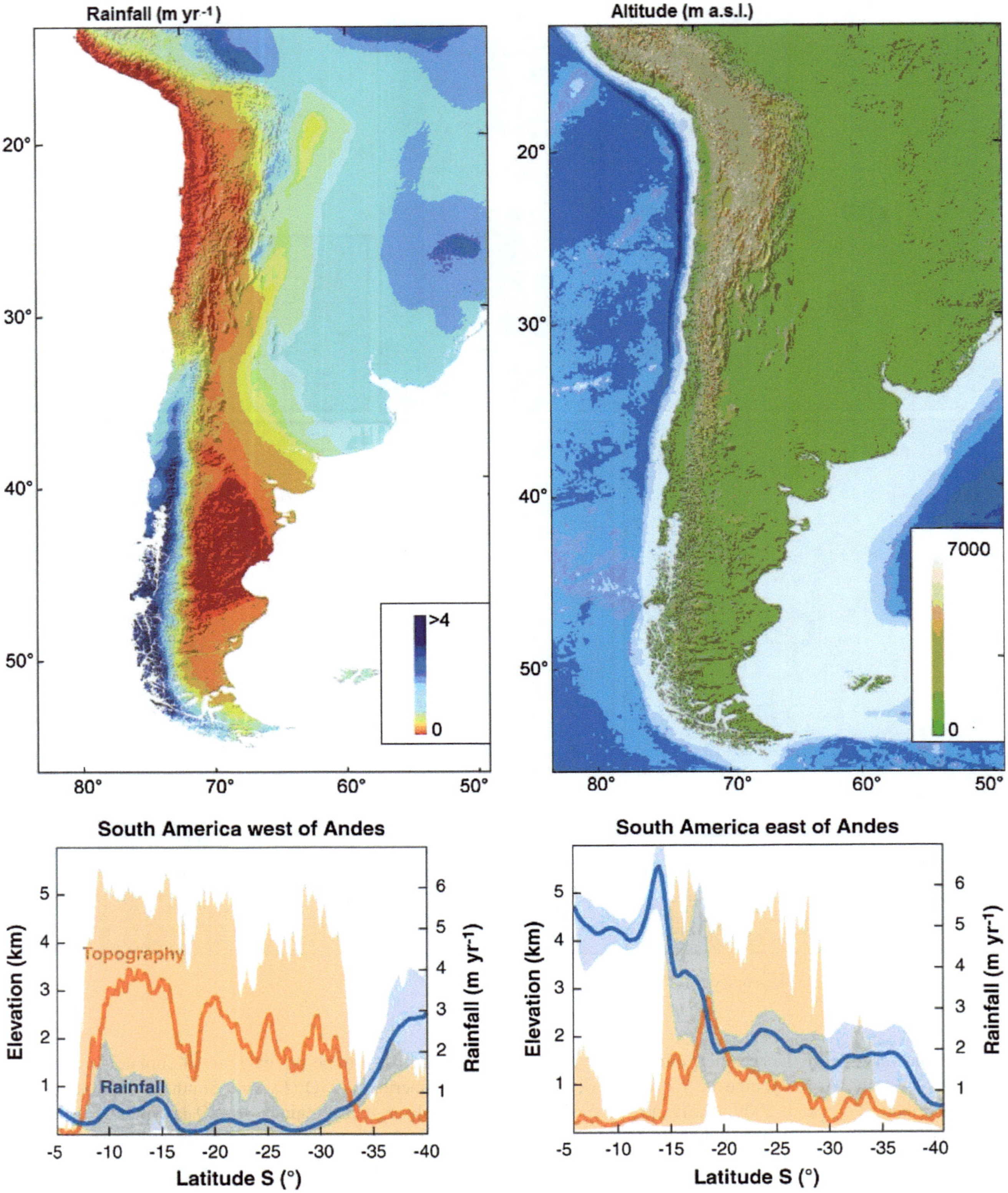

Fig. 1.10 *Above* rainfall and relief at central South America. *Below* rainfall and topographical profiles (250 km wide) on the west and the east side of the Andes (partly from Strecker et al. 2007)

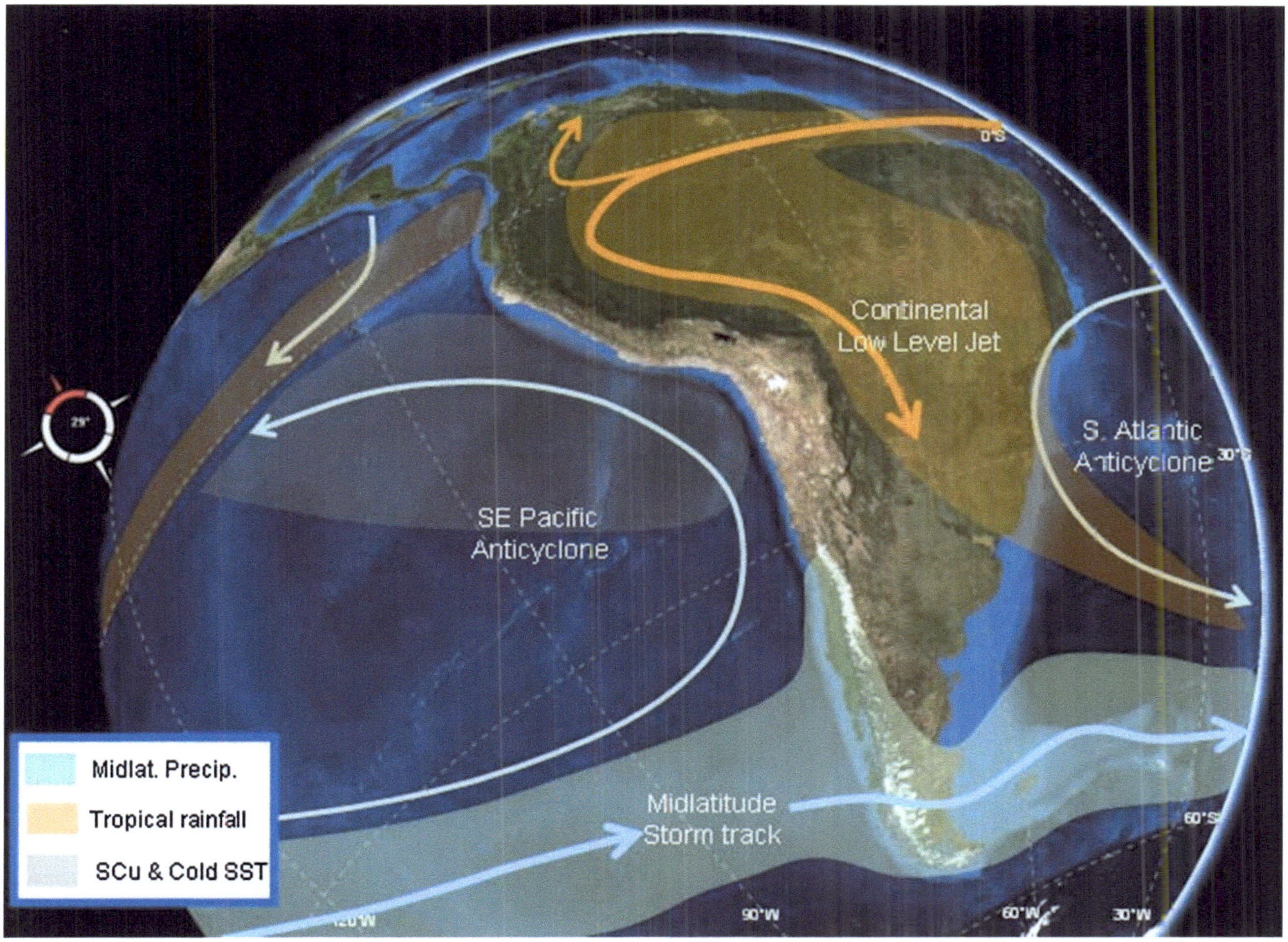

Fig. 1.11 Schematics of the low-level atmospheric flow (roughly from surface to about 1.5 km a.s.l.) around the Andes Range. Major climate features of South America are also depicted (Garreaud 2009)

controlled by the influence of the Pacific Ocean and Antarctic territory. Movement of Antarctic and sub-Antarctic water currents (the Perú–Chile Current) and masses of polar air influence the whole country. The topography of the country strongly controls temperature and rainfall patterns. The Coastal Range and the Andes constitute geographical barriers that block the maritime influence of the Pacific Ocean on their eastern slopes and in the Central Valley. Atmospheric circulation mainly involves the impact of the South Pacific anticyclone, located in front of the Chilean coast, normally between 20 and 40°S. The occurrence of long droughts or large floods is strongly controlled by the location and persistence of that anticyclone and also by *El Niño* currents.

The northern part of Chile (17–27°S), mostly a desert, shows extremely low rainfall (<50 mm yr^{-1}). A recent document (UNESCO, 2010) showing an aridity map for Chile (Fig. 1.12) concluded that 52 % of its territory (northern) is characterised by xeric, hyperarid, arid and semiarid regimes. The Atacama Desert, for instance, has a variation in rainfall from 0 mm yr^{-1} at approx. 2,400 m a.s.l. to 200 mm yr^{-1} at 4,000 m a.s.l. Over the Altiplano Plateau there is some rainfall (200–300 mm yr^{-1}) from December to March, called the *Bolivian Winter*.

According to Houston and Hartley (2003) and Houston (2006), precipitation between 18 and 27°S is dominated by summer convective activity from Amazonia, and data analysis shows that the increase in precipitation with elevation due to the rainshadow effect best fits an exponential correlation. Coupling with limited data from high elevations suggests that the correlation is accurate to 4,500 m a.s.l. and perhaps to 5,500 m a.s.l., suggesting that increased precipitation goes unrecorded over the peaks of the Western Cordillera. Figure 1.13 shows a general climate map with ombrothermic diagrams for different locations in Chile. Insets show mean standardised monthly precipitation for stations dominated by summer and winter rainfall, taken from Houston and Hartley (2003).

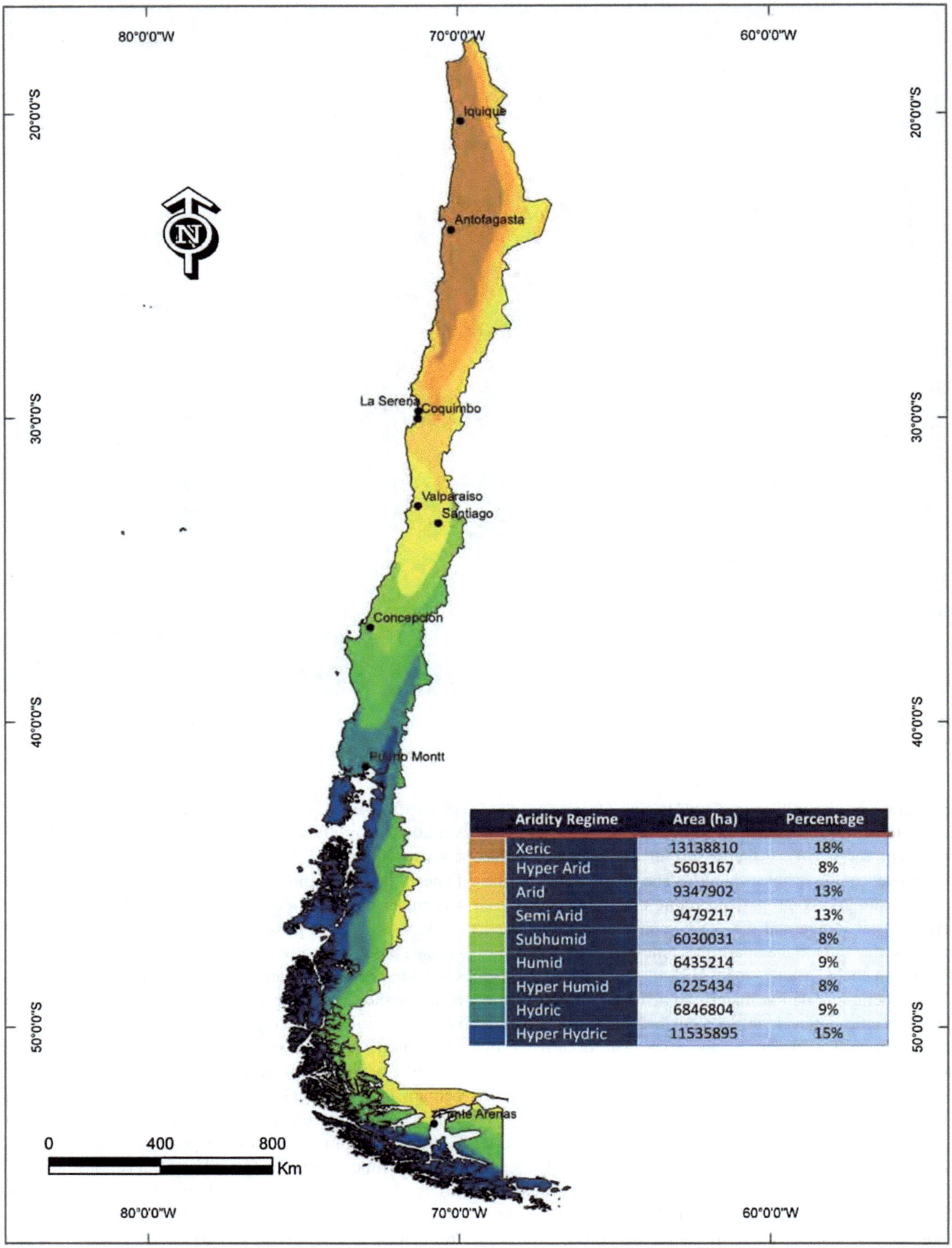

Fig. 1.12 Aridity map of Chile (UNESCO 2010)

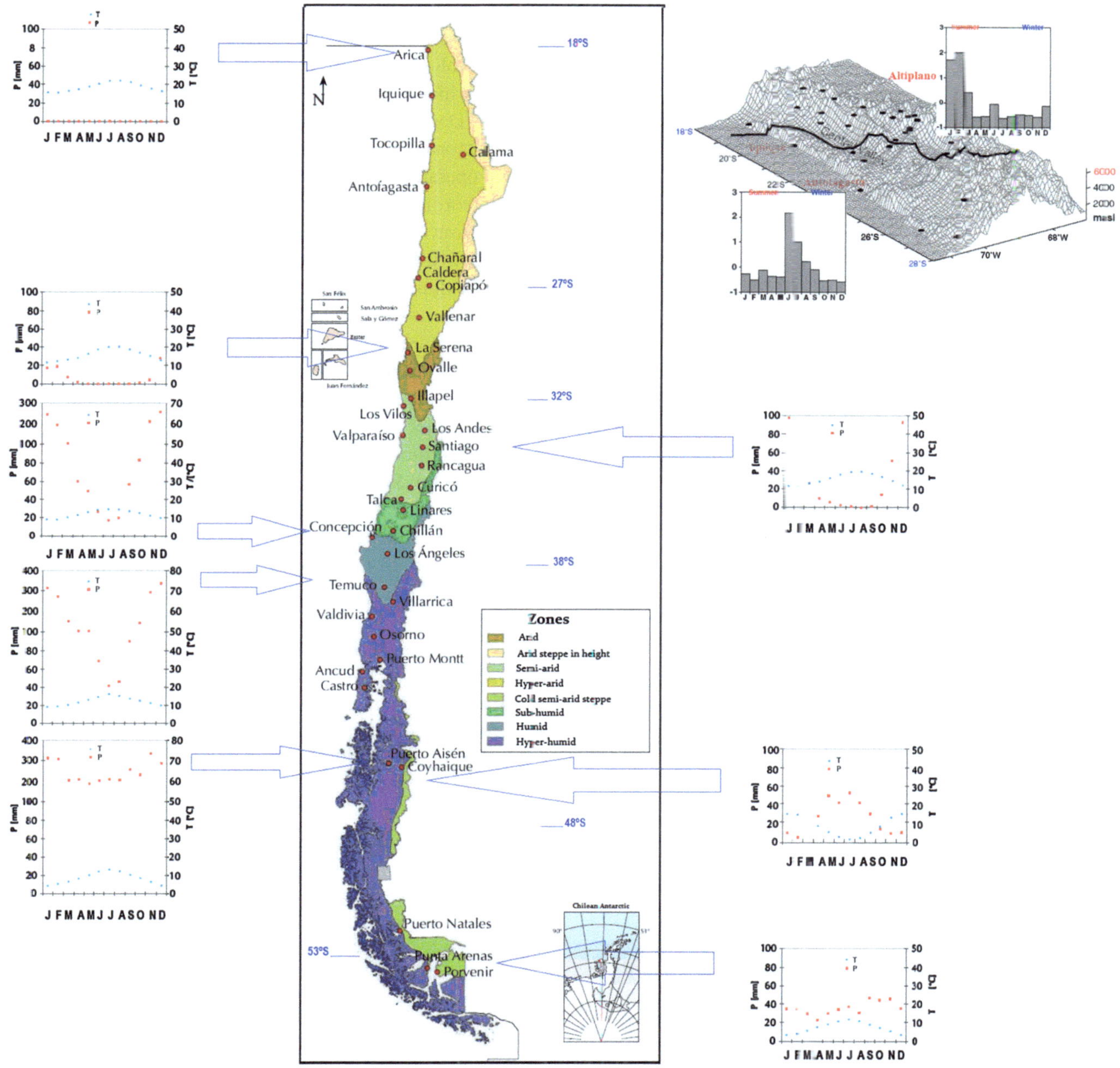

Fig. 1.13 Climate map of Chile (P in *red dots*: rainfall; T in *blue dots*: temperature)

Further south (27–32°S) the climate is semiarid, with scant winter rainfall dominated by winter frontal sources and showing no well-defined relationship with elevation. Between 32 and 38°S, the Mediterranean climate is characterised by rainfall during the winter season (50–1,000 mm yr^{-1}) and a dry summer season.

Even further south (38–42°S) the climate becomes temperate and with increasing rainfall. Between 42 and 46°S, it is very cold and humid, with snow and rainfall of over 3,000 mm yr^{-1}. Due to the predominant humid western winds Western Patagonia as a whole experiences high precipitation. At latitude 40°S, the western slope of the Coastal Range may receive up to 4,000 mm annually. In addition, it is particularly prone to becoming immersed in waves of marine fog. Towards the east, precipitation drops in the Central Valley to 1,000–2,000 mm yr^{-1}, and it increases again on the west-facing slopes of the Andes. There is a narrow zone of transition between 39°S and approximately latitude 47°S, characteristic of most of Eastern Patagonia, that receives about 400 mm yr^{-1} rainfall, beyond which it decreases to 200 mm yr^{-1} or less.

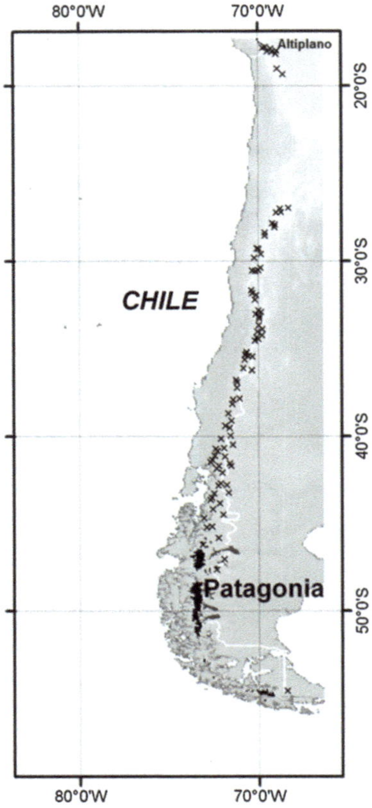

Fig. 1.14 Glaciers along Chile. Single glaciers are shown as crosses, while glacier bodies (ice fields) are shown as black areas (Casassa et al. 2007)

In terms of agrometeorological information, Brunini et al. (2010) report that Chile is known for its use and forecasting of meteorological information for decision support in agriculture, which is either centralised in the Chilean Meteorological Department or in the Regional Centres of Agrometeorological Information (CRIA), where farmers can access operational information, agricultural forecasts and agrometeorological updates in real time. This information ranges from weekly analyses of agrometeorological perspectives to trends and bulletins and effects on different crops and agricultural practices. Issues regarding agrometeorological forecasts for different regions and weekly updates of cold periods are also covered. Agroclimatic risk assessment is made according to the probability of the occurrence of adverse events. Chile is also an active participant in the CYFEN project (Climatic Information Applied to Coping with Agricultural Risk in Andean Countries).

Glaciers in South America have experienced a strong generalised retreat and thinning especially in recent years (Bown et al. 2008) in parallel with the regional and global warming trend. In the Central Chilean Andes, nearly 1,600 glaciers with a total ice area of around 1,300 km^2 have experienced a total volume loss, due to thinning and retreat, of 46 ± 17 km^3 of water equivalent between 1945 and 1996 (Rivera et al. 2002, Casassa et al. 2006). Between 1955 and 2007, a mean frontal rate of -22 m yr^{-1} on the Chilean side of the Andes was estimated by Le Quesne et al. (2009).

Fig. 1.15 Trees of *Prosopis tamarugo* at Zapiga saltlake (Luzio et al. 2010)

Fig. 1.16 Blooming desert (*left*) in Region III and relict southern forest (*right*) in Region IV of Chile

Fig. 1.17 High altitude wetland at Chilean Altiplano (Caquena, Region XV)

Chile has only a few glaciers in its northernmost corner along the border with Bolivia in the Andes that can be considered tropical in the broadest sense, but from 30°S there is an ice-covered area of approximately 27,500 km^{-2}, which contains around 90 % of all glaciers located in the Andes (Fig. 1.14). The Patagonian ice fields are the largest temperate ice masses on the Earth and their outlet glaciers are also some of the most dynamic. Maintained by the southern westerlies, the ice fields have undergone considerable fluctuations throughout the Quaternary and over the past few decades.

1.3 Vegetation

The principal vegetation formations and their main characteristics are briefly described here. This description is based on Moreira-Muñoz (2011), who provided a complete and comprehensive review of Chilean vegetation, including the latitudinal and altitudinal distribution of vegetation formations.

(a) The hyperarid desert formation or desert core, which extends between 18 and 24°S along the coast and interior zones, but vegetation is restricted to the deep agricultural valleys (Lluta, Azapa and Camarones). At the heart of the Atacama desert, vegetation is almost completely lacking, but there are stands of natural and planted forests of *Prosopis tamarugo*, including on five salt lakes (Zapiga, Huara-Pozo Almonte, Pintados, Bellavista and Llamará) with subterranean water dams (Fig. 1.15). Towards the Andes, the vegetation consists mainly of low scrub and in a very thin belt between 2,000 and 2,800 m a.s.l., large cacti of *Browningia candelaris* mark the landscape, responding to constraints described recently by Guerrero et al. (2011).

(b) Sparse coastal shrub vegetation, highly dependent on fog and humidity. The desert coastal scrub extends from 24 to 32°S, generating a transition zone from the desert towards the Mediterranean Central Chile. It encompasses open low scrub that at 30°S gradually changes to xerophytic scrub. This zone harbours two nature phenomena, the blooming desert with only a few mm of rainfall and the isolated forest of Fray Jorge with southern floristic elements on the upper coastal slopes, under the direct influence of maritime fogs (Fig. 1.16).

(c) Andean vegetation, the formation occupying extreme, high environments, ranging from 17°30′S to approx.

Fig. 1.18 Typical high altitude vegetation at Chilean Altiplano (XV Region)

Fig. 1.19 Typical Mediterranean vegetation (*Acacia caven*) at Santiago in Central Chile

Fig. 1.20 Araucaria trees (*Araucaria araucana*) in the Andes (Conguillío Park, Region IX)

40°S along the western Andean slope. This wide latitudinal extension encompasses a very different composition along the North–South profile and over altitude. The intermediate altitudinal belts show the structurally most developed vegetation, as the lowest belts are affected by aridity and the highest by low temperature. These intermediate belts are composed of cushion grasses. Above 4,700 m a.s.l., diverse high Andean wetlands or peatlands (Fig. 1.17) have been characterised in detail (Squeo et al. 2006; Ahumada and Faúndez 2009). Approaching the most arid part of the Atacama Desert there are shrubs in the lower belts. Towards the south, sparse vegetation is composed of cushion grasses.

The treeline changes constantly along the latitude gradient: in the north it is composed of *Polylepis tarapacana* and *P. rugulosa* (Fig. 1.18), the tree that lives at the highest altitudes in the world, while in the central-north it is replaced by shrubs and between 31 and 34°S the treeline reappears. At this latitudinal range, there are Andean scrub and cushion grass vegetation. From 32°60 ′S to the South, at the lower limit of the Andean formation, conifers appear but disappear at around 37°S, to be replaced by deciduous forests.

(d) Entering into the Mediterranean zone, the vegetation changes to a sclerophyllous high scrub (Fig. 1.19). On favourable south-facing slopes, this scrub shows characteristics of woodland, with trees reaching 20–25 m tall. In the more humid stands, there appears a more hygrophilous forest. In contrast, the most exposed and plain areas contain a woody savannah mainly composed of *Acacia caven* and *Prosopis chilensis*. North-facing slopes show a rich array of annual species and characteristic bromeliads together with the cactus *Trichocereus chiloensis*.

(e) Around 33°S the Coastal Range reaches far inside the continent, and above 1,200 m a.s.l., the sclerophyllous woodland makes way for a deciduous forest composed

Fig. 1.21 Vegetation changes within Region XI, west Patagonia (Coyhaique city, *above*), low Andes (Cerro Castillo village, *middle*) and eastern Patagonia (road Puerto Ibañez-Chile Chico, *below*)

of deciduous *Nothofagus* species. The northernmost populations at 33°S seem to be remnants of an ancient distribution of the genus.

Deciduous forests dominate along the Andes and the coast towards the south, surrounding the Central Valley. The core of the deciduous forest between 35 and 36°S is a mesic forest type, dominated by the two broadleaved deciduous species *Nothofagus alessandrii* and *N. glauca*. At around 38°S, this forest shows signs of the transition towards a temperate macrobioclimate, with the remarkable presence of the resinous or conifer forests of *Araucaria araucana* at the coast (Nahuelbuta) and in the Andes (Fig. 1.20). Deciduous forests often transition into stands of *Nothofagus antarctica*, while *N. pumilio* composes the treeline along the Andes all the way to Cape Horn.

(f) Located well into the temperate macrobioclimate, and related to high precipitation levels (>2,000 mm yr^{-1}) is the broad-leaved forest (Valdivian forest). It forms a U-shape, with a coastal and an Andean leg between 39 and 42°S.

(g) At around 41°S on the Andes and 41°30′S on the coast, broad-leaved forests are replaced by an evergreen North Patagonian rainforest mainly composed of large trees belonging to the Nothofagaceae. These rainforests are intermingled with the conifer evergreen forests which dominate the coast and interior, being replaced at altitude by deciduous forest.

(h) As the landscape becomes increasingly fragmented into fjords and little islands south of 47°S and the precipitation exceeds 4,000 mm yr^{-1}, the vegetation adopts the low physiognomy of moorlands. Towards the east, the moorlands are less humid and are dominated by the moss *Sphagnum magellanicum* (Arroyo et al. 2005). The wetlands of Torres del Paine National Park in Patagonia have been studied by Clausen et al. (2006), who determined their variety and type. Most of the interior of Patagonia is covered by the two wide ice fields and to the south of these a deciduous forest of *Nothofagus* reappears, together with the sub-Antarctic evergreen rainforest. In association with the marked precipitation gradient ranging from 4,000 mm at the western side to 300 mm at the eastern side of the low Andes in southern Patagonia and *Tierra del Fuego*, a gramineous steppe of *Festuca* spp. dominates the landscape (Fig. 1.21).

1.4 Land Use

Agriculture and forestry patterns of Chile today reflect great differences in the country's natural environments, the influence of international markets, the impacts of national

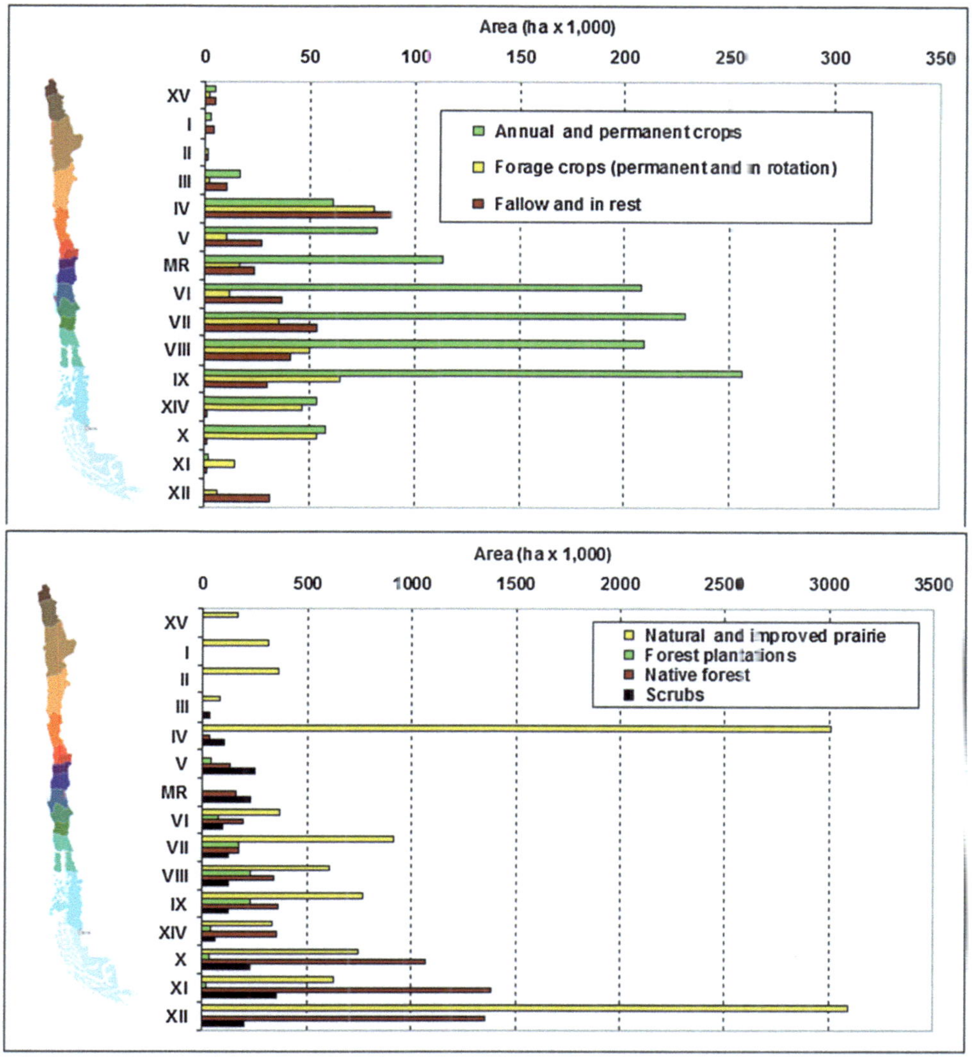

Fig. 1.22 Arable (*above*) and non-arable (*below*) land in the Chilean regions

policies, the imprints of settlement patterns and cultural preferences. The wide range of climates in Chile allows a great variety of crops, vegetables, fruit trees, flowers and grains to be cultivated. Important agriculture is only noticeably absent in rugged portions of the Altiplano and desert landscapes along the Pacific coast of northern Chile. The country invests considerable resources in protecting its privileged phyto- and zoo-sanitary conditions. Agricultural production has adapted well to free markets and the gradual lifting of protectionist barriers and is now modern and efficient.

Land use in Chilean territory according to the last Agriculture and Forestry Census of Chile (INE 2007) is shown in Fig. 1.22 in terms of and non-arable land in each region. The most important annual crops (cereals and potatoes) and perennial species (table grapes and apple) in 2010 occupied almost 600,000 and 88,000 ha of the territory, respectively (Fig. 1.23).

A large number of estates, especially those planted with trees, were auctioned off to large financial groups in the 1970s. By the early 1980s, a few major financial groups controlled 80 % of pine plantations and 100 % of the cellulose and paper-pulp industry. Chilean Forestry Law, Decree 701, provided state subsidies that covered 75 % of the cost of forestation for companies planting pine and eucalyptus in particular (Fig. 1.24). These companies, unlike landowners, encountered few social or political obstacles in their efforts to redraw the ecological landscape of Southern Chile. Throughout the 1970 and 1980s, with subsidies from the state and stimulated by growing markets

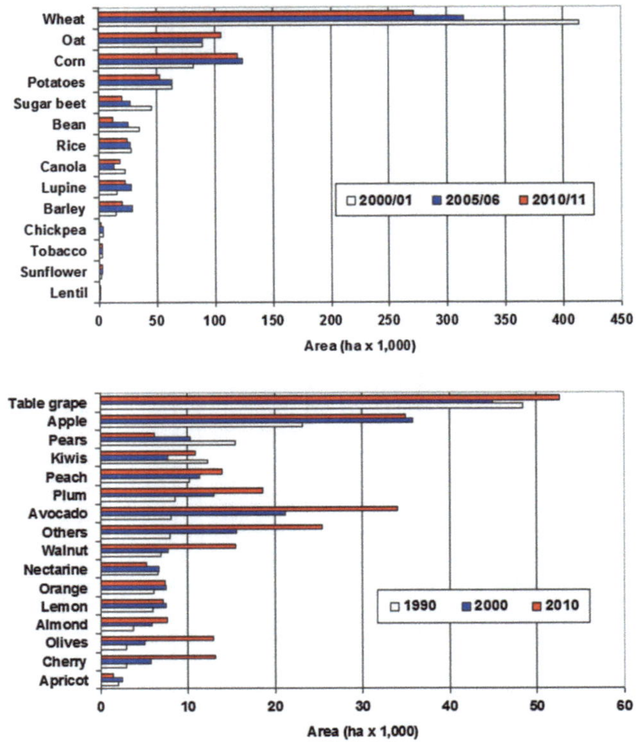

Fig. 1.23 Changes in the area of principal annual crops and fruit trees in Chile over the past decade

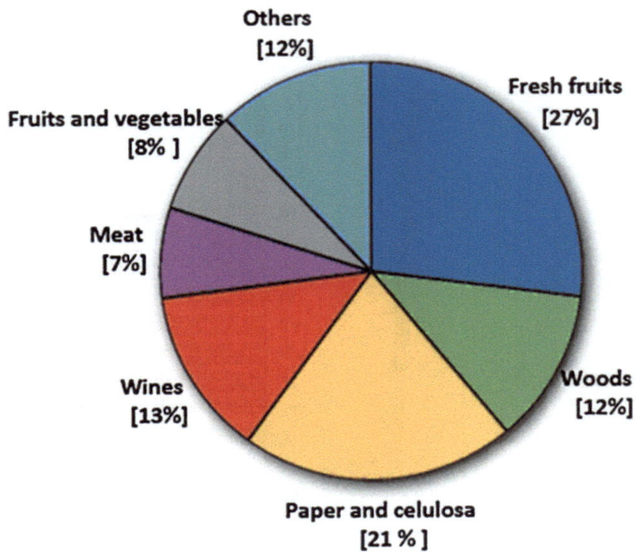

Fig. 1.25 Proportion of exports from different agriculture and forestry sectors in Chile in 2010 (http://www.odepa.cl. Accessed 9 June 2011)

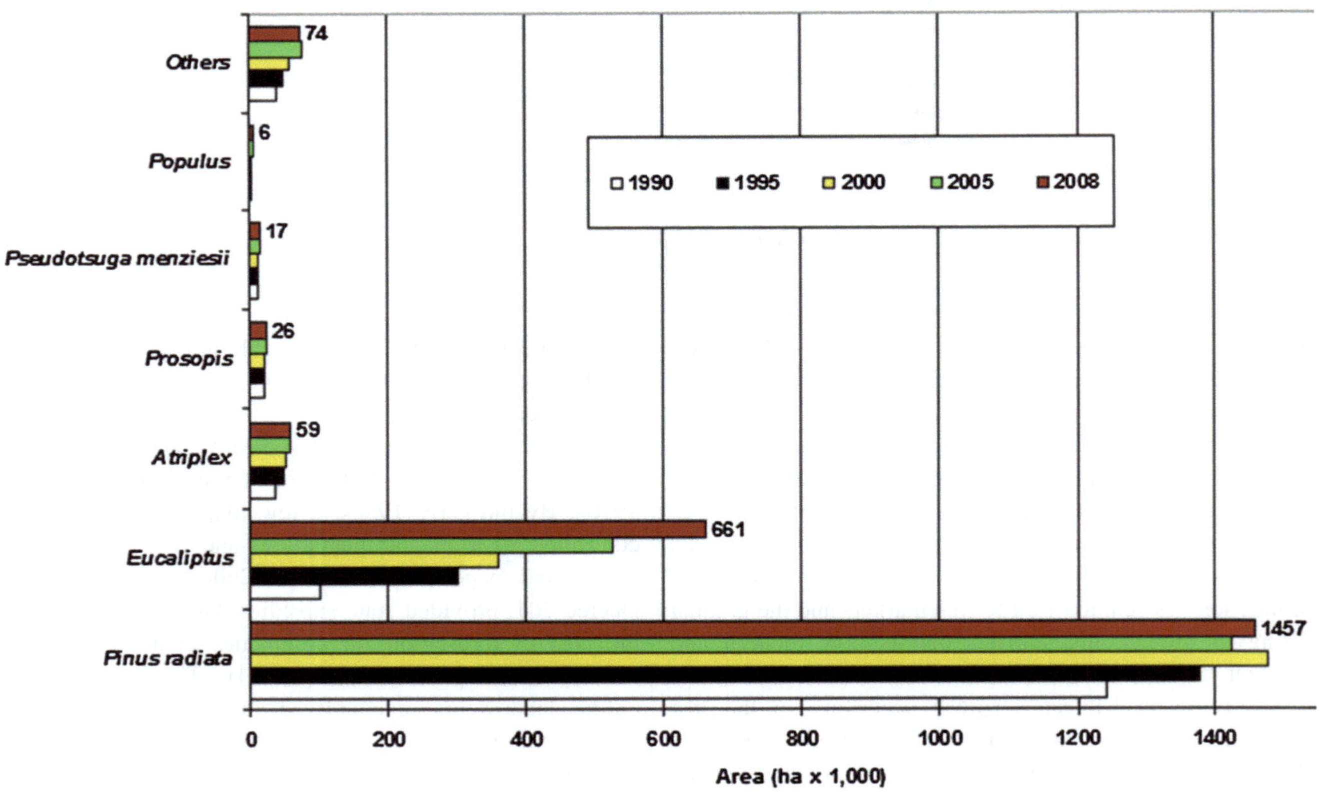

Fig. 1.24 Changes in the area of principal forest trees in Chile in recent decades

Fig. 1.26 Soil (Andic Haploxeroll) with agriculture vocation planted with forest trees in Chile (Region VII)

abroad for cellulose, pulp and wood chips, they proceeded to replace native rainforests with pine plantations.

From an international point of view, fresh fruits, wines, meats and timber were amongst the largest exports (US$ 12.2 billion) from Chile in 2010 (Fig. 1.25).

Chile's economy is one of the most natural resource dependent in the world. According to Altieri and Rojas (1999), over 87 % of Chilean exports are based on only four natural resource sectors, with a quite worrisome ecological footprint of modern agriculture. More recent information (López and Miller 2008) shows that on average, natural resource exports comprised over 40 % of total exports in the period 1990–2004 and that the estimated contribution of natural resource-dependent industries to gross domestic product (GDP) was more than 20 % over the same period.

During recent decades in Chile, the initial agrarian reform was rolled back, restoring some land to the former owners of agrarian estates. Moreover, there is a trend for using soils with agricultural vocation (arable lands) for forestry activities, i.e. soils in Land capability classes (LCC) I to IV are being planted with forest trees (Fig. 1.26).

Therefore, with an economy moving along an eventually non-sustainable path, achieving future food security for all Chilean citizens will depend on conserving soil, water, energy and biological resources. Indeed, knowledge and careful management of all of these vital resources deserves high priority to ensure the effective protection of national agricultural and natural ecosystems.

References

Ahumada M, Faúndez L (2009) Descriptive guide of wetland azonal vegetation systems of the altiplane ecorregion. Santiago, Chile (in Spanish)

Allmendinger RW, González G (2010) Invited review paper: neogene to quaternary tectonics of the coastal Cordillera, Northern Chile Tectonophysics 495:93–110

Altieri MA, Rojas A (1999) Ecological impacts of Chile's neoliberal policies, with special emphasis on agroecosystems. Environ Dev Sustain 1:55–72

Araya-Vergara J (2010) Chile In: Bird ECF (ed) Encyclopedia of the world's coastal landforms. Springer Science + Business Media B.V.

Arroyo MTK, Pliscoff P, Mihoc M, Arroyo-Kalin M (2005) The Magellanic moorland. In: Fraser LH, Keddy PA (eds) The world's largest wetlands: ecology and conservation. Cambridge University Press, London, UK

Bown F, Rivera A, Acuña C (2008) Recent glacier variations at the Aconcagua basin, central Chilean Andes. Ann Glaciol 48:43–48

Brunini O, Pedro Jr MJ, De Oliveira D, Paes De Camargo MB, De Souza Rolim G, Caramori PH (2010) Examples of agrometeorological decision support developed and used in South America. In: Stigter K (ed) Applied agrometeorology. Springer, Berlin

Calderón M, Hervé F, Massonne HJ, Tassinari CG, Pankhurst RJ, Godoy E, Theye T (2007) Petrogenesis of the Puerto Edén Igneous and Metamorphic Complex Magallanes, Chile: late Jurassic syndeformational anatexis of metapelites and granitoid magma genesis. Lithos 93:17–38

Casassa G, Rivera A, Haeberli b W, Jones G, Kaser G, Ribstein P, Schneider C (2007) Current status of Andean glaciers. Global Planet Change 59:1–9

Casassa G, Rivera A, Schwikowski M (2006) Glacier mass balance data for southern South America (30°S–56°S).' Knight PG (ed) Glacier science and environmental change. Blackwell, Oxford

Cembrano J, Lara L (2009) The link between volcanism and tectonics in the southern volcanic zone of the Chilean Andes: a review. Tectonophysics 471:96–113

Clausen JC, Ortega IM, Glaude CM, Relyea RA, Garay G, Guineo O (2006) Classification of wetlands in a Patagonian National Park, Chile. Wetlands 26(1):217–229

Farías M (2007) Tectonic and erosion in the Chilean central Andes relief evolution during Neogen. Dissertation, University of Chile (in Spanish with French summary)

Garreaud RD (2009) The Andes climate and weather. Adv Geosci 22:3–11

Garreaud RD, Vuille M, Compagnucci R, Marengo J (2009) Present-day South American climate. Palaeogeogr Palaeoclimatol Palaeoecol 281:180–195

Gioncada A, González-Ferran O, Lezzerini M, Mazzuoli R, Bisson M, Rapu S (2010) The volcanic rocks of Easter Island (Chile) and their use for the Moai sculptures. Eur J Mineral 22:855–867

Guerrero PC, Durán AP, Walter HE (2011) Latitudinal and altitudinal patterns of the endemic cacti from the Atacama Desert to Mediterranean Chile. J Arid Environ 75:991–997

Houston J (2006) Variability of precipitation in the Atacama Desert: Its causes and hydrological impact. Int J Climatol 26:2181–2198

Houston J, Hartley A (2003) The central andean west-slope rainshadow and its potential contribution to the origin of hyper-aridity in the Atacama Desert. Int J Climatol 23:1453–1464

INE (2007) VII Agriculture and Forestry Census. National Institute of Statistics, Chile (http://www.censoagropecuario.cl/. Accessed 20 June 2011)

Lara L (2010) The oceanic islands of Chile. In: Díaz-Naveas J, Frutos J (eds) Marine geology of Chile. National Oceanographic Committee, Chile (in Spanish)

Le Quesne C, Acuña C, Boninsegna JA, Rivera A, Barichivich J (2009) Long-term glacier variations in the Central Andes of Argentina and Chile, inferred from historical records and tree-ring reconstructed precipitation. Palaeogeogr Palaeoclimatol Palaeoecol 281:334–344

López R, Miller SJ (2008) Chile: the unbearable burden of inequality. World Dev 36(12):2679–2695

Luzio W, Casanova M, Seguel O (2010) Soils of Chile. In: Luzio W (ed) University of Chile. Maval Press, Santiago (in Spanish)

Moreira-Muñoz A (2011) Plant geography of Chile. In: Werger MJA (Ser. ed) Plant and vegetation, vol 5. Springer, Berlin

Muñoz N, Charrier R, Jordan T (2002) Interactions between basement and cover during the evolution of the Salar de Atacama Basin, Northern Chile. Rev Geol Chile 29(1):55–80

Pankhurst RJ, Hervé F (2007) Introduction and overview. In: Moreno T, Gibbon W (eds) The geology of Chile. The Geological Society, London

Paskoff R, Araya-Vergara J (2010) Coastal geomorphology. In: Díaz-Naveas J, Frutos J (eds) Marine geology of Chile. National Oceanographic Committee, Chile (in Spanish)

Rehak K, Bookhagen B, Strecker MR, Echtler HP (2010) The topographic imprint of a transient climate episode: the western Andean flank between 15.5° and 41.5°S. Earth Surf Process Landforms 35:1516–1534

Rivera A, Acuña C, Casassa G, Bown F (2002) Use of remotely sensed and field data to estimate the contribution of Chilean glaciers to eustatic sea-level rise. Ann Glaciol 34:367–372

Segerstrom K (1964) Quaternary geology of Chile. Brief outline. Geol Soc Am Bull 75:157–170

SERNAGEOMIN (2003) Geologic map of Chile: digital version. National Geology and Mining Service. Digital Geologic Publication N°4 (CD-ROM, version 1.0) Santiago-Chile (in Spanish)

Squeo F, Warner BG, Aravena R, Espinoza D (2006) Bofedales: high altitude peatlands of the central Andes. Rev Chil Hist Nat 79:245–255

Strecker MR, Alonso RN, Bookhagen B, Carrapa B, Hilley GE, Sobel ER, Trauth MH (2007) Tectonics and climate of the Southern Central Andes. Annu Rev Earth Planet Sci 35:747–787

UNESCO (2010) Atlas of Arid Zones in Latin America and the Caribbean. CAZALAC. Technical Documents PHI-LAC, No 8

2 Main Features of Chilean Soils

Chilean soil scientists have combined their efforts for a long time to produce a soil map of the country and special mention should be made of the work of Robert and Díaz who compiled a first map of the main Chilean soil groups (Fig. 2.1) following similar criteria used by the Dokuchaev classification system. Three decades later, Luzio and Alcayaga (1992) published a complete map showing soil associations, classified at Great Group level of USDA Soil Taxonomy (Fig. 2.2). Both works give a clear notion of the soil spatial variability along Chilean territory, which reflects its geology, geomorphology and climate diversity. Luzio and Alcayaga (1992) report in the hyper-arid zone only Orthids and Orthents without argillic horizons (due to extreme aridity), Argids which are more frequent in the arid zone, Ochrepts, Xerolls and Xeralfs in the central zone, besides some weakly developed volcanic soils (Xerands), while more developed volcanic ash soils (Udands and Aquands) together with organic soils (Hemists and Saprists) could be the most important soils in extreme southern zones. It is important to note that information on soils in extreme zones was scarce and the Andisol concept was still in the definition process (Beinroth et al. 1985), so most volcanic soils were classified as Inceptisols at the time of the Luzio and Alcayaga (1992) report.

Soil Taxonomy (USDA) is the classification system used in all soil surveys of Chile, while for World Reference Base (WRB) only some correlations have been developed. Salazar et al. (2005) observed a 100 % correlation between Andosols and Andisols in Region X, with Vitric Andosol associated to the Andean Mountains, Silandic Andosol to the Central Valley and the Aluandic Andosol to Chilcé Island. The Umbrisols associated to the Andean Mountains and also to the Coastal Range are correlated to Inceptisols. Cambisols, Acrisols and Luvisols are acceptably correlated to Inceptisols, Ultisols and Alfisols, respectively. Casanova et al. (2007) identified 11 reference soil groups from WRB in VII Region, located mainly in the Central Valley of Chile. They described Andosols towards the Andean Mountains and Lixisols mainly towards the Coastal Cordillera. Lixisols and Alfisols, Umbrisols, Cambisols and Inceptisols, Phaeozems and Mollisols and finally Andosols and Andisols were described as the best soil-correlated reference soil groups and soil orders.

For readers looking for more detailed information on the soils of Patagonia and who are familiar with the WRB classification, Fig. 2.3 includes a partial old map of Chile reported by Gut (2008).

Chilean Natural Resources Information Centre (CIREN) is the official repository institution of soil surveys in Chile, which offers products as paper orthophotos or a digital file containing the image raster georeferenced area of the orthophoto for any size and a vector digital file with the boundaries of the property. Delivered on CD-ROM as a georeferenced TIF file and a vector file, it includes separate layers of farm boundaries, soil cartography and Land capability classes (LCC).

Around 65 % of Chilean territory has been covered by detailed and semi-detailed soil cartography (Fig. 2.4), with soil cartographic units (phases of soil series and soil associations in remote areas) traced over the above-mentioned orthophotos (Fig. 2.5) and at agrological scale, normally 1:20,000 and in more remote areas 1:500,000 (Table 2.1).

The standard notation used for map units includes three letters corresponding to soil series name (examples in Fig. 2.5, RLV: Rinconada Lo Vial soil series; PUD: Pudahuel soil series; MPC: Mapocho soil series), followed by an Arabic numeral denoting different soil phases, which are described in references to Table 2.1. The Roman numeral and minor case letter represent LCC and sub-class, respectively.

2.1 Soil Formation in Chile

As regards geological and geomorphologic patterns in Chile (see Sect. 1.1), the soil parent material is varied, including old rocks, volcanic ash, fluvial and/or glacial deposits and colluvial and/or alluvial deposits. Residual and colluvial

M. Casanova et al., *The Soils of Chile*, World Soils Book Series, DOI: 10.1007/978-94-007-5949-7_2,

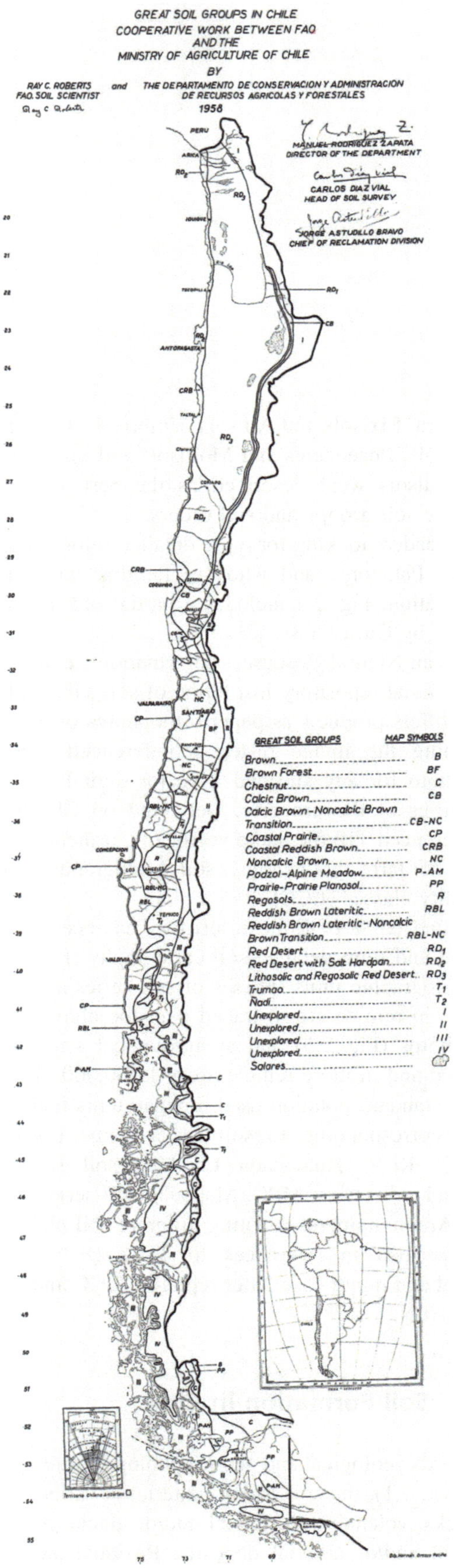

◀ **Fig. 2.1** Soil map of Chilean soils (http://eusoils.jrc.ec.europa.eu/esdb_archive/eudasm/latinamerica/lists/ccl.htm). Accessed 31 March 2012

cessation of leaching losses, silicate transformation and biological activity and key processes in pedogenesis almost everywhere on Earth. When long-term hyperaridity minimises these processes, the result is soils composed primarily of atmospheric salts and dust, which physically expand the landscape as they form (Ewing et al. 2006).

On the other hand, topographical position results in younger soils deposition in the Central Valley of Chile being in an active process of construction, while some old alluvial soils often show the development of a cambic horizon. Watersheds with opposing slope aspect in the central Coastal Range show diverse soil development, as noted by Casanova et al. (2000) when soil hydraulic conductivity was measured and morphological features described (Fig. 2.10). In fact, Badano et al. (2005) in the Mediterranean zone of central Chile observed that slope aspect strongly influences mesoclimate conditions. In an altitudinal sense, soil diversity observed by Cavieres et al. (2000) in central Chile and Squeo et al. (1993) in the Andes of the Hyper-arid zone of Chile determines a certain vegetation zonation in response not only to temperature and radiation gradients, but also changes in soil chemical properties. Physical soil properties such as soil water content must also be included as a variable with altitude.

Douglass and Bockheim (2006) used 34 pedons on 4 moraine groups to investigate mechanisms and rates of soil development in Argentinian Patagonia with coarse-loamy, mesic, Typic Haplocalcids or Calcic Haploxerolls occurring under short grass-shrub steppe in a semi-arid climate. For similar soils observed in Chilean Patagonia, dominant soil-forming processes reported in this environment are melanisation, calcification and the accumulation of clay-sized particles. Organic carbon (OC) accumulates rapidly in these soils, but significantly higher amounts in the oldest two moraine groups are most likely the result of slight differences in soil-forming environment or grazing practices. Accumulation rates of carbonate and clay decrease with age, suggesting either decreased influx in the earliest part of the record or attainment of equilibrium between influx and loss. There are no changes in soil redness, and preservation of weatherable minerals in the oldest soils indicates there is little chemical weathering in this environment. Measured dust input explains the accumulation of both clay and carbonate. The authors present a carbonate cycling model that describes potential sources and Ca mobility in this environment. Calibration of rates of soil formation creates a powerful correlation tool for comparing other glacial deposits in Argentina to the well-dated moraines (Fig. 2.11) at Lake Buenos Aires (Lake General Carrera, in Chilean territory).

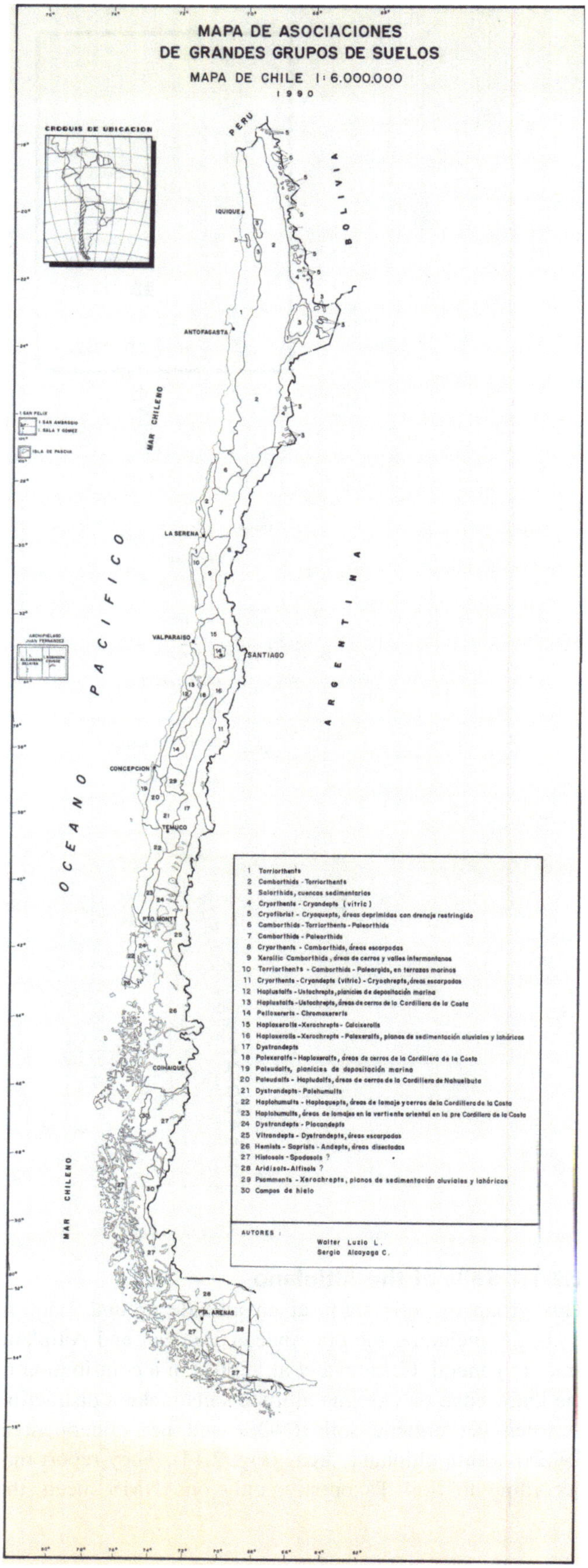

Fig. 2.2 Soil map of Chilean soils, after Luzio and Alcayaga (1992)

2.2 Major Soil Zones

Using soil surveys along Chile, the various soil formation factors and diverse soil formation processes, Luzio et al. (2010) defined in detail eight so-called edaphic zones of Chile (Table 2.2). However, these are characterised and synthesised here as four major soil zones (Table 2.3 and Fig. 2.12).

2.2.1 Soils of the Hyper-Arid to Semi-Arid Zone

Soils of this zone are associated with diverse geomorphic features along nearly 1,500 km of Chilean territory, where it is possible to differentiate soils of the Altiplano, soils of the Longitudinal Central Valley, soils of the Coastal Range, soils of marine terraces, soils of valleys and soils of *serranías* (Fig. 2.13).

According to the age of geological materials in the Hyper-arid sector, advanced pedogenic evolution should not be expected and even less translocation of components within the profile, because of the climate characteristics, a factor that creates a weak organic regime, at least in the northern sector of this zone. Many workers (Risacher et al. 2003; Berger and Cooke 1997) report that crusts and salars are characteristic features not linked to any particular geomorphic position within this zone.

Duricrusts are a common feature of the Central Valley and on the inland slopes of the Coastal Range. Salts occur either as fabric-preserving cement, consisting of original depositional textures of alluvium or colluvium or as fabric destructive nodules and masses. Where extensive precipitation of salts has destroyed the original fabric, the coarse fraction appears to have been expelled to the surface. Because dissection is so limited, duricrusts are best exposed in workings for nitrates and other salts or on dissected slopes where erosion has exposed slope duricrusts. Sulphate duricrusts are found on the inland slopes of the Coastal Range and in the Central Valley. In the latter, they may be impregnated by exotic salts such as nitrates, perchlorates and iodates. The presence of calcium and sulphate in the soils is due to rock weathering and groundwater remobilisation weathering, rather than accession as cyclic salts, followed by aeolian redistribution in the landscape. Sulphate duricrusts are absent from the comparatively higher rainfall areas of the pre-Andean sector. Further south, towards Copiapó city, sulphate duricrusts are replaced by calcretes. The origin of the nitrate salts has been the subject of much debate, but the most recent geochemical studies appear to confirm deposition of atmospheric nitrate as the source (Prellwitz 2007; Mishalski et al. 2004).

Northern Chile is also characterised by a succession of north–south trending ranges and basins occupied by

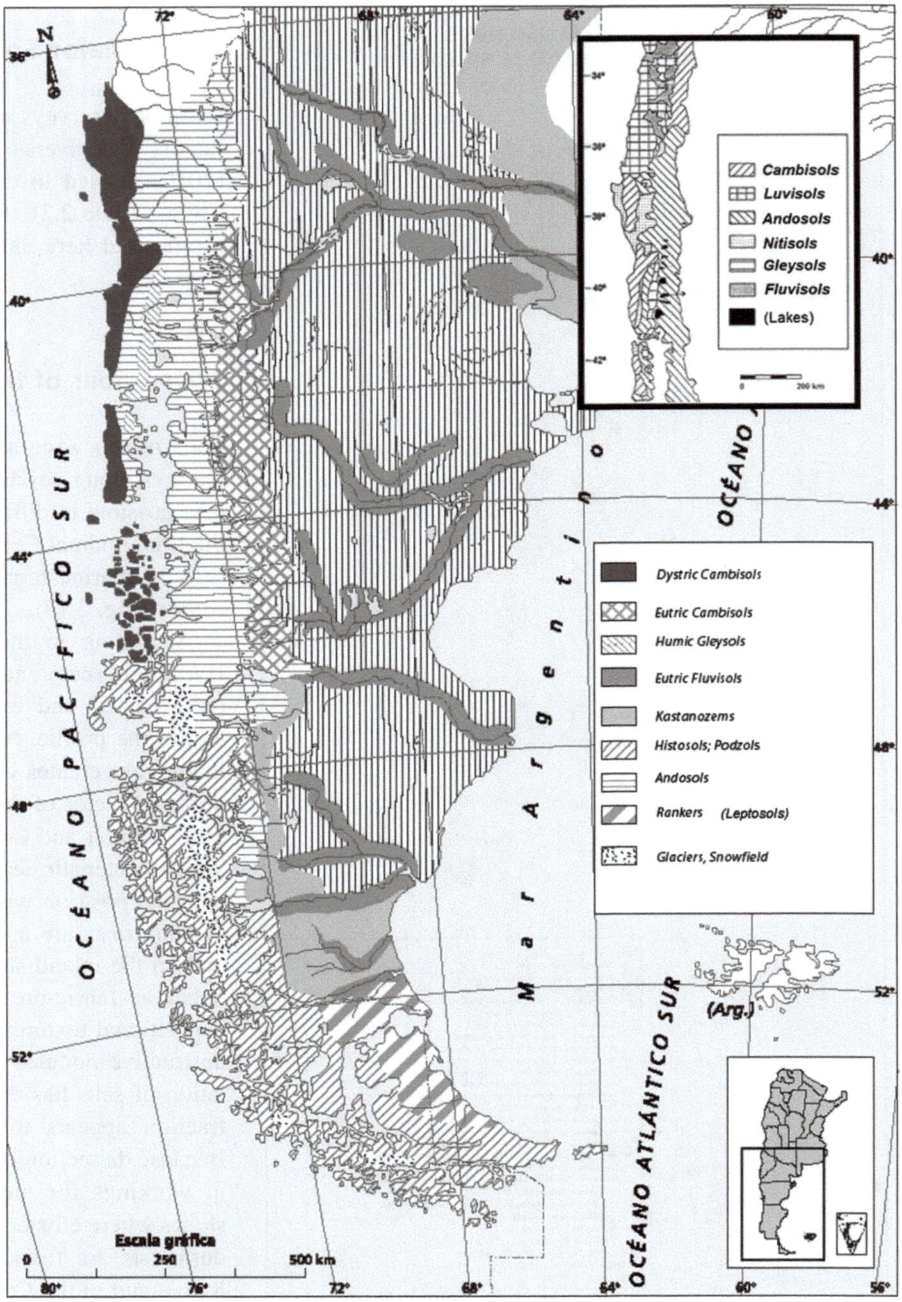

Fig. 2.3 Generalised WRB system map of southern South America and Chile. Adapted from Gut (2008) and Aravena et al. (1993)

numerous saline lakes and salt crusts, collectively called *salares* (salars). Fossil salt crusts are found to the west in the extremely arid Central Valley, while active salars receiving permanent inflows fill many intravolcanic basins to the east in the semi-arid Coastal Range. Sea salts and desert dust are blown eastward over this range, where they constitute an appreciable fraction of the solute load of very dilute waters (salt content <0.1 g L^{-1}).

2.2.1.1 Soils of the Altiplano

This group of soils starts at an altitude around 1,500 m a.s.l., i.e. including the pre-Andean, Andean and Altiplano area. In general, Casanova et al. (2008) in a contribution to the knowledge on extreme altitude soils make a distinction between net organic soils (NOS) and net mineral soils (NMS) within altiplanic areas (Fig. 2.14). They report that according to Soil Taxonomy, only one NMS meets the

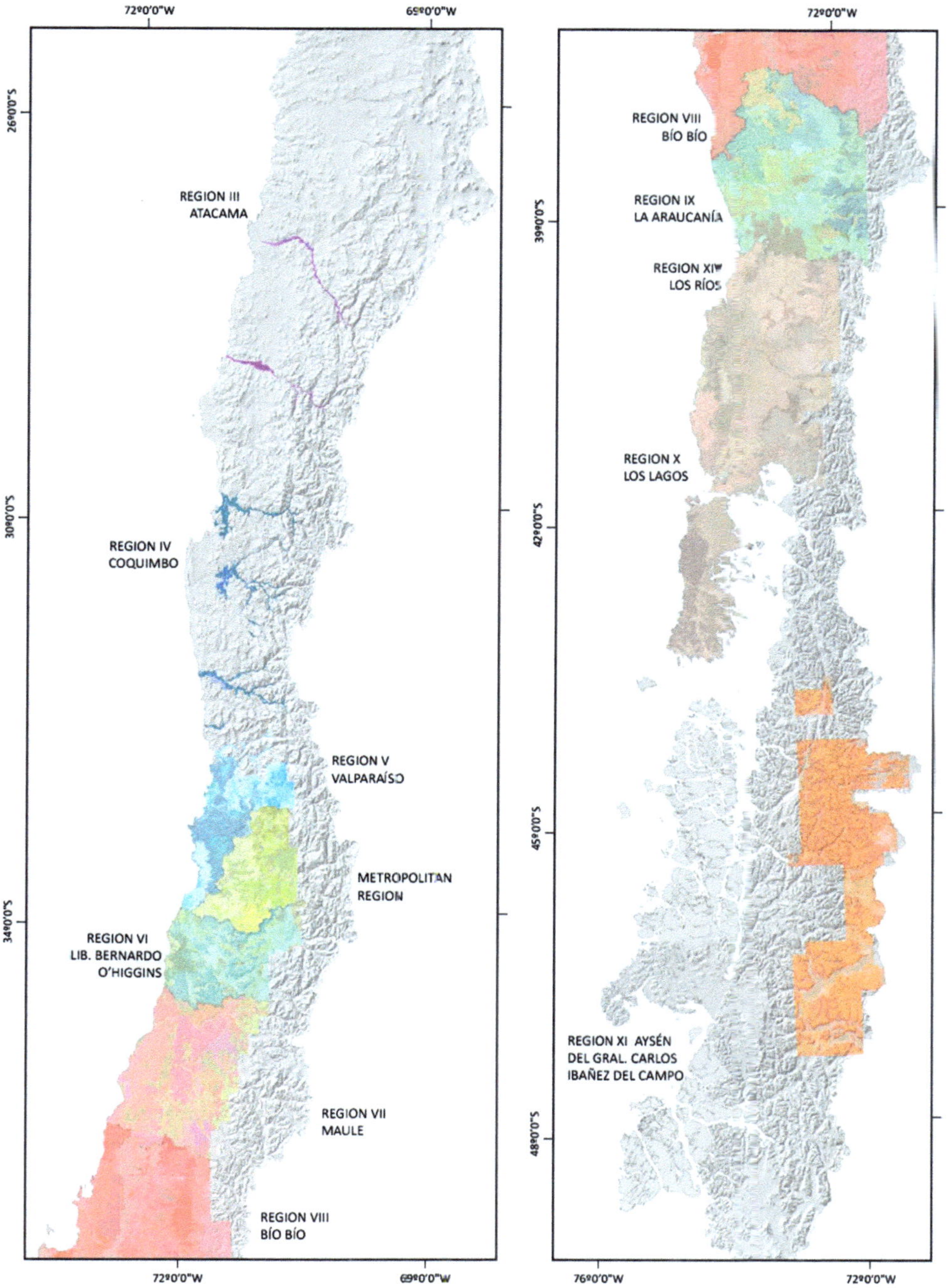

Fig. 2.4 Map showing soil survey cover in Chile, Region III–XI (see Table 2.1)

requirements for Andisols, while six of eight pedons meet the requirements for Andosols according to WRB, the others corresponding to Leptosols and Regosols. With regard to NOS, one profile is classified as a Histosol (Soil Taxonomy), while both profiles analysed meet the requirements to be considered as Histosols (WRB).

Later, Luzio et al. (2010) distinguished soils with skeletal morphology, soils with incipient pedogenesis and soils on depressive basins.

Skeletal Soils

These incipient soils are developed only with primary pedogenesis and from diverse parent materials, and it is possible to identify a common undeveloped coarse-textured and eventually medium-textured A horizon (ocric epipedon). Weak subangular block structures are sometimes observed, but massive or single-grain conditions are more frequent. With a typical A–R profile (A-Cr, when some weathering is possible) these soils normally reach

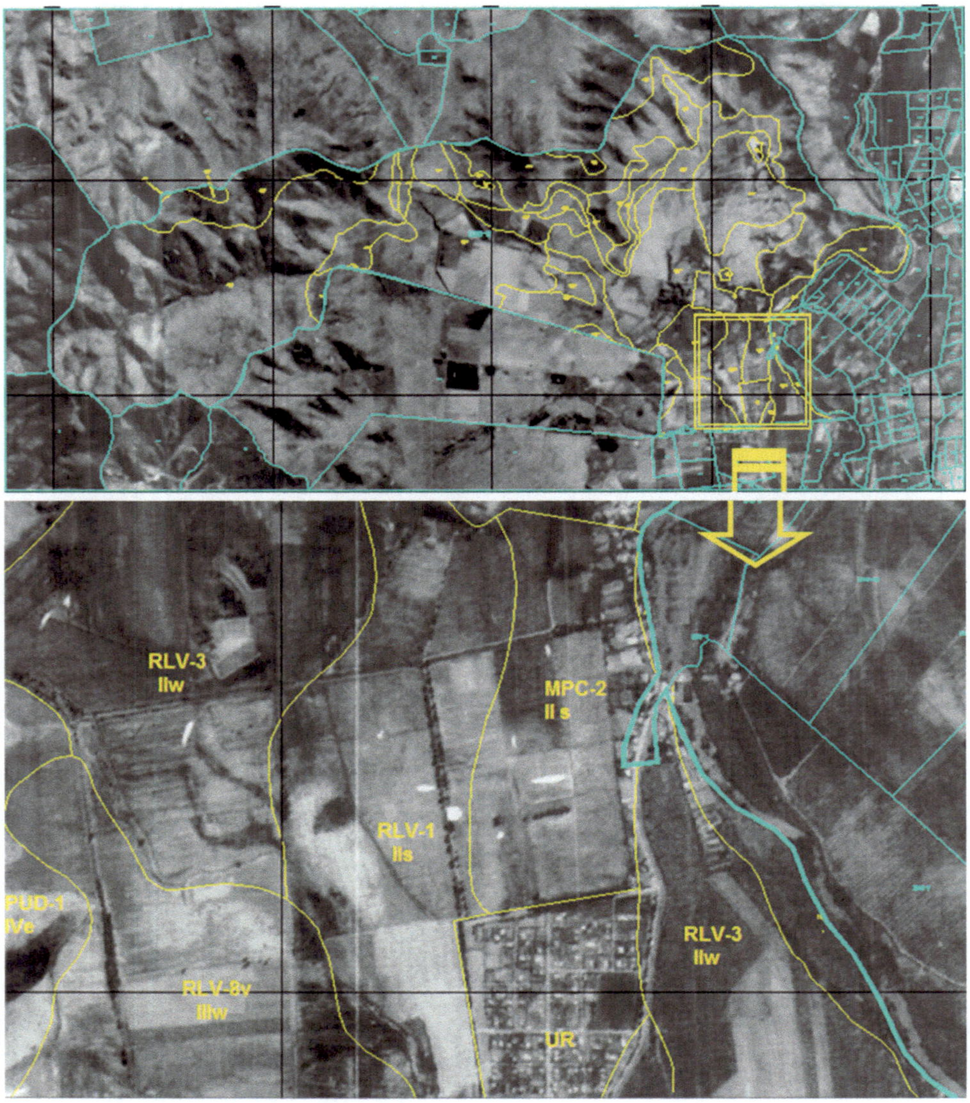

Fig. 2.5 Example of predial window with digital soil cartography information. *Above* cartographic units (phase of soil series) in *yellow* within Experimental Agronomic Station (Germán Greve Silva) belonging to Faculty of Agronomic Sciences, University of Chile. *Below* zoom to *upper yellow inset* (elaborated from www.ciren.cl)

Table 2.1 Soil survey scale and cover in Chile

Region	Cartographic scale	References
III	1:20,000 (valley)	CIREN (2007)
IV	1:20,000 (valleys)	CIREN (2005a)
V	1:20,000 and 1:50,000 (valleys); 1: 100,000 (hills)	CIREN (1997a)
Metropolitan	1:50,000 (valleys); 1: 100,000 (hills)	CIREN (1996a)
VI	1:20,000 and 1:50,000 (valleys); 1: 100,000 (hills)	CIREN (1996b)
VII	1:20,000 and 1:50,000 (valleys); 1: 100,000 (hills)	CIREN (1997b)
VIII	1:20,000 and 1:50,000 (valleys); 1: 100,000 (hills)	CIREN (1999)
IX	1:100,000	CIREN (2002)
X and XIV	1:50,000 (valleys); 1: 100,000 (hills)	CIREN (2003)
XI	1:20,000 and 1:60,000 (valleys); 1:250,000 and 1:500,000 (hills)	CIREN (2005b)

Fig. 2.6 Coastal Range residual igneous soil (Ultic Palexeralf) observed in Regions VII and VIII of Chile

Fig. 2.7 Coastal Range residual metamorphic soil (Typic Rhodoxeralf) observed in Region VIII of Chile

Fig. 2.8 Soil (Vitrandic Durixerept) derived from acid volcanic sands and pumice. Metropolitan Region of Chile

40–60 cm depth. Volcanic coarse fragments are normally present at the soil surface in 5–50 % abundance, with sizes from coarse angular gravel to pebbles. In addition, a thin (1–10 cm) sandy surface coating is observed, with its origins in aeolian transport processes. In IREN (1976), andesitic, rhyolitic and basaltic gravels are characterised, while sands correspond to pumice. The very low organic matter content (OC < 1 %) decreases with depth in most soils, except in those with more stratified profiles. These soils have largely been classified as Lithic Usthorthents.

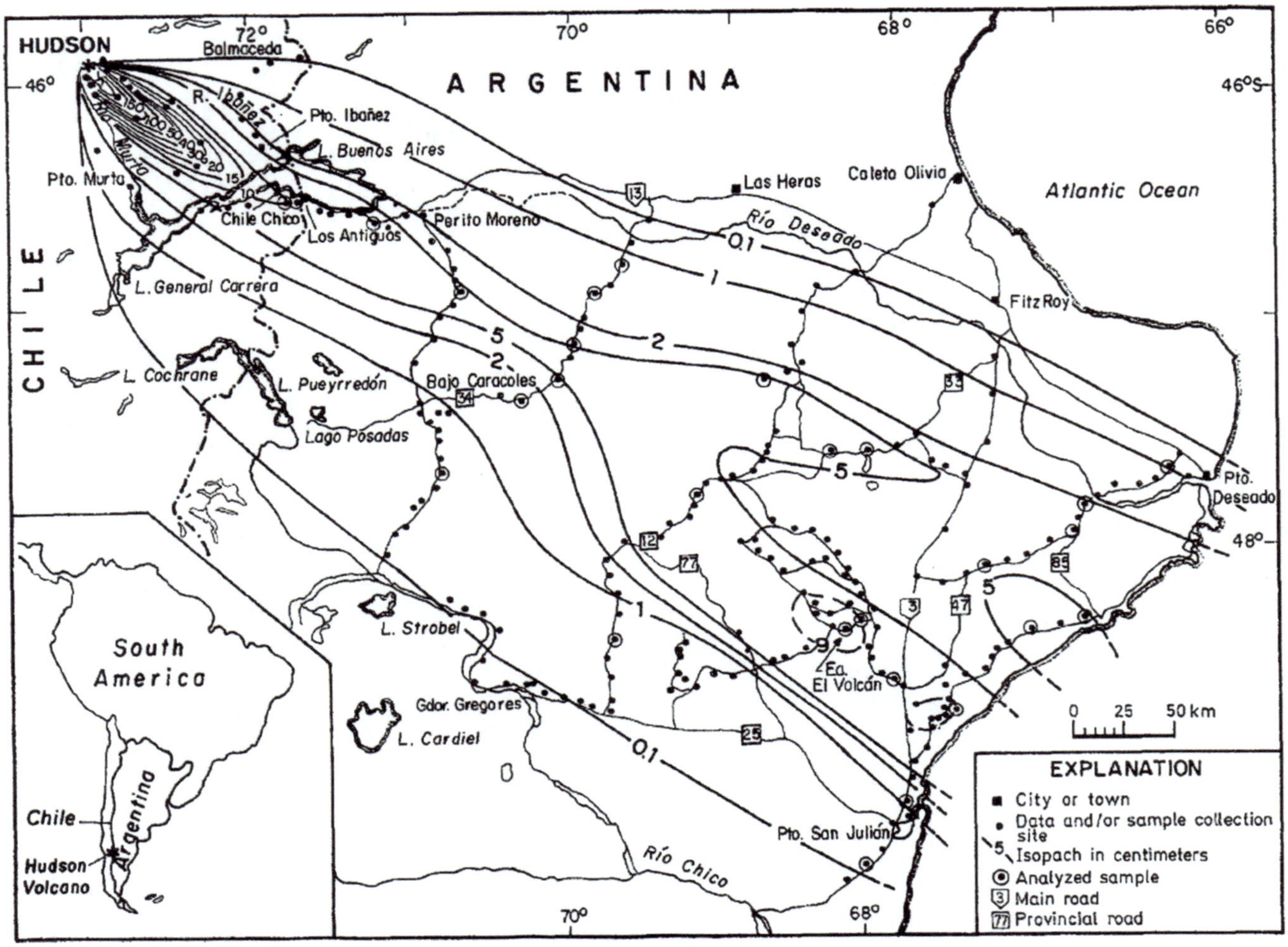

Fig. 2.9 Location and isopach map of primary eruptive deposits (cm) in Region XI of Chile (Scasso et al. 1994)

Fig. 2.10 Slope aspect differences within a Coastal Range small watershed (Metropolitan region)

Fig. 2.11 Glacial features around Lake General Carrera (Chile, Region XI) or Lake Buenos Aires (Argentina)

Table 2.2 Edaphic zones of Chile (Luzio et al. 2010)

Edaphic zone	N limit	S limit
Desert zone	18°00′S	29°00′S
Arid and Semi-arid zone	29°00′S	32°00′S
Arid Mediterranean zone	32°00′S	37°45′S
Wet Mediterranean zone	37°45′S	43°00′S
Rainy zone	43°00′S	50°00′S
Magellan zone	50°00′S	55°00′S
Antarctic zone	62°00′S	63°00′S
Easter and Juan Fernández Islands	–	–

Table 2.3 Major soil zones of Chile

Soil zone	Subdivisions	Soils
Hyper-arid to Semi-arid zone	Altiplano	Skeletal soils
(18–32°S)		Soils with incipient pedogenesis
		Soils in depressive basins.
	Longitudinal Central Valley	Soils of depositional plains and Andean piedmonts
		Salty soils
	Valleys	Soils of intermittent rivers
		Soils of permanent watercourses
	Coastal range	
	Coastal plains	Undeveloped soils
		Indurate soils
		Soils with natric horizons
		Sedimentary soils
	Serranías	
Mediterranean zone	Andean mountains	Non volcanic soils
(32–43°S)		Volcanic soils
	Pre-Andean mountain	Non volcanic soils
		Volcanic soils
	Longitudinal Central Valley	Soils with high base saturation
		Soils with argillic horizons
		Soils with cambic horizons
		Typical Andisols (trumaos)
		Aquic Andisols (ñadis)
		Holocene sandy soils (arenales)
	Coastal range	Igneous parent materials
		Metamorphic parent materials
	Coastal plains	
Rainy and Patagonian zone	West side of North Patagonia	Andisols
(43–56°S)		Other soil orders
	East side of North Patagonia	Soils with high base saturation
		Soils with low base saturation
	South Patagonia (Magallanes)	Organic soils
		Mineral soils
Insular and Antarctic zone	Easter Island and Juan Fernández Archipelago	
	Antarctic	Mineral soils
		Ornithogenic soils

Figures 2.15 and 2.16 illustrate a typical profile and landscape, respectively.

Soils with Incipient Pedogenesis

Figure 2.17 shows an example of soils developed from stratified tephras between Calama and Ollagüe, with an A–C profile. Other shallow soils from volcanic materials (pumice fragments throughout profile) can also be observed between Putre and Parinacota (Fig. 2.18). Textural classes are medium to moderately fine at the surface and coarse sandy with pumiceous gravel (almost 70 % with diameter <4 cm) to 120 cm depth. Although blocky structure is well-developed near the surface, below 20 cm depth massive conditions are generally seen.

Exceptionally, more intense pedogenetic processes are observed in some highland soils which are not occupying a

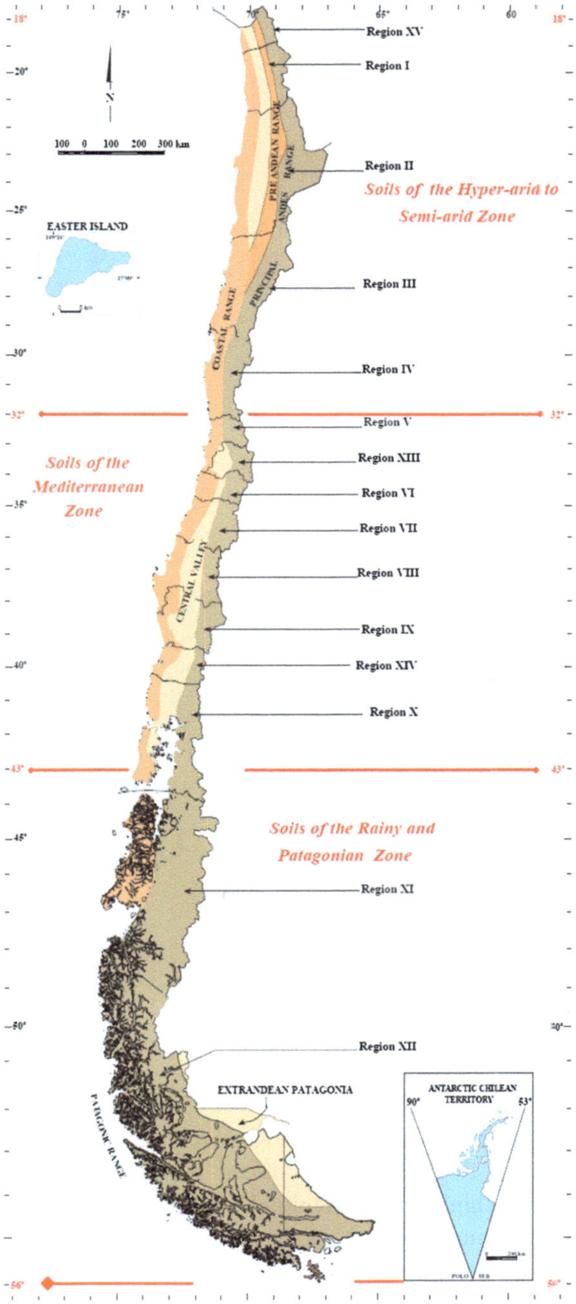

Fig. 2.12 Major soil zones of Chile

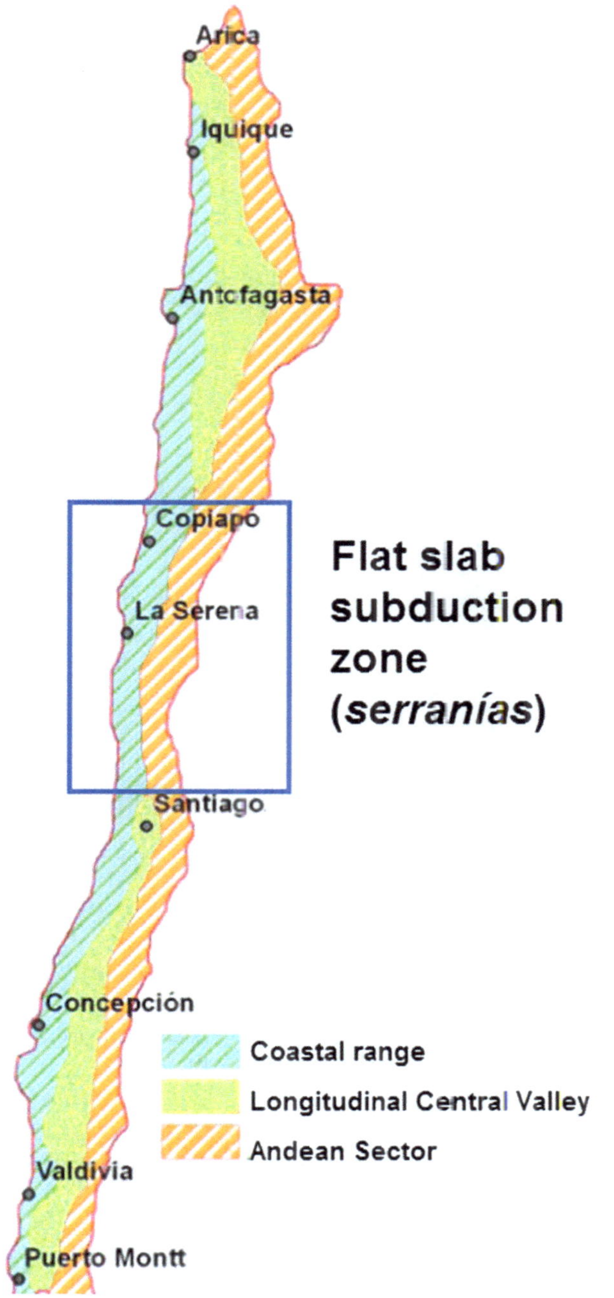

Fig. 2.13 Main physiographical changes along the Hyper-arid to Mediterranean zone of Chile

defined geomorphic position and show evidence of cambic horizon formation (B_w). Fine and moderately fine-textured B horizons with angular or subangular blocks are frequently observed but, occasionally, a strong coarse prismatic structure has been reported. Some C horizons are strongly compacted and present abundant pumice gravel or tuff, with different degrees of weathering. Significant lithologic discontinuities have been formed by frequent and successive petrographically diverse alluvial, aeolian or volcanic materials, transported/deposited by flooding. The clay fraction in their profiles must be responsible for the higher soil water retention capacity and cation exchange capacity (CEC), because as in skeletal soils, the organic matter (OM) content is low.

Considering previous antecedents, normally some of these incipient soils are very close to meeting the minimum requirements to be classified as Andisols. Therefore,

Fig. 2.14 Sites where 10 Altiplanic net organic (NOS) and net mineral (NMS) soils occur. *Inset*, *above* an example of NOS (CQN, Caquena soil) and *below* a NMS (GTR, Guallatire soil)

Usthorthent denomination is appropriate in some cases, but when a cambic B horizon is developed the soil is classified as a Haplocambid (Soil Survey Staff 2006).

Soils on Depressed Basins

Common features of these soils or wetlands are constrained drainage, high organic matter content and accumulation of salts (Table 2.4). In a depressed landscape associated with inundation sectors of streams, poor drainage sites rich in plants are formed which are locally named as *vegas*, are mostly characterised as mineral soils rich in organic matter. However, other organic soils are formed in closed basins with very poor drainage, which are known as *bofedales*.

Ahumada and Faúndez (2009) classified these soils according to their salinity levels, vegetation characteristics and drainage conditions, including an additional soil organic type named *pajonal* (Fig. 2.19). The same authors reported that all these soils occupy a total area of 43,234 ha (48 % in Region XV, 21 % in Region I, 22 % in Region II and 9 % in Region III). Cepeda and Novoa (2006) studied soils in the highlands of Region IV and reported that where water accumulates or the watertable is at the soil surface, sites with characteristics of wetlands/wet pastures are found, locally named *vegas* or *veranadas*. In ecological terms, these are the richest most diverse and productive sites in the highlands, making possible seasonal livestock (normally goats) transhumance (Fig. 2.20).

Luzio et al. (2002a) reported some soil characteristics in two *bofedales*, classified as Haplofibrist and Haplohemist, normally with fibric surface horizons and sapric at depth (Fig. 2.21). Other highly saline pedons have been classified as Endoaquents and Halaquepts/Endoaquepts.

2.2.1.2 Soils of the Longitudinal Central Valley

The Longitudinal Central Valley with 40–55 km wide interfluves (pampas) that correspond to a plateau with altitude

Fig. 2.15 Skeletal soil (Lithic Ustorthent) of 20 cm depth over liparite stone, Region I (Luzio et al. 2010)

increasing from 500 to 1,000 m a.s.l. in the west to 1,900–2,300 m a.s.l. in the east, resulting in a slope of 1.5–2.0°W. The pampas are separated by the deep gorges (1,000 m) of a few rivers coming from the Andes. In physiographic terms this area is gently undulating and deflated, apart from hard crust sectors with salts or silica. The Central Valley is a thick fill of detrital and lacustrine sediments (gravel, sand, silt and clay) of the middle Tertiary to Holocene age, often covered with salt crust remnants of ancient salars. Roughly between 22 and 25°S lies what is known as the Atacama Desert, the oldest continuously arid region in the world. Within this desert, Quillagua (21°39′S–69°32′W, mean rainfall 0.05 mm) can lay claim to being the driest place on Earth.

The Longitudinal Central Valley becomes increasingly discontinuous south of about 26°S (see Fig. 2.13), although the general features of the Coastal Range and pre-Andean ranges remain in place, and disappear entirely near 27°S. The topography to the south of 27°S is continuously hilly to mountainous from the ocean to the Andes (*serranías*).

As shown in Sect. 1.1, the Central Valley is bordered by a steep, <2,000 m high escarpment that rises from the Pacific Ocean to the west, and by the high Andes (>5,000 m) to the east, but its southern boundary is subject to debate. Rundell et al. (1991) suggest that the present-day desert border lies at Elqui Valley (~30°S) based on a shift in vegetation pattern, but the nearly lifeless zone for which the Atacama is renowned ends north of the first transverse Copiapó Valley (Chañaral, ~26°S).

Large, low-angle alluvial fans fill the Central Valley and converge onto the broad pediment and pediplain surfaces. Extensive ephemeral, dendritic drains join axial channels at valley centres, particularly to the north and east. Ewing et al. (2006) and Amundson (2003) reported some developed soils in the zone and described three profiles corresponding to Aridisols (Fig. 2.22), either on alluvial fan (Yungay and Copiapó pedons) or stream terrace (Altamira pedon) positions.

Soils in Depositional Plains and Andean Piedmonts

These soils cover an area between 18 and 29°S, with altitude from 1,000 to 2,800 m a.s.l., i.e. an area laid out by the Andes (and pre-Andes) and the Coastal Range. Clarke (2006) in a detailed description reports that the extensively deflated surface consists of fragments of gypsum duricrust, andesitic pebbles and a lag of lithic sand. Stone circles and polygons form a network of stones enclosing areas of bare ground, being locally common in a number of areas. The soils are characterised by very weak profile development and a

Fig. 2.16 The Atacama Desert. **a** The western Coastal Range. **b** The eastern Coastal Range. **c** Hillslope in the longitudinal Central Valley. **d** Boulder field in the absolute desert, where skeletal soils are present (partially from Placzek et al. 2010)

mixture of alluvial and colluvial sediments, which may be saline, also including soils occurring in the Andean piedmont.

These regular plains, including soils with different salinity levels, were dissected mainly by alluvial processes, probably more intensely at the beginning of the Quaternary. As a consequence, some hilly (slope gradient <10 %) landscape can be observed, with remnant plains (called *pampas*) separated by valleys and streams of different sizes. Soils from main valleys (Lluta, Azapa, Camarones, etc.) are described later.

According to Luzio et al. (2010), these extensive plain sectors contain what could be considered true desert soils, as described by Díaz and Wright (1965) and the most representative Atacama Desert soils, Aridisols. Abundant (>40 %) surface angular coarse fragments are observed, with a mixed acid and basic composition. Like an extensive pavement, similar to a stone desert, armoured surfaces comprise intricate mosaics of coarse particles, where deflation may be a relatively unimportant process of pavement formation.

Shallow soils developed on rhyolitic tephras, coarse-textured at the surface and with a single-grain structure, are common, but in some cases it is possible to find somewhat evolved soils in terms of a weakly expressed B horizon and moderately coarse-textured and subangular blocks structure. Calcareous coatings occur at a cracked R layer, expressing some translocation through the soil profile.

In some planar depositional basins, stratified soils (sandy and sandy loam) are observed (Fig. 2.23) in sectors with accentuated microrelief, but some fine-textured layers (silty and silty loam) can be included in some cases. Although excluded from a salar concept, irregular EC values at depths (2–30 dS m^{-1}) have been measured. The $CaCO_3$ content is also variable (1–6 %).

Salty Soils

The pre-Andean Depression, also called the salar de Atacama Depression, is a large intramontane basin at 2,500 m elevation, filled with Tertiary to Holocene clastic and evaporitic sediments of continental origin. According to

Fig. 2.17 Stratified volcanic soil with abrupt tephras (ashes and lapillis) layers, located between Calama and Ollagüe, Region II (Luzio et al. 2010)

Hardie and Eugster (1970), in most saline lakes worldwide, three main groups of brines may be generated by evaporation and chemical division: alkaline brines (Na/HCO_3–CO_3–Cl, pH > 9, traces of Ca and Mg), sulphate-rich brines (Na/SO_4–Cl, low Ca, pH < 9) and Ca-rich brines (Na–Ca/Cl, low SO_4, pH < 9). In northern Chile, most belong to the SO_4-rich and Ca-rich groups, and alkaline salars are almost completely lacking, due to the abundance of native sulphur in the Andes and atmospheric deposition of gypsum-rich desert dust from the Central Valley (Bao et al. 2004; Risacher et al. 2003).

Fig. 2.18 Volcanic origin soil profile (*above*) in sedimentation plain (*below*), located between Putre and Parinacota. Acid pumice ash is apparent in the subsoil, Region XV (Luzio et al. 2010)

Soils developed in these evaporite deposits represent an extreme in terms of characteristics and properties. Although these soils have a surface extremely hard crust (which in some cases can be up to 1 m thick), they consist of salts where vegetation can eventually thrive in good conditions (see Fig. 2.24).

Fig. 2.19 Characteristic vegetation on altiplanic soil types in depressed geomorphic positions, after Ahumada and Faúndez (2009). *Bofedal* (*above*), *pajonal* (*middle*) and *vega* (*below*). Saline (*left*) and no saline (*right*) types

2.2.1.3 Soils of the Valleys

A number of valleys and streams cut across northern Chile from the Andes Mountains to the Pacific Ocean where alluvial soils have developed on fluvial deposits, while between the rivers (interfluves) the soils are dry and some are infertile.

Responding to intense mining activity and an exponential increase in tourism during recent decades, high capacity infrastructure (roads, gas and water pipes and power lines) have been built in the northern Hyper-arid to Semi-arid zone to facilitate the exchange of goods and energy, and to supply urban and industrial centres located in the absolute desert with food and water. Therefore, a high demand for water for industrial and urban purposes is creating severe conflicts with natural ecosystems, farmers and indigenous cultures. In this regard, Lagos and Blanco (2010) recently claimed that the coexistence of high income levels and inequalities in the zone is not the result of insufficient resources, but of a lack of commitment to development by the main regional stakeholders.

Soils of Intermittent Rivers on Desert Valleys

In the northern Hyper-arid to Semi-arid zone only intermittent rivers or dry valleys are present in a superimposed network. The former corresponds to the active drainage basin of the Lluta River, San José River (Azapa Valley) and Camarones Valley (Fig. 2.25). Parallel valleys are very well-preserved in the interfluve zones of the pre-Andean sector.

The rivers differ greatly from one another in amount of water, seasonality of flow, fluctuation from year to year and character of accessible irrigable lands. Maximum flow

Fig. 2.20 Surire *bofedal* at Surire salt lake, Region XV

Fig. 2.21 Caquena *bofedal*, Region XV. Landscape (*left*) and soil profile (*right*)

Table 2.4 Altiplanic soil types in depressed geomorphic positions. Adapted from Ahumada and Faúndez (2009)

Type		Water supply	Vegetation architecture	SOM contents	Salts outcrop cover (%)
Bofedal	Saline	Permanent from lagoons, surface runoff and high water table, with permanent soil wet	A cushion, low grass with a growth very compact	High	>5
	Non saline	Permanent from lagoons, surface runoff and high water table			<5
Pajonal	Saline	Lagoons and surface runoff. More restricted summer soil saturation	Caespitose plants, >40 cm tall	Moderate	>30
	Non saline	Lagoons and surface runoff. Summer soil saturation			<30
Vega	Saline	High plasticity, with soil low saturated to completely saturated	Non caespitose and rhizomatic plants, <40 cm tall	Very variable	>20
	Non saline	Lagoons and surface runoff. Soils at least in field capacity during summer			<20

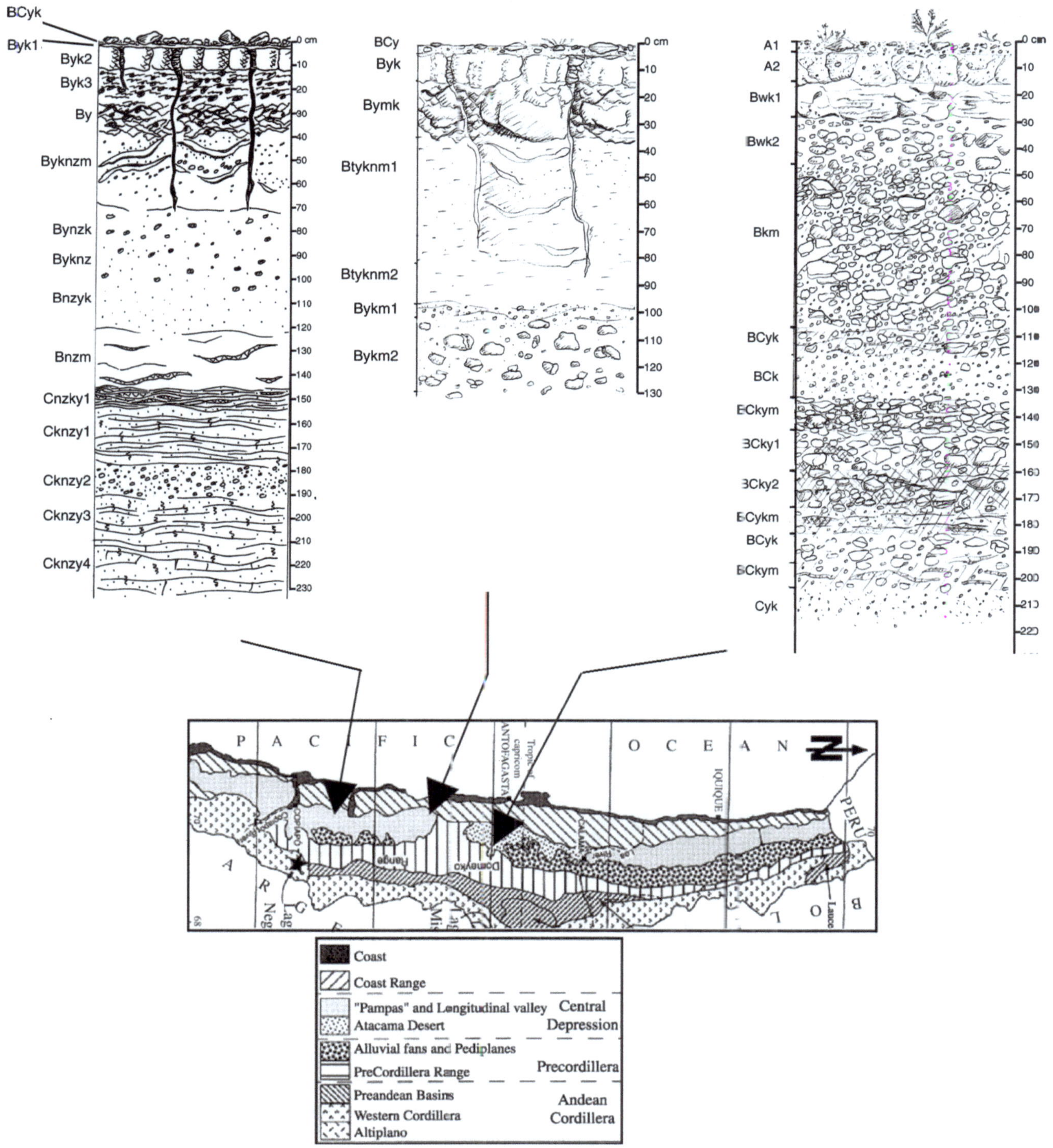

Fig. 2.22 Aridisol profiles from the Atacama Desert (see characteristics in Appendix, page 160) after Ewing et al. (2006) and Amundson (2003). Above, from left to right: Yungay, Copiapó and Altamira pedons

occurs during the southern summer (October–April), and many of the streams dry up completely during the winter. The water provided by these rivers, directly or through the intermediary of groundwater, supports irrigation agriculture along the valley floors, the deltas by the coast or the alluvial fans where the streams emerge from the mountains (Meigs 1966).

Although most soils in these valleys are alluvial, a few colluvial soils or soils with colluvial influence pedons are present. Apart from the Chacalluta Valley near Arica city

Fig. 2.23 Characteristic soil of depositional plains very stratified and with an A–C profile (Luzio et al. 2010)

(SAG 2002), there are only some old soil surveys such as for the Lluta River (Díaz 1958), Camarones Valley (Díaz and Meléndez 1958) and Azapa Valley (Wright and Meléndez 1961). From very general soil descriptions of these surveys, it is possible to conclude that stratified, poorly developed but deep soils occupy several alluvial terraces, with variable salinity and diverse drainage classes, where agriculture is practised.

After *Pampa del Tamarugal*, a continuous desert plateau (crossed only by the Tarapacá basin) in the longitudinal valley (19–22°S), 30–60 km wide and with elevations of 900–1,200 m, and the Loa River emerge in the desert. Occupying 2,430 km^2 in a conspicuous embayment along the Andean front, the Salado River basin is an important source of surface water and groundwater and is the main perennial tributary of the Loa River, the only river in northern Chile to start in the Andes and actually reach the coast.

In general, at the Andes mountain level, narrow valleys with slopes of up to 100 % are found (Fig. 2.26). Most valley bottoms are composed of coarse sediments such as rounded and/or faceted gravel. There are sectors where there has been possible alluvial terrace formation, constituted by sands in a clear graded stratification. In those cases, where slight pedogenesis exists, an A horizon is formed by organic matter accumulation. Indeed, these soils reach the minimum to be classified as Entisols (Ustorthent Great Group). However, at the margin of water streams, common paludisation processes determine the existence of organic or mineral soils (rich in organic matter), despite the thin and poorly developed soil profiles.

Muñoz et al. (2007) found Histosols, Mollisols and Entisols in the Tarapacá basin, the Entisols being derived from coarse sediments with high salt content and elevated pH. Peaty soils and the presence of salts in soil at depth have been reported north of the Loa basin, but to the west sandy

Fig. 2.24 Soil profile at the Pintados salar. Under a 30 cm thick crust, plant roots can be observed (Luzio et al. 2010)

loams to sandy-textured soils with a silty loam layer at depth can be found.

Soils of Valleys with Permanent Watercourses

The exorheic basins begin near 27°S (Region III) in Chile, where rivers of pluvial and snowy regimes run with a general East–West orientation (Fig. 2.27). At this southern portion of the Hyper-arid to Semi-arid zone, more detailed soil surveys have been carried out for main valleys corresponding to Region III (Copiapó and Huasco rivers; CIREN 2007) and Region IV (Elqui, Limarí and Choapa rivers; CIREN 2005a).

The Quaternary continental and coastal deposits along the Copiapó River, the first river of this group, are associated with morphostratigraphic units such as alluvial fans, dunes, fluvial and marine terraces (Fig. 2.28). According to CIREN (2007), along the Copiapó Valley, the main soil series are classified principally as Typic Haplocambids, Typic Haplocalcids, Typic Haplosalids and Sodic Haplocambids.

An incipient geomorphic development can be observed at the upper and medium portions of this valley with abrupt hillsides and without meanders and dominated by coarse materials (fine sands to medium gravel), although locally fine sediments have been deposited. Apacheta (piedmonts) and Amolana (terraces) soil series (Typic Haplocambids) characterise these sectors (Fig. 2.29). Other less developed colluvium soils coexist in this sector, with those used for table grape production showing weak soil structure, abundant coarse fragments and some profile HCl reaction.

The landscape in the lower section of this valley, i.e. from Copiapó city to the coast, is open, with fine-textured soils in remnant terrace positions, which are rich in soluble salts (EC can reach 85 dS m^{-1}) and carbonates (pH 8.0–8.8) throughout their deep and stratified profiles. Most soils present a darker Ap horizon, where the highest values of chemical properties are expressed, and then decrease below ~20 cm depth. As in the upper part of the valley, less developed colluvial soils are found and are considered skeletal. A representative soil in this sector is Ramadilla soil series (Sodic Haplocambid).

Therefore, considering the presence of coarse sediments in the upper part of the valley and fine materials in the lower part, the principal constraints of these Aridisols are stoniness, drainage in some cases, salinity, alkalinity and sodicity. However, most soils are included in I–IV LCC, where table grape production is an important activity.

The Huasco River runs through the second most important permanent valley of the Hyper-arid to Semi-arid zone, where the main soil series are classified as Typic Torriorthents, Aquic Torriorthents, Xerollic Haplargids, Xerollic Camborthids, Xerollic Torriorthents, Typic Natrargids and Xerollic Calciorthids (CIREN 2007). Like the Copiapó Valley, the Huasco Valley opens (~1.5 km) near Vallenar city (~28°S) showing at both river margins alluvial terraces, differentiated in at least three levels (higher, intermediate and lower recent terraces). In addition, some soils in the piedmont position are present at intermediate and high levels of this valley with stony soil profiles, coarse-textured and resting over an alluvio-colluvial substratum. Variable but not excessive slope gradients are observed.

Very stratified and poorly developed soils occupy the lower plain alluvial terraces of this valley. These coarse to medium-textured stony soils have drainage and salinity constraints and are classified as Aquic Torriorthents (soil series Paona and Bellavista).

Medium or intermediate terraces include more productive soils of this valley, with good drainage, and also stratified and medium to fine-texture classes. According to

Fig. 2.25 *Left* valleys with intermittent rivers in the extreme north of Chile. *Triangles and circles* represent volcanoes of different ages (partially from Schlunegger et al. 2010). *Right* fog or *camanchaca* in Lluta Valley

Fig. 2.26 Lluta River and eight sections (*left*) along the watercourse, described by Kober et al. (2009)

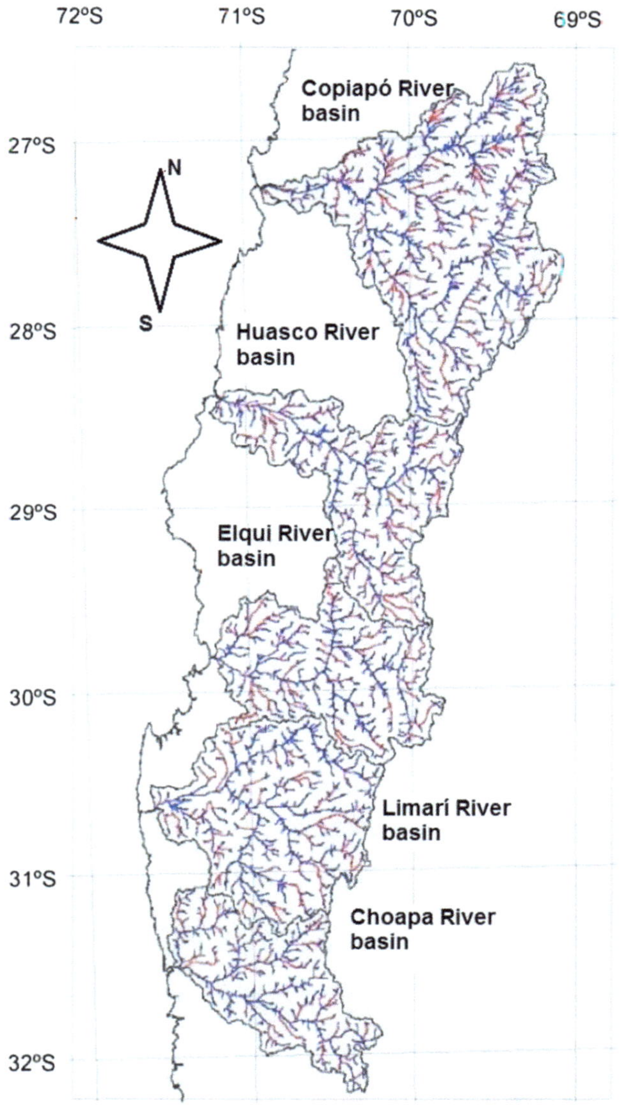

Fig. 2.27 Principal valleys with permanent watercourses. Regions III and IV of Chile

CNR (1993), the Chanchoquín soil series (Xerollic Camborthids) is very representative of these soils.

The highest level of these alluvial terraces shows at least three different types of soil profile. Those with exiguous development, well-drained plains and coarse-textured over a coarse alluvial substrate are represented mainly by the soil series Tatara (Typic Torriorthents). Some soil profiles include a shallow calcareous hardpan (locally named tertel) or only $CaCO_3$ in the case of Xerollic Calciorthids (soil series Ventana). Other calcareous soil profiles, but without tertel, coexist in these high level terraces, e.g. the Cavancha soil series.

Figure 2.30 displays characteristic soil profiles of the three pedons selected as examples here.

The Elqui River represents the next important permanent watercourse in this zone. A nival regime drives this basin and it is expected to be among the ecosystems most affected by climate change. Glacial or fluvio-glacial processes determine, in the highest portion, deep and narrow valley formation, which added to strong winds and reduced vegetal soil cover commonly, triggers landslides, debris flows and avalanches. However, organic wetlands (*vegas*) are frequent in less inclined sites. Alluvial soils on recent and remnant terrace positions (3–6 km wide) are the most representative soil in the lower and medium portions of this valley, respectively (Fig. 2.31).

More developed soils on older terraces have been also described, which are associated with slopes of the valley and can be confused with piedmonts and colluviums, despite the alluvial characteristics present in their profiles. The common A–C horizon sequence can reach a certain development expressed by a cambic horizon within the soil profile (A–B–C).

At the lower terraces, the soils are thin (<40 cm depth), loamy to loamy sand at the surface (A horizons) and abruptly changing to a C horizon constituted by rounded gravel and stones (~80 %), interstitial sands and fine gravels. In general, soils with a well drained and plain topography are included in II and III LCC, although some soil phases can be slightly undulating and others stony and thin and classified as VI LCC. These poorly developed soils are taxonomically classified as Typic Torriorthents (Chapilca soil series) and Typic Torrifluvents according to CIREN (2005a).

However, more developed Typic Haplocambids can be found in similar geomorphic positions with an A–B–C profile and 60–90 cm depth over a sandy and very gravelly alluvial substrate. These are mostly classified as II and III and some cases as IV when imperfect drainage is verified.

In the coastal sector, alluvial soils on marine terraces and in the valley bases have evolved from both continental and marine sediments with both terrace types and a long beach framing the bay of Coquimbo.

The Limarí River drains into a basin along a stretch between 30°S and 31°30′S where, apart from the A–C and A–B–C soil horizon sequences observed in other valleys, A–B_t–C and vertic B_{ss} horizons have been described (Aburto et al. 2008a, b).

Soils with A–C profiles are present in alluvial terraces or occupying colluvial positions with variable slope gradient (1–3 % to 15–20 %). Slightly deep (<60 cm) and well drained, these soils rest over a substrate of angular and subangular gravel and stones (~40 % in volume) with a fine-textured matrix. Surface horizons range between sandy clay loam and clay loam, with a subangular blocky structure in contact with the substratum (massive). Organic carbon content is <2 % at surface and ~0.2 % at depth. The topographical variability determines that these soils be

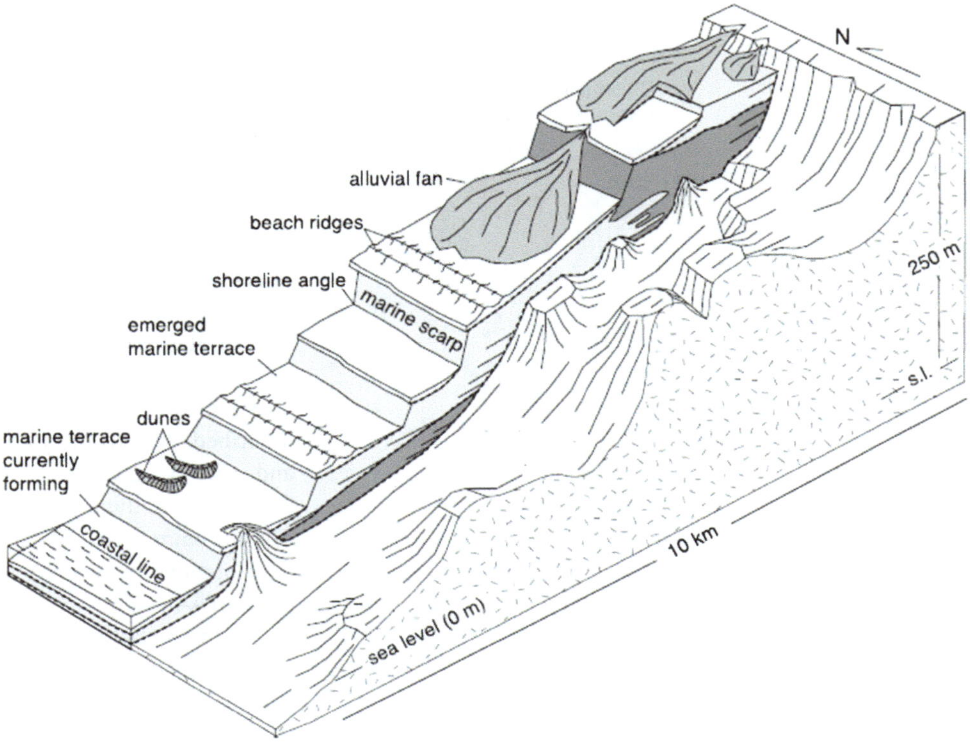

Fig. 2.28 Schematic sedimentary units along the Copiapó Valley (Marquardt et al. 2004)

Fig. 2.29 Characteristic pedon (Typic Haplocambids) in the upper part of the Copiapó Valley. Apacheta soil series (Luzio et al. 2010)

included in III–VII LCC. The pedons classified (see Appendix, Huatulame soil series) clearly represent the characteristics of this soil group.

Soils with B horizons (generally Haplocambids) are found on slightly sloping alluvial terraces (2–5 %). Also slightly deep, these soils rest over a gravelly substratum with sandy to sandy loam matrix. A strong blocky structure dominates, OC content is <2 % at the surface, decreasing regularly with depth. Some profiles (Huamalata soil series) have EC < 5 dS m^{-1} at the surface. LCC varies between II and IV, mainly with drainage conditions (imperfect to poor).

Other soils with a depth of around 90 cm are present in well-drained colluvial and piedmont positions, with slope gradients from 5–8 % up to 20–40 %. Below the strongly structured clay loam-textured A horizons, the clay content increases significantly, showing advanced pedogenetic evolution, as evidenced by development of an argillic B_t horizon. Due to topography, these are classified in III–VII LCC. Taxonomically, the Typic Haplargid (Serón soil series) represent good example of this group of soils.

Coexisting with the soils described above, it is possible to find deep Vertisols in high alluvial terraces of the Limarí Valley, differentiated as Petrocalcic Calcitorrerts (Tuquí soil series, CIREN 2005a) with a subsoil rich in $CaCO_3$, and Sodic Haplotorrerts (San Julian soil series, Aburto et al. 2008a) with elevated exchangeable Na and EC values and gypsum

Fig. 2.30 Characteristic pedons on alluvial terraces of the Huasco Valley (CIREN 2007). **a** Paona soil series (Aquic Torriorthent), **b** Chanchoquín soil series (Xerollic Camborthid) and **c** Cavancha soil series (Xerollic Haplargid)

content. Cutans and slickensides are common in both soils, which are well drained and have been included in III–VI LCC, depending on soil profile depth and/or slope (Fig. 2.32).

The Choapa basin (including the Choapa and Illapel rivers), like the above valleys, also includes alluvial soils in terrace positions and colluvial pedons in piedmont positions (Fig. 2.33). The age of terraces varies and colluvial material derives from granitic rocks of different degrees of weathering. Compared with the other valleys analysed, all pedons observed here present an A–B–C horizon sequence with cambic horizons, denoting greater pedogenic maturity.

Alluvial soils can exceptionally reach slope gradients of 5–8 % and commonly <3 %. A gravelly substratum with coarse interstitial material is frequent and, although surface horizons are fine-textured (silty clay loam and sandy clay loam), subsoil can be finer (clay loam, silty clay or clayey). The well-structured subsoil can reach depths between 55 and 110 cm. Organic carbon content is <2 % in A horizons, decreasing regularly with depth (~0.3 %). Only the Tunga soil series contains free $CaCO_3$ (2.9 % at the surface and 9.3 % at depth). Drainage classes and surface coarse fragments represent criteria to classify soil phases into I–IV LCC. However, all this group of soils (Choapa, Huintil, Illapel, Quelén, Tranquilla and Tunga soil series) correspond to Haplocambids (Fig. 2.34).

On the other hand, colluvial soils coexist in the valley with slope gradients between 1–3 % and 2–5 % in plains situations or 15–20 % and 20–30 % in scarped positions. Granite mainly comprises the substrate in different weathering states or granitic sands (maicillo). Clay at depth follows a similar trend to that in alluvial soils and the structured solum varies between 50 and 80 cm. The OC content is higher than in those soils (2 and 4 % in surface), but also undergoes a regular decrease to 0.3 % at depth. Although carbonates are absent, EC varies between 0.4 and 4.9 dS m^{-1}. Due to their topographical position in the valley, all these soils present good drainage and depending on stoniness, rock outcrops and inclination, are classified from III to VII LCC. Taxonomically, they also belong to the Haplocambids except the Tahuinco soil series, which is classified as a Typic Paleargid that includes a B_t horizon. Other soil series of these colluvial soils are Camisas, Cuncumén, Chillepín and El Tambo.

After less prominent rivers (La Ligua, Petorca and Putaendo) at the extreme south of the Hyper-arid to Semi-arid zone, the Aconcagua River is the largest mid-latitude river of Chile, and its characteristics are typical of the temperate latitudes of western South America. Located at the boundary with the Mediterranean zone of Chile, it drains the ice- and snow-fields of the Andes and Coastal Plain. The river basin is located about 50 km north of Santiago and the upper watershed has several peaks above 5,000 m a.s.l. and vegetation is sparse. In the lower watershed, intensive agricultural production depending almost entirely on irrigation takes place. The Aconcagua River crosses important cities at Region V of Chile (Los Andes, San Felipe and Quillota), ending in the Pacific Ocean near Valparaíso.

Soils in recent and remnant alluvial terrace, colluvium (piedmont) and sedimentation basin positions have been described. The former are less developed than other soils in the valley; coarse to medium-textured, stony along soil profiles and resting over a substratum with rounded coarse fragments. However, according to CNR (1993), the soil series most representative are Chagres and Hualcapo, both

Fig. 2.31 Highest sectors of the Elqui Valley, with intense agriculture use (Luzio et al. 2010)

Fig. 2.32 Vertisols in alluvial terraces of the Limarí Valley, San Julián soil series (Sodic Haplotorrert) and Tuquí soil series (Petrocalcic Calcitorrert)

classified by CIREN (1997a) as Mollisols. Most of these soils are included in III and IV LCC due to coarse fragment amounts, depth and drainage.

On the other hand, remnant terraces host the best and more productive soils in the valleys formed on alluvial plains or mudflows reshaped by watercourses. Deep, well-structured and drained fine-textured profiles are characteristic. The Pocuro soil series (Fig. 2.35) is a clear example of these soils, also classified as Fluventic Haploxerolls but included in I and II LCC, although it is frequently possible to observe a plough pan.

Colluvial soils in piedmont positions present a variable slope, stoniness and depth, but are well drained (LCC: III and IV). Mainly classified as Inceptisols, some pedons are included in order Mollisol (CIREN 1997a). As an example, only the Ocoa soil series is included here (see Appendix).

Some soils in the valley are derived from fine sediments evolving under excessive soil water contents (lacustrine

Fig. 2.33 Illapel River descending from the Andes Mountains, Region IV of Chile. Alluvial terraces and piedmonts are apparent

Fig. 2.34 A Typic Haplocambid (Quelén soil series) in the Choapa Valley, Region IV of Chile (Luzio et al. 2010)

conditions). Situated in low and depressed landscape positions, they are fine-textured and show imperfect to very poor drainage and elevated $CaCO_3$ contents. In general, taxonomically, these soils have been classified as Inceptisols, Mollisols or Histosols (Typic Xerumbrepts, Petrocalcic Palexerolls, Typic Calcixerolls, Entic Haploxerolls, Typic Medihemists or Fluvaquentic Endoaquolls) and the Palomar soil series is included here as an example (see Appendix).

2.2.1.4 Soils of the Coastal Range

The Coastal Range or Coastal Cordillera in northern Chile, which is 10–50 km wide, is a prominent forearc, a morphological feature in northern Chile that runs parallel to the coast. The most remarkable feature is a very steep, nearly 1,000–2,000 m, coastal cliff that plunges 45° into the sea in the extreme north. Cut by a series of faults, it is characterised by high relief that contrasts with the smooth, duricrusted surfaces further inland, which is probably due to strong salt weathering under the influence of periodic coastal fogs (*camanchaca*). According to Juez-Larré et al. (2010), uplift and delineation of the range as a separate morphotectonic unit (i.e. distinct from the adjacent Central Valley) is considered to have taken place in the Oligocene and continues to the present day.

The western margin of the range comprises the coastal scarp—a prominent break in slope—that extends for 900 km along the coastline. Although the origin of the scarp has been the subject of some debate, the absence of active fault scarps and the irregular morphology of the scarp

Fig. 2.35 Pocuro soil series, a Fluventic Haploxeroll (CIREN 1997a), near Los Andes city, Region V of Chile

suggest that it represents a weathered paleo-cliff line. This plunging paleo-cliff most likely developed through marine erosion during post-mid-Miocene uplift. Onlap of Middle to Upper Miocene marine sediments onto the scarp can be observed in a number of places between 21 and 27°S in northern Chile.

The top of the Coastal Range is an ancient, well-preserved erosive surface and testifies to the extremely arid conditions that prevailed in the Atacama Desert since the Early Pliocene. Such a surface is believed to result from intense erosion of the northern Chilean margin during past time periods. The eastern side of the Coastal Range is a transitional zone where a smooth topography marks the gradual boundary of the Coastal Range with the Central Valley.

Along the Coastal Range of the Hyper-arid to Semi-arid zone, soils can be separated into two broad types according to development of soil profiles: A–C–R and A–B–C. The former are skeletal sandy or sandy loam soils, in hillside positions, stony at the surface and with frequent rock outcrops, over fresh or weathered rock at least 50 cm deep. Soil organic matter content in the A horizon is <1 %, base saturation (BS) is low (<8 %) and pH varies between 6.2 and 7.0. Soil salinity is also low (EC < 1 dS m^{-1}). These Torriorthents (Luzio et al. 2010) are frequently included within LCC IV and VI.

Luzio et al. (2010) describe a Coastal Range soil at 25°S (Taltal, Region II) in a scarped landscape without terrace transition, where the slope gradient is over 20 % with abundant coarse angular fragments (~80 %) at the surface and at depth, and common rock outcrops (Fig. 2.36). The thin superficial A horizon shows a slight accumulation of carbonates and an abrupt boundary with underlying B or C horizons. A coarse-textured pedon is reported by these authors with high amounts of organic matter and soluble salts in surface layers.

In contrast, in Region IV of Chile, some soils of the Coastal Range (slope 6–12 %) have developed cambic and argillic horizons, although rock outcrops and surface stones are common. Despite being granitic residual soils, the complete profiles described are probably influenced by colluvial processes and marine sedimentation. Sandy loam and sandy clay loam textural classes have been described in surface, B_w or B_t horizons and are fine-textured (clay loam to clay) and strongly structured (subangular blocks and prismatic). Soil pH varies between 6.1 and 7.7, OC content reaches 2.0 % at the surface and 0.3 % at depth. Soils with a cambic horizon have been classified as Camborthids or Haplocambids (Mincha soil series) and those with an argillic horizon as Paleargids (Luzio et al. 2010).

In conclusion, in the western Hyper-arid to Semi-arid zone, only two continuous longitudinal morphological units are recognised: the Coastal Plain and the Coastal Range. In the majority of the north of Chile, the Great Coastal Escarpment separates the Coastal Cordillera from the Coastal Plain. The Coastal Plain, with 3 km average width, contains sedimentary deposits and abrasion marine terraces, partially covered by alluvial fans, which extend from the present coastline to 300 m a.s.l. (Paskoff 1989). However, between the cities of Arica (18°S) and Iquique (20°S), virtually no marine terraces are preserved. In many places south of Iquique, there is a narrow coastal plain, 1–3 km wide, that extends south beyond Chañaral (~26°S, Region III). This plain broadens to the north of Antofagasta, at the structural block of the Mejillones Peninsula.

2.2.1.5 Soils of the Coastal Plains

The northern coast of Chile includes the most emerged parts of the Southern Central Andes forearc closest to the Pacific Ocean trench. The pervasive Hyper-arid climate conditions along with the good preservation state of the morphostratigraphic (i.e. marine terraces) and morphostructural (i.e. fault scarp) records allow studies in neotectonics. However, southward of La Serena (Coquimbo), the morphological signs are more difficult to interpret, probably as a

Fig. 2.36 Characteristics of a northern Coastal Range soil profile with colluvial coarse fragments comprising almost 80 % (Luzio et al. 2010)

result of the combined effects of greater complexity and greater dissection by the wetter climate (Regard et al. 2010).

Morphogenetic inheritance of plate tectonics and of the marine cycles associated with the Holocene highstands has been important, generating a coastline of bays which have marine terraces and beach-ridge plains. Paskoff (1989) in a sectorisation of Chilean coastal plains in the Hyper-arid to Semi-arid zone, describes their littoral features from 18 to 25°S and from 25 to 33°S, which are denoted as desert and semi-arid coast, respectively.

The escarpment of the Coastal Range (18–25°S), which is 700 km long and 700 m a.s.l. on average, is largely inactive from Iquique city (20°13′S, Region I) to the south with Pleistocene marine terraces exposed at the foothills attesting to formerly intense marine degradation of the coastal cliff profile, which is the result of marine erosion acting on an actively uplifting coastline since the Pliocene (González et al. 2003). North of Antofagasta (23°38′S, Region II), cliffs are cut into horizontally bedded Tertiary sandstone (Fig. 2.37).

Over the terraces, large alluvial fans were formed during the rainy Quaternary by streams of the Coastal Range. Although the escarpment is from the end of the Miocene, a recess as a result of marine erosion during a major transgression occurred in the Upper Miocene/Lower Pliocene. A narrow continental platform borders an aligned coast with a marked N–S orientation. Only the Mejillones peninsula (23°15′S) which is 55 km in length and 20 km in width breaks this regularity. It consists of N–S orientated uplifted and downfaulted blocks, with a tectonically emerged coastal platform developed on sandstones, shales and coquinas. Flores-Aqueveque et al (2010), characterising wind erosion processes over the flat geomorphology of the northern portion of this peninsula, also observed marine terraces of Pleistocene age (highest terraces) and mineralogical composition of moving particles similar to that of the soils, with quartz, feldspar and calcite as the most important minerals, followed by clay minerals, gypsum and amphibole. The ages reported indicate that substantial uplift has affected the peninsula.

From 25 to 33°S, the coastline is mainly rocky and lower (150–200 m a.s.l.), behind which extend abrasion marine terraces, stepped up to the first spurs of the Coastal Range. Resulting from the interference between sea eustatic fluctuations and the upthrust trend of the coast, there is general recognition of the Quaternary marine terraces along 110 km of the coast between 27 and 28°S. Uplift of these terraces, within the Chilean flat-slab sector, has been attributed to subduction of the oceanic Juan Fernández ridge beneath this continental section (Le Roux et al. 2005, 2006).

Detailed studies by various authors (Martínez et al. 2011; Pfeiffer et al. 2011; Saillard 2008; Quezada 2007; Marquardt et al. 2004) have been carried out in bays such as Caldera (Region III), Coquimbo-La Serena and Tongoy (Region IV) and Con Con (Region V). Talinay and Huentelauquen, both in Region IV of Chile, are the locations where different levels of marine terraces (Fig. 2.38) have been studied (Saillard et al. 2009; Aburto et al. 2008b; Le Roux et al. 2006).

Pedological studies on marine terraces are concentrated mainly in Region IV (Luzio et al. 2010), where it is possible to establish some general differences between soils. Lower and intermediate marine terraces include (a) soils without profile development, (b) indurate soils (tertel and petrocalcic horizon), (c) soils with natric horizons and (d) sedimentary soils. More pedogenetic development is observed at some intermediate and higher terraces, as evidenced by cambic horizons, and in some cases, argillic horizons.

Undeveloped Soils

Although dunes are present along the coast of Chile, between 30°00′ and 31°45′S (Los Vilos) and near La

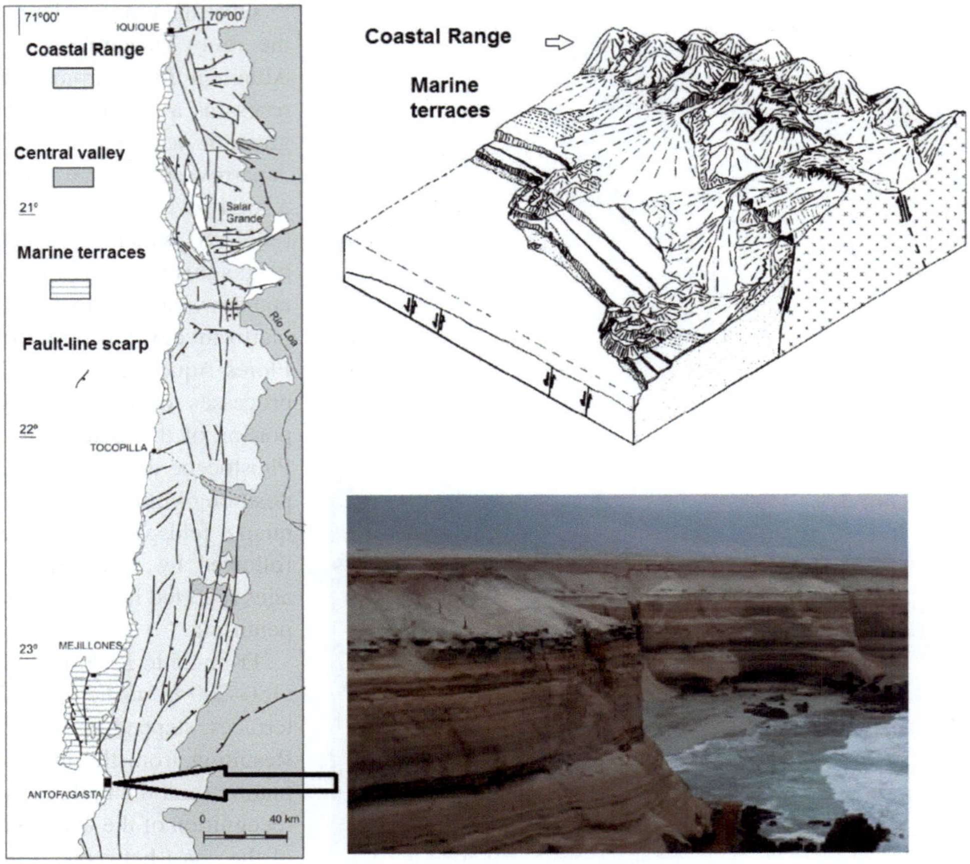

Fig. 2.37 Marine terraces and coastal cliffs in Region II of Chile

Serena city, stabilised dunes that have resulted in very scarce soil development can be distinguished by the absence of diagnostic horizons. Slope gradient is <6 % with some rock outcrops, but without stoniness at the surface or in the profile. The A–C horizon sequence is coarse-textured (sandy, loamy sand until sandy loam) on deep profiles (~ 150 cm) and is common, in some cases, over marine terraces (Aburto et al. 2008b). With an assumed granitic origin, these sandy soils are characterised by a pH between 6.3 and 8.0 and surface organic matter content <2 %, which suggests soil stability. Some pedons show relict redoximorphic features (iron nodules and laminar concentrations) formed on depressive positions or by probably more humid conditions than today's climate (Fig. 2.39). Classified as Torripsamments, LCC varies between II and III, but some soil phases correspond to Class IV due to complex slopes (5–8 %). A characteristic pedon of this soil group found near La Serena city is La Compañía soil series.

However, in the southern part of this Hyper-arid to Semi-arid zone, other stabilised dunes (paleodunes) displaying more advanced development occupy the highest marine terraces (CIREN 1997a). The Longotoma soil series, an Entic Haploxeroll (32°23′S), is a good example of this pedogenic evolution.

Indurate Soils

These limitative soils can be found on marine terraces and at colluvial positions and coastal piedmonts. They are slightly deep and thin soils, with a substrate consisting of a layer of hard/extremely hard consistency which in some cases consists almost entirely of carbonates and in others material, but the cementing agent is a mixture of silica and manganese without reaction to HCl, so that locally in both cases, a *tertel* denomination is used.

Nevertheless, for soils with calcareous tertel it is necessary to distinguish between those with a calcareous substratum pedogenically not related to formed soil and soils with a petrocalcic horizon (e.g. B_{km} or C_{km}) directly vinculated to soil formation processes. Soil pH is high (>8) along profiles, OC decreases regularly with depth (1.5–0.3 %) and EC is maintained below 1 dS m^{-1}. These soils belong to LCC, commonly III and IV, but exceptionally are classified as VI or VII. The Cerrillos de Elqui (Calcid Petrocalcid) soil series includes pedons representative of this soil group.

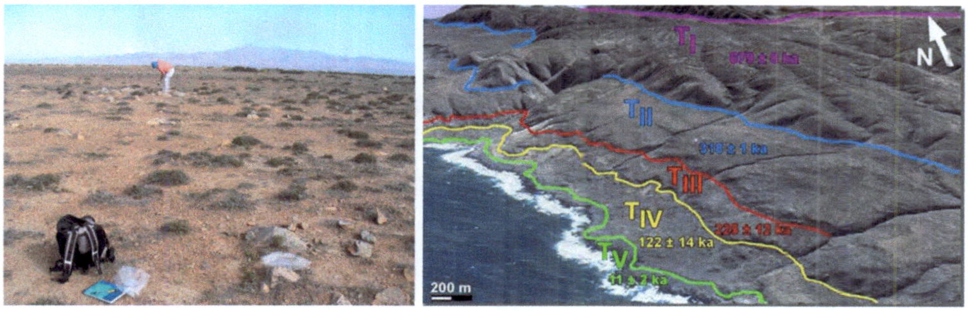

Fig. 2.38 Distribution of abrasion marine terrace levels at the coast of Region IV of Chile (Saillard 2008)

Fig. 2.39 Stabilised dune (Typic Xeropsamment) with certain pedogenic development at Los Vilos (Region IV)

Recently, Pfeiffer et al. (2011) described an aeolian sand cover deposit that also underwent pedogenic processes overlying calcrete. Aburto et al. (2008a) reported similar soils in the same paleobay (Tongoy), but classified these as Xeric Petrocalcids (Fig. 2.40).

Soils with Natric Horizons

Natrargids, characterised by the occurrence of a natric horizon but lacking a duripan or petrocalcic horizon within a metre of the ground surface, have been observed over marine terraces at Huentelauquén (31°34′S). These soils are present from a slight slope gradient (2–5 %) to gently rolling topography (5–8 %), where sodicity features include a strong prismatic or columnar soil structure and surficial spots of dispersed organic matter (Fig. 2.41). Fine-textured soil profiles can show cutans and redoximorphic features (Fe–Mn concretions). Within the natric horizons, exchangeable sodium percent (ESP) varies between 21 and 65 %. Pedon Huentelauquén, classified as a Typic Natrargid (CIREN 2005a), represent this group but other profiles have been characterised later as Xeric Natrargids (Aburto et al. 2008a).

Sedimentary Soils

These soils on marine terrace positions occur at the transition of the Hyper-arid to Semi-arid zone with the Mediterranean zone, at Region V. In general, they are classified as Alfisols and are coarse-textured at the surface and fine-textured at depth. Most soils show a drainage class depending on topography and resting over different substrates (granite/quartzite sandstones cemented or partially cemented with silica, or gravels with a clayey matrix). The Tabolango soil series is an example of this soil group. Other members of this group have been classified as an Ochreptic Haploxeralf (Mantagua soil series), Udollic Albaqualf (Catapilco soil series) and Ochreptic Haploxeralf (Chilicauquén soil series).

2.2.1.6 Soils of the *Serranías*

Serranías is the local physiographical denomination for a portion of Chilean territory where the characteristic geomorphic continuity (Andes Range—Longitudinal Central Valley—Coastal Range) is broken and active volcanism does not exist, and which coincides with the tectonic Chilean flat-slab sector.

Knowledge of soils in this sector is very limited, because most soil surveys have been conducted in valleys where agriculture activities first developed, but today are occupied by intensive fruit production (Fig. 2.42). The *serranías* sector, 800–1,000 m a.s.l., is interrupted by some plains of rolling topography and by some of those valleys with permanent watercourses (principally Copiapó, Elqui, Limarí, Choapa and Aconcagua) described above.

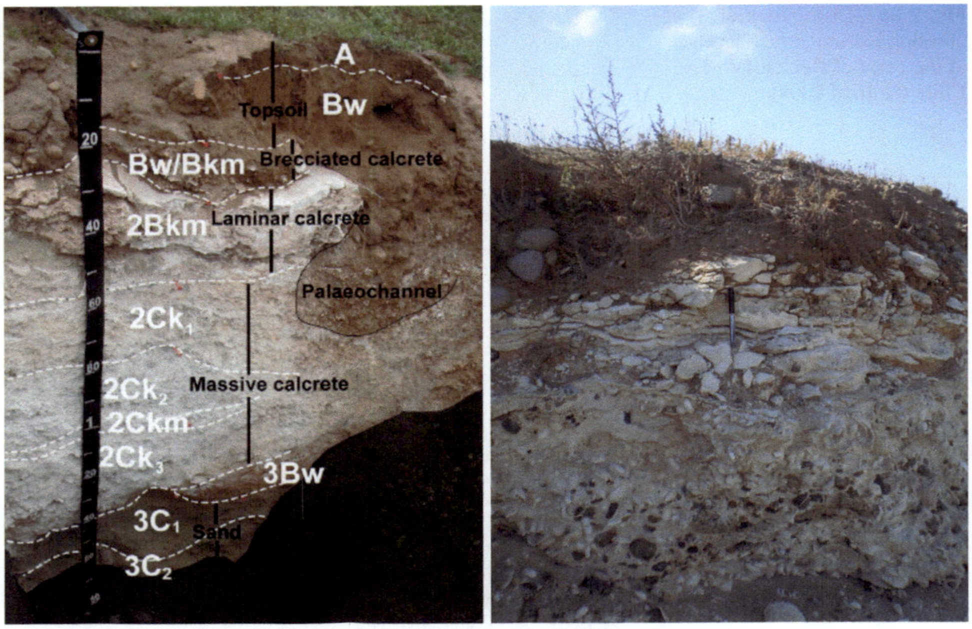

Fig. 2.40 Pedons on marine terraces at Tongoy paleobay (Pfeiffer et al. 2011; Aburto et al. 2008a), Region IV of Chile

Characteristic relief includes moderately hilly (8–15 %) and hilly (20–30 %) sectors, among which depressions are not enough to constitute valleys in a geographical-geomorphologic sense.

Residual soils from granite with diverse degrees of weathering are normally developed. Although it is common to find cambic horizons, few profiles develop argillic horizons and it is possible to observe soils derived from calcareous materials only locally (A–C_K profiles).

Granite-derived soils reach between 60 and 130 cm depth, depending on erosion process rates, which in these sectors are moderate to severe. Upper horizons present a variable textural class and varying gravel content, but clay content increases with depth, so clay loam or clayey B horizons have been described (Fig. 2.43). However, this clay content increase is associated with in situ transformation without illuviation processes. Most soils described in these situations present a cambic horizon (B_w), so the soil group has been classified as Haplocambids, and a pedon of the El Tambo soil series is considered representative.

More developed soils, classified within the Haplargid and Paleargid Great Groups, coexist with less developed soils. Piedmonts or alluvio-colluvial positions, with slope gradients from 2–5 % up to 8–15 %, and reaching between 50 and 90 cm in depth, rest over gravel to boulder substratum or granite weathered to different degrees. Illuviation features (cutans) and the presence of B_t horizons indicate advanced pedogenesis. Morphologically, these constitute a very homogeneous soil group, because they are fine- and very fine-textured, with strong and coarse blocks or prisms. The Combarbalá soil series is characterised here (see Appendix) as a representative pedon of this soil group. Figure 2.44 shows the landscape and soil profile of the Pintacura soil series, which also belongs to this soil group.

Another group of soils in this sector is those overlying calcareous materials (Fig. 2.45). Due to the absence of pedogenic studies, it is not possible to accurately define the processes that have occurred in these soils, and thus determine whether subsurface calcareous materials have served as parent material. These soils are observed at slope gradients that can reach 30 –50 %, particularly in places such as Vicuña (30°02′S) and Punitaqui (30°50′S). Very incipient soil profile development is observed and generally A–C (C_K) horizons have been described with only in few cases a B_w horizon. These Typic Haplocalcids have medium-textured profiles and are not suitable for agriculture. The Marquesa soil series is presented as an example of this soil group.

Southward of this wrinkled (*serranías*) landscape, morphological and geological features change abruptly at Santiago city (33°27′S). As the dipping of the subduction zone again reaches higher values, volcanism and forearc alluvial basins (the Chilean Central Longitudinal Valley) reappear and extend to latitude 41°28′S at Puerto Montt city, where the sea has entered into the Central Valley.

2.2.2 Soils of the Mediterranean Zone

During the Quaternary, the relief was defined in the current zone, disappearing the transverse valleys and defining the physiography in a north–south direction, from the Andes, the Longitudinal Central Valley, the Coastal Range and the Coastal Plains in an east–west direction.

Fig. 2.41 Sodicity features of the Huentelauquén pedon on marine terraces of Region IV

Fig. 2.42 *Serranías* in Region V (La Ligua, *left*) and Region IV (San Lorenzo, *right*)

The crestal elevation of the Andes and Coastal Range in this zone decreases significantly. The latter reveals a gradual elevation decrease from approximately 5,000 m at 33°S to only 2,000 m at 42°S. The materials brought from the mountains, either by ice, lahars or alluvial fans, were deposited mainly in the Central Valley, defining broad valleys where the current rivers flowed out and giving rise to the parent materials of soils. The geology of the Coastal Range south of 33°S is marked by abundant, primarily low-grade metamorphic rocks (especially from ~38°S to the south).

Fig. 2.43 Landscape and pedon of granite-derived soil in the inner *serranías* (Region IV). A B_w horizon developed without illuviation can be seen (Luzio et al. 2010)

2.2.2.1 Soils of the Andean Mountains

Geographically, this sector comprises an area bounded by an imaginary line from north to south, roughly along the 72° meridian. There are few references of these soils, usually due to low agricultural use and the limited accessibility of most sectors.

Non Volcanic Soils

North of parallel 36°S there have been no detailed soil surveys, but overall pedogenic processes have not been strong enough to give rise to soils with a clear differentiation of horizons. Coarse-textured skeletal soils with abundant angular stones and an andesitic and basaltic origin dominate. The best taxonomic approach to these soils is to consider them Entisols (Luzio et al. 2010).

The steep slopes (>30 %), the low vegetation cover (dominated by *Stipa* spp. and *Festuca* spp.) and the irregular regime of rainfall lead to these soils being considered to fall within VIII LCC.

Volcanic Soils

According to Moreno and Varela (1985) there was intense volcanic activity in the Andes of Chile during the Quaternary, with considerable volumes of materials ejected at the surface (Fig. 2.46). Much of the finer pyroclasts (volcanic ash and dust) were transported to Argentinian territory by the action of winds from the west (Bertrand and Fagel 2008). However, it is necessary to consider that this area has a large influence of easterly winds (locally called *Puelche*), which had the effect of depositing tephras in the valleys and

Fig. 2.44 The Pintacura soil series, a Typic Haplargid, in Region IV of Chile (Luzio et al. 2010)

Fig. 2.45 Soil over limestone near Punitaqui, Region IV (Luzio et al. 2010)

flanks of the Chilean Andes, from where the deposits extended to the Central Valley in the south.

From, approximately, Chillán city to the south (37°S), between 600 and 1,400 m a.s.l., there is a dominance of soils derived from coarse and medium grain size tephras (pumice, cinder and fragmental material), with ages ranging between 600 and 14,000 years (Pino et al. 2004). These soils are deep and moderately deep, with textural class varying between coarse sandy loam and silty loam. At the surface the colour is often brown, of 10YR hue, and can reach a yellowish brown of 7.5YR hue at depth. Most of the soils have scoriaceous volcanic gravel as substrate and an abrupt boundary, with particle size varying between 2 and 6 cm in diameter. The slope is between 30 and 50 %. Permeability is moderate to fast and drainage is good to excessive.

Even though not all the soils derived from volcanic materials have evolved to Andisols, the 36°S parallel marks a natural limit beyond which rainfall has allowed this evolution. In southern parts of xeric-thermic and udic-mesic areas, the evolution of the volcanic material has been enough to qualify some of these soils as Andisols, with bulk density (Db) <0.9 Mg m^{-3}, P retention >85 % and $[Al_{ox} + ½ Fe_{ox}] > 2$ %. The OC content is elevated at the surface, reaching values of 9.5 %, with a gradual decrease to approximately 2.3 % at 120 cm depth. In some soils, the effective cation exchange capacity (ECEC) is <2 $cmol_c$ kg^{-1}, indicating a low availability of exchangeable cations and a low amount of Al^{3+} (KCl).

The use of these soils is limited due to P retention and slope (20–30 %), even though the soil may be deep (Fig. 2.46). Their agricultural potential is limited by susceptibility to

Fig. 2.46 Sequence of paleo-soils in a deep and steeply sloping deposit of volcanic ash close to the Lanin volcano

erosion, being classified within IV, VI and VII LCC, predominantly suitable for forestry, especially native species. Areas with slope gradient >50 % are considered within VIII Class. Los Nevados soil association (Acrudoxic Hapludand), is considered representative of the soils in this area.

In udic-isomesic areas, it is possible to find a first group of soils (among 900–1,200 m a.s.l.) with steep relief and coarse particle size distribution, all belonging to cinder and pumice families, which are alternating layers of pumice and sand. Most of these soils, because of the steep slopes, are only suitable for forestry and present a dense cover of native forest.

A second group of soils, also in steep positions, are characterised by their marked thixotropy. The tephras have been deposited in discordance on basaltic rocks without weathering (Fig. 2.47). With a particular disruption, such as an earthquake, the soil behaves like a liquid, causing landslides that are facilitated by the steep slopes and the discordant contact with the underlying rock.

When landslides occur due to thixotropy, vegetation cover is not able to sustain the soil, which is considered extremely susceptible to erosion and classified in VII and VIII LCC. Los Riscos soil series (Acrudoxic Hydrudand) is representative of this group of soils.

A third group of soils in the udic-isomesic area of the Andes Mountain are Andisols without thixotropy. These are moderately deep to deep soils, some of them over pumice substrate with different degrees of weathering. All are well drained, with slopes varying from 15 to 30 %. Textural classes are medium and fine, but some soils show stratification with coarser materials, as a result of successive tephra deposition. They correspond to Hapludands, with ECEC <2 $cmol_c\ kg^{-1}$, which are identified into the Acrudoxic sub-group. The main limitation of these soils is the steep slope, so the LCC may vary between IV and VII.

2.2.2.2 Soils of the Pre-Andean Mountains

Like soils of the Andean Mountains, the soils in this sector can be classified as not volcanic and volcanic, the former predominantly Mollisols (36°S to the north) and the latter sufficiently developed to satisfy the central concept of Andisols.

Non Volcanic Soils

These correspond to soils situated on hills, piedmonts and alluvial-colluvial deposits with pedogenic development which allowed the formation of subsurface diagnostic horizons and mollic epipedons. Slopes vary from 2 to over 30 % (Fig. 2.48), with substrates constituted by andesitic and dioritic rocks and depths that can exceed 1 m. Textural classes are highly variable, and OC content can reach values of 3 % in the surface horizon.

The deeper soils located at low slope can be classified as I and II LCC, but those less than 50 cm deep and at slopes over 30 % are classified as VI and VII. Challay soil association is considered representative of this group of soils.

From 35°S, the soils are located in high terraces and undulating hills of the pre-Andean Mountains which derive from Holocenic volcanic ash. The slopes vary between 1 and 15 % and the soils are deep, loam to silty loam, with surface OC content close to 5 %, pH between 6.0 and 6.5 and BS < 50 %.

Recent studies (Stolpe et al. 2008) concluded that these soils have evolved to such a level that they cannot be considered Andisols, since the bulk density exceeds the boundary value of 0.9 $Mg\ m^{-3}$ and the mineralogy is dominated by halloysite, which is the result of loss of silicon in allophane (Tan 2000). Bramadero soil series (Andic Haploxeroll) is characteristic of this group of soils, originally classified as Humic Haploxerand by CIREN (1997b).

Soil physical properties and the well-drained conditions give these soils a high productive capacity that is only limited by the slope, with LCC between I and III. In some cases (udic-isomesic zone), it is possible to identify a group of soils located between 300 and 900 m a.s.l. with weak

Fig. 2.47 Lithic contact between soil and basaltic rock in an abrupt limit (*left*). Landslide processes favoured by steep relief and thixotropic behaviour (*right*)

profile development, moderately deep, in hill positions and with slopes that vary from 15 to 50 %. These soils are derived from volcanic ash with varying grain size, which were deposited on coarse sand and cinder sediments, mostly of basaltic origin. Although the parent material is volcanic, these soils have not developed andic properties in their fullness.

Because of the coarse grain size dominance, the position in the Andes Mountains and the steep slopes, it can be assumed that the absence of andic properties results from lack of evolution of the original tephra to short-range order minerals (MORC), being classified as Andic Dystrudepts and within VI and VII LCC. Choshuenco soil series (Andic Dystrudept) represent this group of soils.

Volcanic Soils

Closer to the Central Valley, but always within the concept of pre-Andean Mountains, these soils reach the Andisol central concept, being classified as Haploxerands and Hapludands in the southern part of the xeric-thermic zone and the udic-mesic zone, and as Fulvudands in the udic-isomesic zone. They are deep and moderately deep soils located on alluvial terraces as a result of the alluvial re-transport of volcanic ash, on fluvio-glacial remnant terraces or on compacted and cemented volcanic conglomerates.

The soils located on alluvial terraces present a substrate of gravel and usually rounded and subrounded stones at depths of around 100 cm, with a soil matrix that can range from sandy to clay loam. When the slope gradient is <3 %, the soil may have a moderate to imperfect drainage class, even when it has moderately rapid permeability. Slightly undulating (2–5 % slope) soil phases generally have good drainage.

In the case of soils classified as Haploxerands, the OC content ranges between 3.6 and 7.7 % at the surface and 0.7 % at 100 cm depth, but is very variable through stratification. In the case of Hapludands, the OC content is high (>5.5 %) in the surface horizon and decreases regularly with depth, except for those soils with buried soil below 50 cm depth, where the OC content increases again. Other soils may have higher OC content in the surface horizons (10–14 %), where a melanic or similar epipedon is dominant.

The pH is in the range from weakly acidic to strongly acidic, ECEC remains between 3 and 10 $cmol_c\ kg^{-1}$ and Al saturation is <5 % for most soils in the sector. The water content retained at 33 kPa is maintained above 50 % for all horizons of most soils and, in some cases, can reach 80 %, which is consistent with andic properties.

Soils located closer to the Central Valley have the best productive potential, mainly because of more appropriate topographic position, for example alluvial terraces, remnant terraces or fluvio-glacial deposits with low slope gradient. It is possible to find soils in II and III LCC, with 1–3 % slope, such as the Cunco soil series (Acrudoxic Hapludand), representative of the soils in this area.

Fulvudands are characterised by their very dark brown to black colour (Munsell chroma <2) in deep deposits, which can be up to 2 m in depth above the substrate. These soils are formed from the weathering of fine tephras deposited on flat lacustrine conditions or alluvial fans, with gravelly and rounded stony substrate. The upper horizon has OC content >6 % and textural class ranges from loam to silty loam and silty clay loam, with well-developed structure and excellent radical penetration. Despite the typical constraints of the Andisols and BS usually <10 %, they are very productive soils with good agricultural potential and can have II LCC. Their main constraints are the imperfect drainage and the slope, making them slightly to moderately undulating.

A special case of volcanic soils are those derived from volcanic sands on alluvial positions in the pre-Andean Mountains of Region VIII, which have a fine fraction that

Fig. 2.48 General view of landscape in the pre-Andean Mountains at Santiago and soil profiles of the Challay soil association (Lithic Haploxeroll, *left*) and the Santa Sara soil series (Haplic Palexeroll, *right*), after Valle (2012)

has andic properties. These soils are slightly deeper, up to 75 cm, resting on a substrate of sand and gravel of volcanic origin. Textural class ranges from sandy loam to silty loam, but the substrate is always sandy and structureless. The pH is maintained close to 6.0 and BS is <40 %. See in Appendix, the properties of the Antuco soil series (Humic Vitrixerand).

The soils are well drained, but of limited agricultural suitability, ranging between IV and VII LCC depending on depth, gravel content and textural class.

2.2.2.3 Soils of the Longitudinal Central Valley

Geographically, this valley covers a large area located between the western flank of the Andes Mountains and the eastern flank of the Coastal Range, ranging between 50 and 100 km. However, these limits are somewhat fuzzy, due to the intervention of landscape-altering agents such as alluvial sediments, aeolian volcanic ash, alluvial lahar flows and some glacial sediment all of which generated deposits of up to 500 m thick (Araneda et al. 2000). The Central Valley is crossed by numerous water courses in an east–west direction, which is of great agricultural importance and has resulted in intense productive activity.

In this whole area, between 50 and 400 m a.s.l., there are differences in topography, ranging from alluvial terraces with flat or nearly flat surfaces to rolling hills with up to 15–20 % slope gradient. This variability gives rise to a wide range of soils with different characteristics, although most soil surveys have established that there is a predominance of soils derived from Holocene and Pleistocene tephra, so most of the soils would have volcanic sediments originating from different geological times.

Soils are associated here into six groups, according to their soil moisture and temperature regimes, considering morphological properties, geomorphic features and pedogenic development. The first group, in the xeric–thermic zone, includes soils with BS > 50 % throughout the soil profile and dark-coloured epipedons, mostly classified as Mollisols, and in few cases as Inceptisols or Vertisols. A second soil group is of higher age (Pleistocene) and more profile development, showing an illuviation horizon (B_t). A third soil group has BS < 50 % and less evolution, although it is possible to distinguish a B horizon with no evidence of translocation of components (B or B_w). The fourth group is represented by Holocene volcanic soils with umbric or melanic epipedons, which correspond to the typical Andisols and are locally called *trumaos*. The fifth group includes poorly drained Andisols in an aquic soil moisture regime, locally called *ñadis*. The last group, concentrated in Region VIII and also found in alluvial terraces of the Coastal Range, is a group of holocenic sandy soils of volcanic origin, locally called *arenales*.

Soils with High Base Saturation

This group includes a large number of soils with variations in origin, parent material and transport agent, but the common feature is their intense productive activity. The soils are in positions of alluvial terraces, piedmonts and lacustrine sedimentation basins. The landscape is flat to nearly flat, with slopes ranging from 1 to 5 %. The depth and texture class are very variable, but in the case of alluvial soils the substrate is typically rounded stones (Fig. 2.49). Base saturation is greater than 65 %, with some exceptions; the pH varies between 6.8 and 8.2 in the Maipo river basin, due to the contributions of carbonate in irrigation water, but decreases towards the south to between 5.6 and 7.6. See in Appendix the main properties of a profile of the Maipo soil series (Fluventic Haploxeroll).

Most of these soils have good permeability and good drainage, which determines their inclusion in LCC between I and IV, depending on the depth to the rocky substrate. While most of them are Mollisols (Haploxerolls), some cases are classified as Inceptisols (Haploxerept).

Fig. 2.49 The Central Valley of the Mediterranean zone is the main base for agricultural production. Walnut trees on Maipo soil series (Fluventic Haploxeroll)

Another soil group with high BS values originates from lacustrine sediments. The Metropolitan Region (33°S) defines the limit for soils with presence of pedogenic carbonate and high EC values reaching 18 dS m^{-1} (Fig. 2.50). Dominant slopes are in the range 1–3 %, the soil is fine-textured and the depth is limited by a cohesive dense horizon (by clay or cemented by carbonates) or a permanent watertable, the latter imparting black colour in surface horizons and grey colour at depth.

The OC content ranges between 1 and 8 % in the surface horizon and decreases regularly with depth, while depending on carbonates and salinity, the pH ranges between 7.0 and 9.3 around Santiago city and between 5.3 and 7.9 to the south, increasing with depth. The drainage conditions determine II–VI LCC for these soils, but with artificial drainage the soils improve their productive potential. Agua del Gato soil series (Petrocalcic Calciaquoll) is representative of this group.

A special group of soils with high BS are those derived from volcanic materials, but as a consequence of the climate conditions (xeric-thermic soil regime, mainly from Chillán city to the north) did not form Andisols. They are located in flat positions within the landscape or sloping up from 15 to 20 %, depending on transport agent. The soils have a variable depth, usually resting on a cemented or consolidated material. Base saturation increases with depth, the pH ranges between 5.2 and 7.9 and the OC in the surface horizon ranges between 0.6 and 3.5 %. Quillayes soil series (Aquic Haploxerept) is representative of this group.

Because the substrate is cemented or consolidated, drainage restrictions are normal, so it is possible to find common to abundant redoximorphic features. Thus LCC ranges from II to VI, depending on depth and drainage.

In the Central Valley from the udic-mesic zone to the south, it is possible to find a group of soils characterised by high BS. These are classified as Mollisols but with a low areal representation. This soil group includes deep pedons, medium and fine-textured, located in alluvial terraces and flat deposits of fluvio-glacial origin, with topography ranging from flat to gently undulating. The dominant colours are in the 10YR hue, with low chroma value (Munsell), i.e. dark brown, dark greyish brown and black in some cases. The OC content is not particularly high, as would be expected, and in some soils is even close to the lower limit for the presence of a mollic epipedon. The BS is over 50 % across all profiles and the CEC varies from 12 to 27 cmol$_c$ kg^{-1}, which is an important factor in concluding that the soil mineralogy is mixed, with dominance of 1:1 clay minerals. Only a few pedons (Santa Rosario soil series) contain evidence of Holocene volcanic ash inputs, because P retention reaches 78 % and bulk density is less than 1 Mg m^{-3} in the epipedon and about 1 Mg m^{-3} in the whole profile. In general, these soils have good productive potential, with II and III LCC dominating, although some soil phases show low potential due to a watertable close to the surface, excessive slopes and erosion risks.

Soils with Argillic Horizons

These soils are located in hillocks with slopes from 2–5 % to 20–30 % and are moderately deep to deep and medium to fine-textured. There are two dominant soils, those located closer to the Andes Mountains in fluvio-glacial sediments, and those located close to the Coastal Range. The former dominate between 34 and 36°S, approximately, with a highly variable BS (38–83 %), pH between 5.0 and 6.6 and OC values close to 2.2 %. At contact with the fluvio-glacial substrate, soil redoximorphic features are common, classified from II to IV LCC, but with good potential for crops (Fig. 2.51). Talca soil series, a Pachic Palexeroll (Stolpe et al. 2008), represent this soil group.

Close to the Coastal Range, these soils are deep and moderately deep (Fig. 2.52), resting over fluvio-glacial conglomerates, are variable and partially weathered from andesitic-basaltic material or volcanic breccias. Located at

Fig. 2.50 A drained Petrocalcic Calciaquoll (Agua del Gato soil series, *above*) with Ckm horizon and *gley colour* persisting at depth. A Sodic Haploxerert (Batuco soil series, *below*)

positions on hillocks with gentle to moderate undulating topography, with slopes predominantly 8–15 % but 20–30 % in some soil phases. Although slowly permeable, the soils are well-drained and only few soil phases in concave positions may have imperfect drainage. The upper horizons are clay loam or silty clay loam, whereas the subsoil (B horizon) is always clayey; i.e. there is a clear increase in the fine fraction with depth, so that the clay content of B horizons varies between 40 and 65 %. The strong erosion that these soils have suffered because of inappropriate farming on slopes has led to most of the profiles being currently described are incomplete because of the beheading of the surface horizons (Fig. 2.52).

Another feature that distinguishes this soil group is the colour of the horizons, ranging from the hues 2.5 and 5YR, from dark reddish brown to dark red. The clay content and the colours are the reason why these soils are known as *Red Clayey Soils* (*Suelos Rojo Arcillosos*).

Weakly to strongly acidic (6.2 and 5.3) soil pH is observed, with low Al saturation in the exchange complex. At the same time, BS varies between 31 and 65 % through the profile. The highest BS values found in deeper horizons allow these soils to be classified as Alfisols. The CEC is variable among pedons and horizons within the same profile (13–31 $cmol_c\ kg^{-1}$), which may be associated with the original stratification of parent materials, in addition to the translocation of components to which the soils have been subjected.

South of Temuco city (39°S) the andic properties begin to appear, which could be considered evidence that these soils originated from old volcanic ash deposits (Besoaín 1958). The bulk density is usually slightly higher than 1 $Mg\ m^{-3}$ but in some horizons may be less than that, P retention is over 50 % in all soils and can reach 66 % in some horizons, and $[Al_{ox} + \frac{1}{2} Fe_{ox}]$ is always less than 1 %, but close to this value. According to these data, it seems logical to deduce that the original amorphous materials have evolved into some form of crystalline minerals (Besoaín and González 1978) and that it is still possible to find some features that identify them. The productivity of these soils is largely associated with the topography and the current and potential soil erodibility. Slightly undulating soil phases (slope 2–5 %) are classified as III and IV LCC and show slight

Fig. 2.51 General view of the Mariposa soil series (Ochreptic Haploxeralf) and redoximorphic features of the fluvio-glacial deposit

erosion, but phases with 15–20 % slope gradients show severe sheet erosion and frequent gullies, being included on VII LCC and destined for permanent grassland or forest, excluding tillage crops. Representative soils of this group are Collipulli (Typic Rhodoxeralf), Mininco (Typic Rhodudalfs) and Metrenco (Typic Paleudults) soil series.

Soils with Cambic Horizon

There are soils with different origins and positions in the landscape, but that share the common feature of development of a cambic horizon. One group corresponds to deep and moderately deep soils on alluvial terraces, with 1–3 % slope and well drained. Another important group corresponds to thin soils over a duripan formed by volcanic breccias, conglomerates or gravels with varying degrees of weathering, that inhibit rooting. Finally, there are also soils on hillocks and hills with slopes greater than 30 %.

The soils, deep and moderately deep developed on alluvial terraces, have moderately fine to fine textures, silty loam and silty clay loam in the surface horizon, clay loam and clayey in the B horizon. The colours are in the hues 7.5 and 10YR, with low chroma values (Munsell). The normal sequence of horizons for these soils is an umbric epipedon over a cambic horizon with BS < 50 % throughout the profile. In some cases, there may be a duripan under 100 cm. In some soil phases, the redoximorphic features are abundant, showing current conditions of poor and imperfect drainage. These restricted drainage conditions determine that these soils are classified as III and IV LCC.

Fig. 2.52 Eroded (*left*) and non-eroded (*right*) Collipulli soil series (Typic Rhodoxeralf) in Region VIII of Chile

Soil bulk density is over 1.0 Mg m^{-3} in all horizons and all soils, although occasionally some horizon can have a value of 0.9 Mg $m^{-3.}$ The retention of *P* may be unusually high (80 %) for this type of soil, thus indicating some source or contamination with volcanic material. The OC content reaches 4.7 % in the epipedon, decreasing regularly with depth, and soil pH is maintained in the range of weakly acidic throughout the profile. The CEC is very variable from one soil to another and from one horizon to another within the same profile (13–34 $cmol_c$ kg^{-1}). Nueva Imperial soil series (Typic Humaquept) is representative of these soils.

The slightly deep and thin soils are restricted by a duripan at depth, cemented by silica and manganese, always in abrupt boundary with the overlying soil. The boundary layer has been described as lying between 50 and 60 cm depth, restricting the possibilities for use, even though the other properties remain more or less similar. The dominant textural classes are silty loam, clay loam and clayey, with a slight increase in the clay fraction to the subsurface horizons. All these soils occupy alluvial terraces in flat or nearly flat topography, with some phases poorly drained in more depressed positions. They are classified as III and IV LCC.

The soils located on hills with 30–50 % slope are deep, formed from ancient volcanic ash deposited in the Central Valley between 300 and 400 m a.s.l. These soils have a high clay content (40–55 %) with amounts increasing towards the B horizons, are reddish brown and dark reddish brown (5 and 2.5YR), well structured and can rest on a conglomerate very weathered under 120 cm depth. Their origin from ancient tephra is related to some properties associated with andic properties of Holocene volcanic soils, i.e. Db < 1.0 Mg m^{-3} and P retention 60 %, values too high to ignore a possible influence of volcanic deposits. Because of their topography on hillocks and hills and their suitability for forest, these soils are classified within VI and VII LCC.

From the udic-mesic regime zone to the south, it is possible to distinguish two kinds of Andisols through their position in the landscape. These are locally called trumao and ñadi and correspond to the deep and well-drained Hapludands (trumao) in softly-steeply higher positions of landscapes and the moderate to slightly deep and poorly drained Placaquands (ñadi) in flat and depressed positions in the landscape (Fig. 2.53), commonly associated with outwash from glaciers.

Typical Andisols (*Trumaos*)

In the Central Valley from Chillán city to the south, soils derived from Holocene volcanic ashes are widely represented and satisfy all requirements to be considered Andisols. As deep and moderately deep soils, they have no restrictions on rooting in most soil series and their respective soil phases. Located in recent or older remnant alluvial terraces, where the original volcanic ash has been redeposited by water, the substrate of most of these soils is composed of rounded gravel with interstitial sand. In some soils, the substrate can be less than 1 m depth, without limitations to root penetration (Fig. 2.54).

Dominant textural classes are silty loam in the surface and silty clay loam or clay loam in the deeper horizons. The most common colours are hues 7.5 and 10YR, and only in rare cases are there reddish brown and red soils (2.5 and 5YR). The OC content varies between 3 and 11 % in the epipedon, with a gradual decrease with depth, but is not less than 1 % even at 120 cm depth. In soils that have a marked alluvial influence, the OC has an irregular decrease, clearly marking the stratification of materials. The BS is particularly low, <25 % in all soils, reaching 2 % in some horizons, as in the Toltén soil series.

Except for the restriction common to all these soils (high *P* retention), they are very suitable for agriculture, with most soil phases classified into II and III LCC due to slight limitations on drainage in the lowest sectors of the relief. See properties of the Toltén soil series (Acrudoxic Hapludand) in Appendix.

A second group of Andisols includes deep soils, dark brown and black coloured, with a high content of OC throughout the profile (Fig. 2.55), with values between 8 and 6 % to 60 cm, and in some soils, 3 % at 140 cm depth. They commonly contain an umbric epipedon, a melanic epipedon or a horizon that meets the requirements of the melanic epipedon, except for the chroma value (Munsell). The water content can vary from 64 to 80 % at 33 kPa and 38 to 61 % at 1,500 kPa. The CEC can reach values of 57 cmol kg^{-1}, indicating dominance of short-range-order minerals. Like all the soils of this sector, BS remains under 17 % and in some horizons may reach values as low as 2 %.

These soils have good potential from the agricultural point of view and most have been classified as II and III LCC, with only a few soil phases included within IV and VI LCC, mainly due to the increase in slope or drainage restrictions (Osorno and Paillaco soil series, respectively). Corte Alto soil series (Typic Hapludand) is a very deep and stratified Andisol, which occupies remnant terraces and slightly steep positions, with very well-expressed andic properties.

Aquic Andisols (*Ñadis*)

In addition to their andic properties, these soils share the common feature of restricted drainage that in some stages can reach very poor conditions. This constraint is created both by the topography and soil morphological features. Generally, they occupy flat and depressed positions in the landscape (Fig. 2.53) and a fluvio-glacial deposit consisting of gravel and sand has been described as the substrate.

Above this deposit, a duripan and/or a placic horizon (locally named *fierrillo*) is present which prevents the passage of water and plant roots (Luzio et al. 1989). The

Fig. 2.53 In the foreground, a *trumao* and in the low relief, a *ñadi*, volcanic ash-derived soil developed over a glacial substrate with drainage restrictions creating a Bhsm horizon (*red arrow*). *Ñadi* soils can be moderately deep, e.g. the Frutillar soil series (Typic Placaquand, *above*) or thin, e.g. the Alerce soil series (Histic Placaquand, *below*)

Fig. 2.54 Loncoche soil series (Acrudoxic Hapludand) with natural prairie and detail of soil profile

fierrillo is very thick (<2 cm), extremely strong and cemented with Fe and Al, with mixed black (N 2/0) and yellowish red (5YR 5/8) colours. This diagnostic horizon is moderate to strongly acid, with an OC content too high for this type of horizon. Palma (1993) reported OC contents from 1.1 to 5.6 % and found that Fe is the most abundant element in the fierrillo, particularly Fe extracted with oxalate. Al is present too, but to a lesser extent, so Palma (1993) concluded that Fe is an essential element in cementing these horizons.

The depth of these soils is very variable depending on the position of the fierrillo within the profile and they range from very thin to deep soils. The dominant slopes range from nearly flat (1–3 %) to slightly undulating (2–5 %), so surface runoff is slow. In some soils, the watertable can be found near the surface in winter and drops below 1 m in summer, but other soils remain saturated throughout the year. These characteristics are maintained in the soils of Chiloé Archipelago, in particular, and have a wide distribution in the core of the Big Island of Chiloé.

The topsoil is usually an O horizon, consisting of a bed of *Sphagnum* sp., roots and leaves, and depending on its thickness can be a histic epipedon. The most frequent colour is black, from 5 and 10YR. At depth, there is always good development of the profile which implies the presence of a cambic horizon. In field descriptions, it is possible to note an increase in the clay fraction with depth, as the textural class ranges from loam or silty clay loam at the surface to silty clay loam or even clay loam in the B horizon.

In general terms, these soils have potential use for pasture and in most cases there is a dominance of imperfect to very poor drainage. However, some soil phases (Frutillar soil series) are moderately well drained and classified as IIw LCC. All soils are considered Duraquands, Endoaquands or

Fig. 2.55 The Puerto Octay soil series, a Typic Hapludand on morrenic deposits at Llanquihue Lake, and detail of the soil profile

Placaquands. Maullín soil series (Hydric Endoaquand), is an example of this soil group.

Holocene Sandy Soils (*Arenales*)

A special soil group corresponds to those derived from volcanic sands deposited in alluvial fans and terraces of the Laja River (Region VIII), corresponding to basaltic-andesitic materials, coarse textured, stratified and developed in a xeric-thermic regime (Mella and Khüne 1985). Moreover, some of these materials have been mobilised to the valleys of the Coastal Range and deposited in alluvial terraces (Fig. 2.56).

The main geomorphologic positions in which these soils are disposed correspond to alluvial fans, terraces and flat fluvio-glacial deposits. The substrate may be gravel and rocks with a sandy matrix, located deeper than 160 cm. Textural classes are particularly homogeneous, ranging from sandy loam to sandy, as in the Coreo and Arenales soil series.

In general, the soils are deep and very deep (150–160 cm depth) with elevated permeability, but through topographical position they may contain drainage restrictions. The root system is well developed, so roots are reported even up to 160 cm depth.

In the Arenales and Coreo soil series, the OC content is 1.0 and 1.8 %, respectively, and BS increases with depth (from 20 to 85 % in some soils). The pH remains uniform in all soils, with values between 5.4 and 6.7. The LCC varies between IV and VI, mainly as a consequence of the coarse soil texture. Arenales soil series (Dystric Xeropsamment) is considered representative of this group.

2.2.2.4 Soils of the Coastal Range

The delimitation of this sector is based on geological and geomorphologic features, which means the Coastal Range, rising gradually in some sectors and more sharply in others west of the Central Valley, until it reaches 800 m a.s.l. Two main parent materials are recognised, which together with the climate determine the properties of these soils. There is a continuity of the northern Coastal Range to the Mediterranean zone, with igneous (granitic) materials, but about from 37°45′S to the south metamorphic rock begins to appear.

Igneous Parent Materials

Angular gravels, especially of quartz, have been described on all these soils with a dominant hilly topography, with slope gradients from 8–15 % up to >50 % (Fig. 2.57), although some alluvial fans and piedmonts can have lower slopes. The dominant textural classes are medium to fine, but at depth the sand content increases through the presence of quartz. As a result of geomorphic position, all soils are well drained.

The chemical and biological properties correspond to a climate gradient from 33 to 37°S, thus BS ranges from 65 to 95 %, the pH is slightly acid (between 6.0 and 6.6) and OC content can reach 1.5 % in the upper soil horizons at 33°S. However, in soils from 37°S to the south, pH can decrease to 5.0, BS can be lower than 50 % and OC can reach values of 3.0 % in upper soil horizons. The most important constraint is given by the position of hills and steep slopes, so that soils are between IV and VII LCC. Lo Vásquez soil series represent clearly this important group of soils.

Metamorphic Parent Materials

South of 37°45′S, on the eastern side of the Coastal Range, between 250 and 300 m a.s.l., deep soils are developed from volcanic ash deposited on a metamorphic complex.

The most evolved pedogenic soils have developed from metamorphic rocks of the Coastal Range, in hillock and hill positions. Some of them (Correltúe and Araucano soil series) are considered to be derived from old volcanic ash deposits. In the case of the Nahuelbuta (Rhodic Palehudult) soil association, its origin would be directly in the

Fig. 2.56 The Quillón soil series, a Dystric Xerorthent derived from Holocene volcanic sands in alluvial deposits of Region VIII (*left*). From Concepción city to the south, the parent material of the Coastal Range begins to change from igneous (*right* Cauquenes soil association, an Ultic Palexeralf) to metamorphic rock

weathering of metamorphic rocks such as phyllites and mica schists. One characteristic of these soils is that they occupy relief that goes from softly to strongly undulating hills (Fig. 2.58). Andic properties for several of these soils are very close to the limit to be considered Andisols and therefore the influence of volcanic materials in their evolution is undeniable. An example of these soils is Correltúe soil series (Andic Haplohumult), described as resting on mica-schist at 130 cm depth.

Soils derived from metamorphic rock, where no influence of holocenic tephras is detectable, are deep and B_t horizons can reach 150 cm, with dominating fine textural classes (silty clay loam to clay). Although eluviation horizons morphologically are not obvious, either by colour or textural class (very homogeneous), the soils are well evolved because of the presence of argillic horizons. The clay fraction varies between 35 and 73 % in the B_t horizons and soil structure is generally strong, varying from medium and coarse subangular blocks to coarse prisms. The dominant colour varies from dark reddish brown to dark red, in the hues 2.5 and 5YR. Only the high OC content of some surface horizons imparts a brown colour in the hue 10YR.

The CEC, although it is highly variable among soils (14–36 $cmol_c\ kg^{-1}$), shows a decrease with depth, following a trend similar to OC content. In the same way, the BS may vary considerably from one soil to another, but is always <40 %. Only some soil phases with low slope (2–5 % and 5–8 %) are suitable for agriculture, particularly grasslands and cereals, but the best use of these soils is forestry, with VII and VIII LCC.

A second group of soils, located in the areas of flat topography between the metamorphic massifs of the Coastal Range, has been described. These soils have intermediate pedogenic development with a cambic horizon (B_w), mostly occupying recent or remnant alluvial terraces. Materials of these terraces come largely from sediments that have been transported from the highest parts of the relief due to the severe erosion affecting these sectors. Therefore, the soils are rich in mica and quartz minerals from the weathering of metamorphic rocks (Fig. 2.59).

These soils are deep and fine-textured, generally clay loam to clay (27–50 % clay). The dominant colour is dark reddish brown, in 2.5 and 5YR hues, inherited from parent materials. The flat or nearly flat soils (1–3 % slope) have moderately slow permeability and imperfect drainage; the deeper horizons (100 cm) have abundant redoximorphic features (concretions of Fe and Mn and grey colours in 2.5 Y hue).

The OC content is low, not exceeding 2.3 % in the surface horizons, and decreases with depth to 0.1 %. Base saturation remains under 30 % in all horizons and CEC does not exceed 17 $cmol_c\ kg^{-1}$, indicating an abundance of kandites. Soils in depressed positions have imperfect drainage, with concretions and nodules of Fe and Mn occurring in the horizons below 85 cm depth and imparting an olive brown colour in the hue 2.5 Y. Soils are classified according to the imperfect drainage (IIIw LCC) and excessive slopes, as in the case of soils located in colluvial or piedmonts (IVe and VI LCC). Los Copihues soil series (Oxiaquic Dystrudept) is considered representative of this group.

From Valdivia city (~40°S) to the end of the udic-isomesic zone, metamorphic rock is the dominant parent material, determining the existence of Inceptisols with andic properties. In soils that have been described in this formation, it has been shown that the rocks are more weathered, and therefore, have less recognisable petrographic structure. Soil substrate consists of gravel deposits of multicoloured gravels, rounded and subrounded, with different degrees of weathering, very weathered tuff and/or metamorphic phyllites and mica schists, generally very weathered. The depth of the soil is determined by erosion occurrence, giving slight to moderately deep soils (Fig. 2.60).

Complex slopes dominate (20–50 % slope gradient) and only some soil phases show 8–15 % slopes, with moderate

Fig. 2.57 The Lo Vásquez soil series (Ultic Haploxeralf) in a steeply sloping position (*above*). The La Lajuela soil association (Ultic Haploxeralf) with productive activity on the low position of the alluvio-colluvial deposits (*below*). Despite the great depth of weathering of granite, soil depth depends on the degree of erosion to which the soil has been submitted

Fig. 2.58 Landscape of the Correltúe soil series (Andic Haplohumult) and the respective profile

to severe erosion, so they have been assigned to VI and VII LCC, i.e. suitable for forestry. Some sectors (5–15 % slopes) have a better aptitude and are used with annual crops, being classified within II, III and IV LCC. Hueñi soil series (Andic Dystrudept) is very representative of these soils.

2.2.2.5 Soils of the Coastal Plains

This group includes a wide variety of soils not deeply studied, developed on marine terraces, gentle rolling hills, stabilised dunes and remnant terraces, both alluvial and marine (Fig. 2.61). The position in the landscape determines that the slopes are less than those of the Coastal

Fig. 2.59 General view of soil distribution in the Coastal Range of Region IX and detailed profiles of the Los Copihues (*left* Oxiaquic Dystrudept) and Lumaco (*right* Oxic Dystrudept) soil series

Fig. 2.60 General view of a deforested landscape of the La Pelada soil series (Oxic Dystrudepts) and detail of soil profile resting on a weathered fluvio-glacial deposit

Range (>15 %) and the gradient of chemical properties influenced by climate persists as in the case of Coastal Range soils.

The most common substrate is sandstone, often mixed with gravel of mixed composition. Coarse-textured upper horizons are without a definite trend in depth, varying from sandy to clay, but more or less homogeneous in each soil. The depth to the substrate is variable (50–120 cm) and drainage is good to imperfect. As expected, there is a gradient of chemical properties from north to south, with BS between 60 and 90 % in the north and 20–70 % in the south, pH values decreasing from 5.6 to 7.8 in the north down to 4.6–6.5 in the south, and OC content ranging from 1.5 to 3.4 % in the same climate gradient. Matanzas soil association (Oxic Haplustoll) developed from limonite in a marine terrace position, is a good example of these soils.

Although soil analyses and taxonomic classification require updating, in general these soils belong to the orders Alfisols, Mollisols and Inceptisols. Aqueveque (2008) indicates that soils derived from Holocene dunes correspond to Ustipsamments. On the other hand, LCC varies from III to VI, depending on drainage conditions and depth.

2.2.3 Soils of the Rainy and Patagonian Zone

Most of northern Rainy and Patagonian zone has a climate that does not strictly respond to latitude, because of the topographical effects of the Andes Mountains, and then precipitation diminishes towards the Atlantic Ocean. However, in the southern part of this zone the precipitation trend follows latitudinal bands, again diminishing from south to north. This zone includes Region XI (Aysén) and Region XII (Magallanes), where the geomorphology is dominated by overlapped volcanic and glacial processes, which make it complicated to interpret the distribution of soils. Fluvio-glacial deposits have greater representation in the northern sector of Region XI due to the less steep topography of the area, probably by intense glacier activity, leaving large areas of sediments of different natures. In the southern sector of Region XI, it is difficult to distinguish these fluvio-glacial deposits, due to more steep topography and active fluvial erosion processes determined by narrow river valleys.

In the Province of Palena (43°S), glacial valleys are filled with different kinds of materials such as moraines or alluvial deposits. Above them, volcanic materials from the Michimahuida, Apagado, Hornopirén and Chaitén

Fig. 2.61 Soils developed on the coastal plains have an origin that may be alluvial, marine or aeolian, which result in large soil variability, but similar chemical properties (Photo: Aqueveque C)

volcanoes were deposited, which would be the starting materials for most soils of the area (Ahumada et al. 2004).

Díaz et al. (1959–1960) show that during the Lower Cretaceous, the sea invaded the continent gradually, from about 43°S to the south, which resulted in the formation of marine sandstones. Later, during the Tertiary, great volcanic activity began and was responsible for the formation of Paine and Balmaceda clumps. At the end of the Tertiary, there was a new period of volcanic activity that deposited ash, lava and pumice on extensive areas, which were then redistributed by watercourses. During the Quaternary the area was covered by ice which was responsible for major erosive activity and the formation of various channels, lakes and bays. This glacial activity also left a complex system of moraines that act as dams, giving rise to large glacial lakes with varved sediments on the banks of the Strait of Magellan (53°S). At the end of the last glacial period, there was activation of extrusive volcanism, resulting in extensive deposits of basaltic lava and ash.

In the area of Puerto Edén (49°S), Calderón et al. (2007) describe a sequence of schists, gneisses, migmatites and granitoids, biotite and biotite–hornblende granitoid derived from Puerto Edén Igneous and Metamorphic Complex, belonging to the South Patagonian batholith. The authors attribute these formations to the Late Jurassic.

In Magallanes in Region XII (*Tierra del Fuego*), in particular the east sub-Andean foothills, Gerding and Thiers (2002) describe a flat-shaped area with mountain ranges of low altitude, whose peaks rarely exceed 1,200 m a.s.l. There are deep valleys carved by the effect of Quaternary ice and postglacial phenomena, filled with glacial, fluvio-glacial, fluvial and lacustrine deposits and the latest volcanic ash.

Díaz et al. (1959–1960) observed close coincidence between the distribution of Great Groups of soils and systems of moraines that have been described as glacial deposits of Quaternary advances. For example, the substrate of the Castaños soil series is highly weathered and partly disintegrated, while Podzolic soils show a massive clay substrate, known locally as *masacote*, which allows the accumulation of organic matter and the presence of peatlands in the area.

The volcanic deposits correspond to thin materials (ash) and sand-sized particles, but it is also possible to find medium and coarse particles (lapillis). Pumice fragments, most with intermediate degrees of weathering, are widely distributed in soil, either as strata in lithological discontinuities or mixed with the soil matrix. The volcanic materials were deposited covering different types of formations such as alluvial terraces, moraines, colluvial deposits and rocks (Fig. 2.62), so that soils developed from these materials are derived from contemporary events from the same volcanic centres. Therefore, the tephra have led to similar soils, even when the substrate on which they are developed can be of a different nature. Because volcanic events have been sporadic, soils show a marked stratification, and not all possess andic properties, differentiating the soils from Puerto Montt to the north, where the tephras have been more homogeneous in both particle size and composition (Luzio et al. 2010).

As a result of volcanic origin and steep slopes, most soils in this zone are thixotropic. This phenomenon is associated with the presence of gels, in this case derived from volcanic materials (such as allophane, imogolite and ferrihydrite), which change their behaviour from solid to liquid on physical alteration (such as an earthquake). Due to these minerals of low crystallinity, which have a large surface area, the soils have the capacity to absorb an enormous amount of water, which can reach over 100 % at wilting point, so that slopes over 60 % can easily produce landslides.

The forms due to alluvial processes are mainly alluvial terraces, usually 1–3 % and 2–5 % slope, in low positions within the landscape, over rounded gravel substrate and always associated with numerous rivers in the area. Many of

Fig. 2.62 Volcanic soils over different deposits or rocks. Sequence of ash deposited from the Hudson volcano over basaltic rock (*left*), mixed layers in alluvial deposits (*centre*) and volcanic soil over moraine (*right*)

these soils are formed from volcanic ash, with similar properties to the soils of the highest positions, which suggests that the volcanic ash deposits are relatively recent (Holocene). Close to Coyhaique city alluvial terraces can reach great dimensions, constituting potential areas for agriculture and livestock (Fig. 2.63).

Between 42 and 50°S, the soil landscape can be subdivided into two sectors without clearly defined boundaries. The first group corresponds to the Andisols, taxonomically well expressed, but the substrate can be glacial (moraines), fluvio-glacial, alluvial and even rock (Fig. 2.62). These soils are found in the central sector of the zone, approximately between 71°45′ and 72°30′W. On the other hand, there are intermediate developed soils (presence of a cambic horizon) that are found on alluvial terraces or fluvio-glacial sediments, which do not exhibit sufficiently well-expressed andic properties to be considered Andisols.

From 71°45′W to the east (soils of the East of North Patagonia), there is a change in the landscape: forests disappear and are replaced by a dense cover of *Stippa* spp. bunchgrass (Fig. 2.64). The soil properties also change, particularly in relation to BS in the soils described, assuming a xeric soil moisture regime. Agriculture activity is more intense in these areas, including crops impossible to grow in the western sector.

Between 50 and 55°S (South Patagonia) there are few detailed studies of soils. Physiographically, three important forms have been identified in this area: The Magellan Mountains, the pre-Mountains (Precordillera) and the Magellan Pampa. The Magellan Mountains consists of a granitic batholith as a crystalline basement and volcanic rocks of different ages. This formation raises an aspect of wall, with the highest peaks over 3,000 m. Petrographically, schists, phyllites and gneiss of the lower Paleozoic or Precambrian dominate. The present relief is completely influenced by Quaternary glacial action (IREN 1967). The pre-Mountains show a relief softer than the Mountains and are composed of sedimentary and metamorphic rocks such as sandstones and shales, attributed to the Cretaceous and Tertiary. Volcanic materials in contact with the Magellan Mountains have also been described. To the east the landscape is softer, slightly undulating to flat (IREN 1967). The Pampa extends from the foothills (Precordillera) to the Atlantic Ocean with a flat relief composed of Tertiary sediments (sandstones and clay) over Quaternary fluvial-glacial deposits.

Finally, the *mallines* and *vegas* are sectors that generally occupy the lowest positions of the Patagonian relief, and are therefore imperfectly to very poorly drained (Fig. 2.65). Soils in these positions include mineral soils and organic soils, with huge stratiographic variability. In the case of organic soils, the variability is related to the content and degree of humification of organic materials and the depth they achieve. They are found in lake sedimentation basins, terraces and alluvial plains lacking natural drainage. They can also be found in areas of steep slopes breaking from the hills to the alluvial terraces. The *mallines* of Cisne River (Queulat soil series) show moderately flat topography and microrelief (2–5 % slope) and are permanently saturated.

In a broad separation, the area is divided here simply into North and South Patagonia (*Tierra del Fuego*), corresponding in general to Administrative Regions XI and XII respectively.

2.2.3.1 Soils of the West of North Patagonia

Andisols

In this sector, the Holocene tephras have been widely spread, covering large areas of shapes and different origins. This means that the slopes that dominate in each soil are also quite variable, ranging between 1–3 % and 5–8 % in

Fig. 2.63 Sequence of alluvial terraces near Coyhaique city. Volcanic ash was deposited over alluvial substrate, defining the Pollux soil series (Typic Hapludand)

Fig. 2.64 A typical landscape in a possible udic-isomesic area where Andisols are well developed (*left*). From 71°45′W to the east, xeric conditions allow the existence of soils with high base saturation and in some cases with drainage restrictions (*right*)

alluvial terraces, 5–8 % and 8–15 % in moraines and 15–50 % in hills and mountains.

The textural classes of surface horizons are medium and coarse (silt loam, sandy loam and loamy sand) that, practically as a rule, become fine-textured towards the deeper horizons, so there are frequent silty clay and silty clay loam classes. Most of the soils described are deep and moderately deep, with only Murta soil series described in detail as being 67 cm deep. All soils have good root penetration, usually reaching to the substrate.

One characteristic property of these soils is their strong stratification (Fig. 2.66), where strata of pumice lapillis are inserted between layers of ash (Besoaín et al. 2000). Another property common to most of these soils is the thixotropy, which does not manifest itself uniformly throughout the profile, but appears in some layers of ash, which facilitates the occurrence of differential slip processes.

With BS < 50 %, OC content can reach high values in the upper horizons (13 %) followed by an irregular decrease with depth, which is associated with marked stratification. The pH is maintained in the acidic range (6.3 and 4.7) and redoximorphic features are mainly represented by Fe segregations, which have been translocated by organic matter in *trumaos* and *ñadis* (Luzio et al. 1989). The bulk density is low because of the volcanic ash origin and the low intensity of use, and the water retention is highly variable, in some upper horizons exceeding 100 % at field capacity and wilting point (Besoaín et al. 2000). Pollux soil series (Typic Hapludand) a wide area in Region XI.

Because of the slope, stratification and thixotropy phenomena are common in these soils, the VI and VII (VIII in

Fig. 2.65 Depressed sites where mallines and vegas are found, near to Caleta Tortel (Region XI, 47°47′S)

special cases) LCC dominate. The volcanic ash soils in alluvial terrace positions have better suitability because of their lower slope, high depth and good drainage, so they are classified frequently as II and III LCC.

Other Soil Orders

Near Coyhaique city (45°34′S), a deep Andic Distrudept is reported by CIREN (2005b) in an alluvial terrace position with a substratum of glacial clay.

2.2.3.2 Soils of the East of North Patagonia

Although no detailed data are available for moisture and temperature regimes of the soils in this area, it is assumed that they have conditions close to a xeric moisture regime and thermic temperature regime, based on higher agricultural activity and chemical analyses.

A recent study (Pfeiffer et al. 2010) reported relationships between soil Great Groups and the glacial geomorphology of the Pascua River and Baker River basins (Table 2.5), observing only Entisols, Inceptisols and Histosols. Exceptionally, in the east and close to the Andean Mountains, some Andisols are developed (Murta soil series, a Typic Hapludand) according to CIREN (2005b).

A pristine landscape is characteristic of this eastern portion of Chilean Patagonia, where the Chapel of Marble and Baker River emerge as two emblematic pictures (Fig. 2.67).

Soils with High Base Saturation

The most significant feature of this soil group is that in contrast to the other soils of Region XI, BS is usually >50 % throughout the profile.

Most soils in this area are developed in a position of lake terraces, glacial plains, glacial moraines and alluvial deposits (Fig. 2.68). The alluvial areas show gentler slopes, with topography flat and nearly flat (0–1 % and 1–3 %) to gently undulating (5–8 %). In the case of soils developed on plains and glacial moraines, the relief is more proncunced, dominating complex slopes from gently undulating (5–8 %) to strongly undulating (15–20 %).

Soils in this group are moderately deep and deep on a substratum that can be variable: a succession of strata of sand mixed with layers of finer textural classes such as sandy loam, or angular gravel with coarse interstitial material and glacial clays. As might be expected in soils of sedimentary origin, textural class is highly variable, of indistinct origin (alluvial or glacial), but it has been observed that there is some textural uniformity within each profile. The soils are well drained, with the exception of some soil phases of alluvial soils (Chile Chico and Fachinal soil series) that have imperfect drainage, and all have good root penetration to the depth of description.

In almost all soils BS is >85 %, although in the Puesto Viejo soil series it is 67 %. The pH is more acidic in the upper horizons with values between 6.3 and 6.8, whereas in

Fig. 2.66 Glacial landscape and profile of the Pollux soil series (Typic Hapludand). The soil phases are indistinctly located in the base of the valley (2–5 % slope) or on the hills (up to 20–30 % slope). *Right* stratification, with a buried soil at 77 cm depth

Table 2.5 Classified pedons in the Baker and Pascua river basins (Pfeiffer et al. 2010)

Geomorphology	Baker River basin	Pascua River basin
Alluvial	–	Fluvaquentic Dystrudept;Typic Dystrudept
Alluvial fan	Aquic Cryofluvent	Typic Dystrudept
Colluvial	Typic Udorthent	Typic Udorthent; Humic Lithic Dystrudept; Typic Dystrudept; Lithic Udorthent; Lithic Dystrudept; Typic Fragiudept
Fluvio-glacial terrace	Hydric Cryohemist Typic Cryaquent; Typic Cryorthent; Typic Cryosaprist	Aquertic Udifluvent; Fluventic Dystrudept; Humic Psammentic Dystrudept; Oxyaquic Dystrudept; Oxyaquic Udifluvent; Typic Dystrudept; Typic Endoaquent; Typic Fragiudept; Typic Udifluvent; Typic Udipsamment; Typic Udorthent
Fluvial terrace	Typic Cryaquent; Aquic Udipsamment	Typic Dystrudept; Typic Udorthent; Vitrandic Udorthent
Moraine	Fluvaquentic Sphagnofibrist	Lithic Dystrudept; Lithic Udorthent; Typic Dystrudept; Typic Fragiudept
Stoss and lee topography	Hydric Cryofibrist; Lithic Cryofibrist	Humic Dystrudept; Lithic Dystrudept; Lithic Udorthent; Oxyaquic Dystrudept; Typic Dystrudept; Typic Udorthent
Esker	–	Vitrandic Udorthent
Kame terrace	–	Typic Dystrudept
Kettle	–	Terric Sulfihemist; Typic Udifolist
Loess	–	Typic Dystrudept
Outwash plain	–	Typic Dystrudept; Typic Udorthent

the deeper horizons it is more basic (8.3–8.6). Carbonate filaments have been described under 13 cm depth, but only in the Fachinal soil series.

The agricultural aptitude of these soils is rather limited: most of the soil phases are included between IV and VII LCC. Only some soil phases of the Fachinal and Chile Chico soil series are classified as III LCC. From a taxonomic point of view, these soils are classified as Haploxerolls and Palexerolls and only the Puesto Viejo soil series is classified as a Haploxerept. Chile Chico soil series (Oxyaquic Haploxeroll) is considered representative of the soils in this group.

Figure 2.69 shows the landscape at Chile Chico in Region XI (46°34′S).

Soils with Low Base Saturation

These soils, with BS usually <50 %, are distributed in the udic moisture regime, here represented by the Cochrane soil series (Andic Oxyaquic Dystrudept), i.e. a soil with moderate depth in hill positions (20–30 % slope gradients), coarse-textured in upper horizons and fine-textured at depth. Fluvio-glacial deposit is a more common substratum, less frequently they occur over glacial clays, moraine deposits or rocks. Soil permeability is moderate and soils are well

Fig. 2.67 *Chapel of Marble* (*left*) and beginning of the Baker River (*right*) at General Carrera Lake, Region XI of Chile

Fig. 2.68 Landscape of the Puesto Viejo soil series (Typic Haploxerept) and the respective soil profile east of the Andean Mountains (Region XI)

drained. The Land Capability of these soils is limited both by stratification and by slope, classes VI and VII are common. The landscape at Cochrane is shown in Fig. 2.70.

2.2.3.3 Soils of South Patagonia (*Magallanes*)

According to Filipová et al. (2010) the most frequently recorded soil groups here are Histosols and Fluvisols. Gleysols, Vertisols, Regosols, Solonchaks and Solonetzs are detected with much less frequency. There is also considerable variability in soil properties among and within the soil groups. The principal differences between the Histosols and the Fluvisols are the content of organic matter (often peat), pH level (related to the absence/presence of carbonates) and associated soil properties. Fluvisols are more susceptible to salinisation under conditions of aridity, whereas the main threat to Histosols is artificial drainage.

Mineral Soils

The soils described in this area are located on alluvial terraces, with flat to gently undulating relief. The slopes are frequently 1–3 % and 2–5 %, with only few phases of the Última Esperanza soil series being described as having slopes of 5–8 % and exceptionally of 8–15 %. In some soils, the relief is flat to concave which contributes to creating poorly drained soils with a watertable that can fluctuate between 50 and 90 cm depth (Fig. 2.71). In addition, several of these soils have formed organic horizons at the surface (could correspond to histic epipedon) that can reach 25 cm thick.

The soil substrate is variable, even when the soils occupy similar positions in the landscape. Thus, substrates have been described with rounded gravel and interstitial sand, compacted sandstone, cemented clay and a compacted glaciolacustrine material very hard, clayey or silty clay, with some degree of cracking, locally known as *masacote*.

Grey (5Y 5/1) and olive grey (5Y 4/2) colours are common at depth, associated with poor to very poor drainage. The textural class of the surface horizon ranges from silty clay loam to silty loam and in some cases can be fine sandy loam. However, at depth the textural class is highly variable.

Many soils have been described as deep and very deep (100–130 cm); soils with limited depth (thin, 25–50 cm, and slightly deeper, 50–75 cm) correspond to soils with high watertable or its depth was defined by the presence of partially cemented sediments. Drainage is poor to imperfect

Fig. 2.69 Landscape of Chile Chico city (46°34′S, 71°40 W') in Region XI of Chile, on the shores of General Carrera Lake

in all the soils described; the exceptions are the Tres Pasos and Última Esperanza soil series, with slopes ranging from 5 to 8 % and 8 to 15 %. Some soils, such as the Las Chinas soil series, have a high content of Na, with sodium adsorption ratio (SAR) of 59 and 76 for horizons from 0 to 20 and 20 to 50 cm, respectively. LCC is assigned based on drainage, depth and slope, ranging from III to VII, with V being the most frequent.

According to Douglass (2005), who studied glacial deposits from 1 Ma old to the present, OC, CO_3 and clay content increase with the age of soil at a decreasing rate, while the Ca and Mg increase linearly with age. The presence of CO_3 in soils would be the result of both deposition of aeolian dust originating from evaporation from lakes and pedogenic redistribution from parent material.

South of 50°S, in the Province of Magellan, soils have undergone various processes of landscape modelling, so soil substrates are varied, such as glacial and fluvio-glacial sediments, fresh and weathered rounded gravel with sandy matrix and the aforementioned masacote.

Most soils in this area have been described as having abundant gravel in the profile. The dominant structure through the profile is subangular blocky, changing to massive in deeper horizons. In those soils with an organic surface horizon, it is possible to distinguish a granular structure. Depth to the substrate is variable, between 60 and 120 cm. The drainage is related to the position in the landscape soil, so poorly drained soils are associated with flat or nearly flat positions with IV, V and VI LCC; instead soils in gently undulated positions are well drained, with III and IV LCC.

In this area, it is possible to find Podzolic soils (Fig. 2.72) which present a profile with a surface layer formed by peat or a layer rich in organic matter and roots. At depth, there is a mineral horizon, fine sandy loam, sandy silt loam or sandy, grey-coloured (between 2.5Y 7/1 and 10YR 6/1), ending in a silty loam to silty clay loam horizon. The soils are been described as moderately deep to thin over a substrate of glacial sediments with gravel and stones or Cretaceous shale.

However, the current concept of Spodosols (Soil Survey Staff 2006) is based on the presence of an illuviation horizon (B_{hs}), which would not be clearly identified in the descriptions above. Gerding and Thiers (2002), in the area of Bueno River in *Tierra del Fuego*, identified soils to be included in this group, characterised by young soils, thin to very thin, influenced by volcanic activity, topography and climate, pH between 4 and 5.5 with high Al saturation (>60 %). Below the topsoil there is a grey (10YR 5/1), silty loam and structureless horizon, but the formation of this horizon could be influenced by the deposition of volcanic ash and by podsolisation processes.

Certainly, given the continued action of transport agents on the contribution of new parental material, Inceptisols should be an important group, more precisely Dystrustepts. However, soils with coarse textural classes throughout the profile should be considered rather within the concept of Entisols, probably Cryorthents.

Fig. 2.70 Landscape at Cochrane city (47°15′S, 72°34′W) in Region XI of Chile

Organic Soils

Knowledge of organic soils is more limited than that of mineral soils for the area, because there are few studies on recognition and identification of these ecosystems. The few studies that exist have focused on description and identification of plant species that make up each system, but not the soil in which these species are growing.

According to Clausen et al. (2006), there would be nine *major types of wetlands* in Patagonia, considered unique in South America, mainly in terms of plant species that dominate in each wetland. Three of these major types are considered closer to the concept of organic soils.

One of these wetlands is described as *Carex-Nothophagus* (called *mallines*), found mainly in depressions within glacial moraines and slopes that can reach 28 % on hillsides. A second type is similar to peat, described in the vicinity of Grey Lake, and corresponds to the concept of marsh or swamp in which the dominant species is *Sphagnum magellanicum*. The third type of wetland, considered the most important in Magellan, is called *vega*, occupying depressions in the landscape due to glacial processes that have created an impermeable clay layer very close to the surface and slopes lower than 1 % (Fig. 2.73). They are considered more like moist meadows than wetlands, because they constitute an

Fig. 2.71 Landscape near Puerto Natales (52°30′S) city

Fig. 2.72 Soil morphologically similar to a Spodosol. The limited surveying in the area and the lack of analysis result in uncertainty regarding soil classification (Luzio et al. 2010)

essential habitat for the guanaco (*Lama guanicoe*) during spring and summer.

Filipová et al. (2010) identified within the Histosols some Ombric Histosols (Haplohemists), Folic Histosols (Haplohemists) and Episalic Histosols (Sulfihemists).

2.2.4 Soils of the Insular (Easter–Juan Fernández) and Antarctic Zone

2.2.4.1 Insular Territory

Chilean oceanic islands are fragile ecosystems, with few and incomplete soil information available today, that have been modified by humans since prehistoric times and more severely in recent centuries. Therefore, only two inhabited territories in central Chile (Region V) are characterised here: Easter Island and Juan Fernández Archipelago.

Easter Island

An island in the south Pacific Ocean belonging to Chile, Easter Island has different names (Polynesian: *Rapa Nui* or Great Island, *Te Pito te Henua* or navel of the world, *Mata-ki-te-rangi* or the eyes that look to the sky; Spanish: *Isla de Pascua*). It is located 3,515 km west of continental Chile (27°09′S 109°25′W), has an area of 163.6 km^2 and is roughly triangular in shape, with a maximum ground elevation of 560 m a.s.l. It occupies an exceptionally isolated position in the South Pacific Ocean, where the nearest island is Pitcairn Island, located 2,075 km to the west. Only a few soil surveys have been carried out on Easter Island, and although some of these are very old, they represent the only reference available. The origin of the island is volcanic, and it is still possible to see three volcanoes located in each corner of the island (Fig. 2.74): Poike, an older (3 Ma) strato-volcano; Rano Kau, a caldera and the fissure complex of Terevaka and its associated cones (Baker et al. 1974).

Over a small submarine platform of 560 m high, this island presents a hilly landscape, with altitude from sea level to 70 m a.s.l. Average slope ranges between 1 and 15 %, with steep slopes (10–20 %) occurring on volcanic cones. The relief has been described as a sequence of terraces with a nearly level slope (1–3 %) descending down from the volcanic cones, which were probably formed by erosion.

Baker et al. (1974) carried out a comprehensive geological study on Easter Island and found that the lava showed a wide compositional spread, from tholeiites and olivine tholeiites to hawaiites, mugearites, benmoreites, trachytes and rhyolites (comendites). They also determined that hawaiite is by far the most abundant rock type and trachytes and rhyolites are relatively rare, whereas intermediate and acid rocks are concentrated in the south-western part of the island.

The recent geomorphologic evolution of Easter Island has been dominated by volcanic and marine events. The age of the geomorphologic forms defines the interrelationships among these, as is evident in the dense drainage net which is found in Hiva Hiva, which is a maximum of 5,000-years old (Muñoz 2004).

The state of the cliff in coastal areas is directly related to the low weathering processes over the parent material. For instance, there are differences between the north and south side in the fissure complex of Terevaka, but these have been related to differences in height between the cliffs and coastal areas.

In general, soils are very shallow and shallow, frequently with lava outcrops. As the slope increases, the soils are shallower than in flatter areas. Deeper soils are located at the bottom of volcanic cones and in lower areas with sediment accumulations.

The soil factors that determine the soil depth and properties are topography (height and slope) and parent material (type and age) (Wright and Díaz 1962). In the island soil, parent materials are volcanic materials, such as andesitic lavas, volcanic scoria, volcanic tuff, and basaltic and andesitic volcanic ash. Moreover, it is possible to find a mix of old volcanic ash and scoria from the volcanic cones in some areas. These volcanic materials are very fine and the scoria weathered, so therefore it is usually very difficult to identify new and old volcanic materials (Wright 1965). Volcanic eruptions have generated a high amount of basaltic rock, which is named Terevaka basalt, mainly in the central part of the island, covering approximately two-thirds of the total area. Therefore, it is considered that the parent material of the soils is tephras, either volcanic ejecta moved and deposited by wind or lava flows (Louwagie et al. 2006).

Louwagie et al. (2006) noted that in Easter Island the main soil limitations include (i) extensive areas with

Fig. 2.73 General view of landscape with evidence of aeolian erosion and at the *bottom* of the valley, a typical wetland with damp meadows important for feeding auchenids (*Lama guanicoe*)

shallow soils and basaltic rock on the surface; (ii) scarce soil water due to a long dry season with irregular precipitation distribution; (iii) soils excessively drained with low water holding capacity and (iv) soils with low values of available P and K.

Although different authors (Díaz 1949; Wright 1965; Alcayaga and Narbona 1969) agree that the soils in Easter Island are developed from volcanic materials, they disagree on the type of volcanic material and geological processes which created these soils. Díaz (1949) concluded that the soils are derived from intemperised volcanic ash and that the island is composed of andesitic lava over basaltic lava at the bottom. Wright (1965) considered the soils of the island to be tropical volcanic soils, together with nonvolcanic soils, similar to the young latosolic soils described in Western Samoa. Alcayaga and Narbona (1969) concluded that the soils in Easter Island developed from volcanic ash and weathered lava. Mikhailov (1999) added that in addition to Pleistocene volcanic activity in the island, there were intensive hydrothermal processes that generated red-clay deposits in the fault lines, where these clays were deposited in undulating slope positions. Moreover, this author noted that east winds were the source of loess deposits, by transporting fine particles from abrasion zone to plateau, and that the red deposit in the upper soil horizons was caused by hydrothermal metasomatism of volcanic rocks and the only product of soil formation processes would be a humic horizon. Horrocks and Wozniak (2008) claimed that the soils in the island are Oxisols, but they did not report any analytical data or taxonomic characteristics.

On Easter Island the weathering of tephras has generated soils with andic soil properties, with significant amounts of short-range-order compounds, such as allophane, imogolite, ferrihydrite, or aluminium-humus complexes in soils (Louwagie et al. 2006). Thus, soils in the island present low bulk density, high amounts of OM and high P retention. In addition, the soils have loamy textures at the surface and fine textures at the bottom of the profiles, which is similar to volcanic soils in the Mediterranean zone of central continental Chile (Honorato and Cruz 1999).

The main controllers of the soil cover mosaic are altitude, slope, the nature of the parent material (e.g. lava flow or tephra deposit) and the age of these deposits, and only small pockets of arable soils occur, scattered across the rest of the island (Mann et al. 2008). Only soil surveys that include a detailed description are reported here, which in chronological terms include Díaz (1949), Wright (1965) and Alcayaga and Narbona (1969).

Díaz (1949) grouped the soils according to its LCC and identified seven soil series. Although this is a very old soil survey that omits many soil properties included nowadays, it is the primary work used by national and foreign researchers for studies on Easter Island. The author classified the soils in different slope gradients: <6 %, 5–10 %, 10–15 %, and >30 %. The dominant soil colour in surface soils is brown, ranging between light brown, dark reddish brown and light reddish brown with only one white soil, providing details about the parent material, whereas deeper horizons have yellowish brown as the dominant colour. Soil textural class varies between silt loam and sandy loam in all soils. It is assumed that with the actual soil structure types, soil may present no visible structure (massive). In general, the soils are described as deep, ranging from 80 to 200 cm, with a stony surface layer, whereas no stones are found in the soil profiles. Some chemical analyses have been carried out and show that soil pH ranges between 5.5 and 6.6 and CEC between 28 and 60 $cmol_c\ kg^{-1}$.

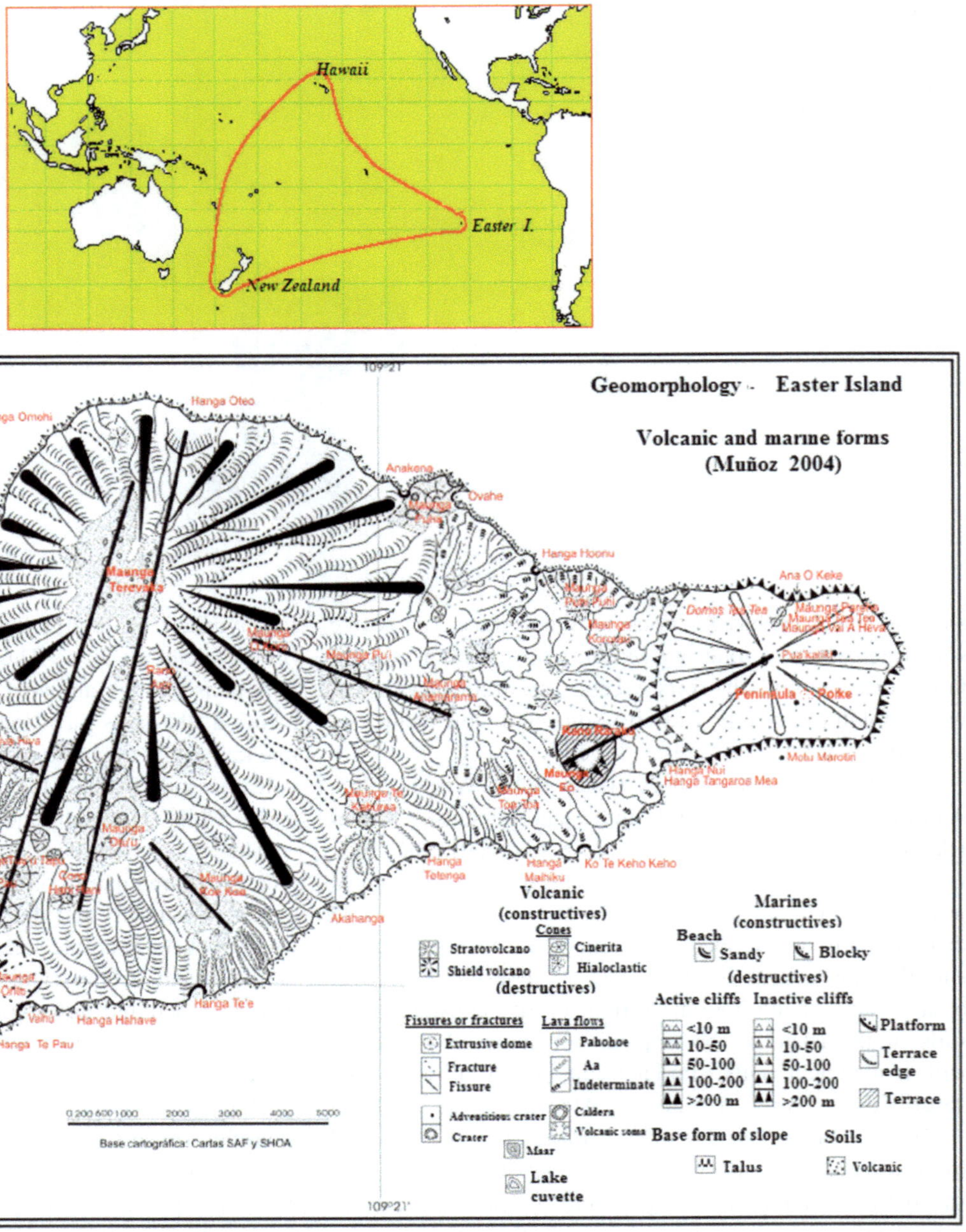

Fig. 2.74 Geomorphology map of Easter Island and its relative location (*below*) in the Pacific Ocean (*above*) modified from Muñoz (2004)

Wright (1965) described the soils on hills, plateaus and gently sloping areas in the volcanoes and grouped these into three soil series, Rano Kao, Vinapu and Poike. Slopes ranged from 7 to 14 %, with the substratum identified as scoria or weathered volcanic tuff. Soils are deep, between 125 and 175 cm to the substratum. The colour in topsoil is dark brown, whereas in deeper horizons it ranges from dark reddish to yellowish red. Textural classes are finer than the soil described by Díaz (1949), such as sandy loam, silty clay loam and silty clay in all horizons. Soil structure is subangular blocky. In that study, the soil pH was found to range between 5.1 and 5.9 at the soil surface, whereas in deeper horizons it was more acid, reaching 4.8. Although unfortunately the soil survey by Wright (1965) did not provide enough data either to compare with those of Díaz (1949) or to classify the soil according Soil Taxonomy, it is clear that the parent materials of the soils of the island are volcanic materials.

Alcayaga and Narbona (1969) carried out a detailed soil survey in the entire island at scale 1:40,000 and described soils on volcano hills, piedmonts and floodplains, generally with slopes level (0–1 %), nearly level (1–3 %), gently (3–8 %) and strongly sloping (15–20 %). In particular, a very stony soil surface was found in soils on volcano hills

Table 2.6 Correlation between World Reference Base and Soil Taxonomy for some soils in Easter Island (Louwagie et al. 2006)

Soil	World Reference Base	Soil Taxonomy
TP3 and VA4	Hyperdystri-SILANDIC (aluandic) Andosol	Typic Hapludand
Akahanga: AK1, AK2 and AK3	(Silti-Pachic) (Silti-Endosonic) (Silti-Hiposodic) Phaeozem	Andic Hapludoll
La Pérouse: LP4	Silti-Pachic Phaeozem	Typic Hapludoll
LP6	Silti-Luvic Phaeozem	Typic Argiudoll
LP1 and VA3	(Hyperdystri-Silandic) (Ferrali-Silandic) Cambisol	Andic Dystrudept
Vaitea: VA2 and LP3	(Hyperdystri-Humic) (Hyperdystri-Umbrihumic) Cambisol	Andic Dystrudept–Humic Pachic Dystrudept
VA8	Humi-Episkeletic Leptosol	Lithic Udorthent

and piedmont positions. The soil substratum was lava flows with different stages of weathering processes, with the exception of the Vaitea soil series, which had a mix of ash and lava flows. Although the authors found that the dominant soil texture class was clay, ranging from dense clay to less dense clay in whole soil horizon, it is not in agreement with volcanic soils (Andisols) in the southern Mediterranean zone. Soil colour in surface soils is generally dark brown to very dark brown, whereas in subsurface horizons it is closer to reddish. The survey found no poorly drained soils, and these soils colours were related to the parent material (tephras). The dominant soil structure is subangular blocky in surface and subsurface horizons, with some cases of no visible structure such as massive at the bottom of the soil profiles. The soils are usually shallow (<50 cm depth), with the exception of the Rano Kao soil series, which is deep (>100 cm depth). All the soils are well drained, with soil restriction to root growth at depth resulting in commonly occurring very fine roots in surface horizons and scarce roots in deeper horizons. Some data showed that SOC at the surface ranges from 1.8 to 4.2 %, decreasing in deeper horizons to values around 0.5 %. A broad range of soil reaction was found in all soil horizons, with soil pH values between 6.3 and 4.9. Cation exchange capacity was found to be medium to high, ranging from 44 $cmol_c\ kg^{-1}$ in surface horizons to 18 $cmol_c\ kg^{-1}$ at the bottom. These authors defined the soils from moderate to extremely limited for agricultural use, with LCC between IV and VI, with the exception of the Orito soil series, which is II.

On the other hand, Louwagie et al. (2006) reported that soils in Easter Island are derived from tephras, clearly show andic soil properties and can be classified as Andisols. Table 2.6 shows the correlation between World Reference Base and Soil Taxonomy for some soils in Easter Island according to Louwagie et al. (2006).

Juan Fernández Archipelago

This archipelago includes raised areas from submarine volcanoes at a depth of 1,000 m bsl, in an extension of 402 km from 74 to 83°W. This submarine volcano series is part of a submarine mountain range that runs north parallel to the continental Chilean coast, related with the Juan Fernández ridge of the Nazca plate. This archipelago is located 670 km west of continental Chile (33°36′S–33°66′S and 80°47′W–78°47′W), and has three main islands: *Más a tierra* or Robinson Crusoe Island, *Más afuera* or Alejandro Selkirk Island and Santa Clara Island. Stuessy et al. (1984), using K–Ar dating of five basalts, found ages of 1.01 ± 0.12 and 2.44 ± 0.14 Ma for Alejandro Selkirk Island, 3.79 ± 0.20 and 4.23 ± 0.16 Ma for Robinson Crusoe Island, and 5.8 ± 2.1 Ma for Santa Clara Island.

Robinson Crusoe Island has an area of 48 km^2 with a maximum ground elevation at 922 m a.s.l. and is long in shape. It has many volcanic centres, which have extended to peripheral structures (Baker et al. 1987). Approximately 180 km west of Robinson Crusoe Island lies Alejandro Selkirk Island, with an area of 52 km^2, which has a cupola shape, very cliffy, with a maximum ground elevation of 1,650 m a.s.l. Alexander Selkirk Island is the youngest and highest of these islands and exhibits strong ecological zonation, with many species associated with particular environments. Santa Clara Island is the smallest, 2.2 km^2, and is located 1.2 km south-west of Robinson Crusoe Island (Fig. 2.75).

The only information available regarding the soil properties of this archipelago were reported by IREN (1982) and updated by Cereceda et al. (1996). Figure 2.76 shows the distribution of three general soil groups according to natural zones defined for Robinson Crusoe and Santa Clara islands. Figure 2.77 shows a more detailed portion of Robinson Crusoe Island.

In general, Rici (2005) reported that soils on all three islands are volcanic in origin, and thus highly susceptible to erosion, particularly in areas where vegetation cover has been overgrazed or removed. There are few permanent surface streams and no lakes, as the soils are very porous and the steep slopes cause rapid runoff. The islands have a maritime climate, with a high average relative humidity of 77 % and mean annual precipitation of 1,181 mm.

Robinson Crusoe Island is composed of basaltic material with different stages of weathering and morphological relationships, where IREN (1982) concluded that cinerite deposits are not related with rock weathering. Similarly, Cereceda et al. (1996) noted that the soils in the islands are volcanic.

IREN (1982) described geomorphologically at least four altitudinal areas: (i) high mountains in the eastern zone of the island, where basaltic basalt dike occupies the top areas; (ii) hilly zones in a medium position in boundaries of watershed in the north side, where slopes range from 35 to 60 %, with continuous outcrops of rocks of different hardness; iii) low hill areas with slopes between 35 and 40 % that have suffered intensive landslide processes due to natural and anthropogenic causes, such as forest clear-cutting and overgrazing, which have formed even depth gullies; and (iv) deposition zones of material transported by alluvial, wind and landslide processes. Alluvial cones and landslides are associated with developed valleys, which are not connected to actual fluvial dynamics, whereas active wind deposition areas have been found around the island but mainly close to the airport in the west zone of the island.

Nowadays, the soils show the effects of different soil formation factors, such as topography, vegetal cover and anthropic activity, but soil erosion has had a great impact on soil characteristics. Thus, IREN (1982) defined three soil types associated with relief:

i. Soils of mountains of new origin and derived from volcanic ash, classified as Andepts of new origin according Cereceda et al. (1996). These are located in the central part of the island covering 1,356 ha, approximately 29 % of the total area of the island. Soils are located between 250 and 300 m a.s.l., where slopes may reach 80 %. In these areas, vegetation includes indigenous forest. Soils are thin and characterised by A–C horizons and have loam to coarse texture, usually with basaltic outcrops. Some measured soil properties show that soil pH is slightly acid to neutral (6.0–6.8), soil organic matter levels are high (17–25 %), and CEC is also high (70–100 $cmol_c\ kg^{-1}$). The Yunque and Chifladores soil series are characteristic of these positions (IREN 1982).
ii. Soils of intermediate level, located in hilly zones in a medium position, where slopes range from 20 to 40 % and colluvial material has been deposited. According to Cereceda et al. (1996), these soils may correspond to more evolved Andepts. They occupy 705 ha, approximately 16 % of the total area of the island. These soils are more evolved than the Andepts of new origin and present an A–B–C soil horizon distribution, while in some soils there is an argillic horizon (B_t). Some measured soil properties show that soil texture class ranges from loam to clay, usually with fine gravel, soil organic matter levels between 4 and 9 %, and that CEC

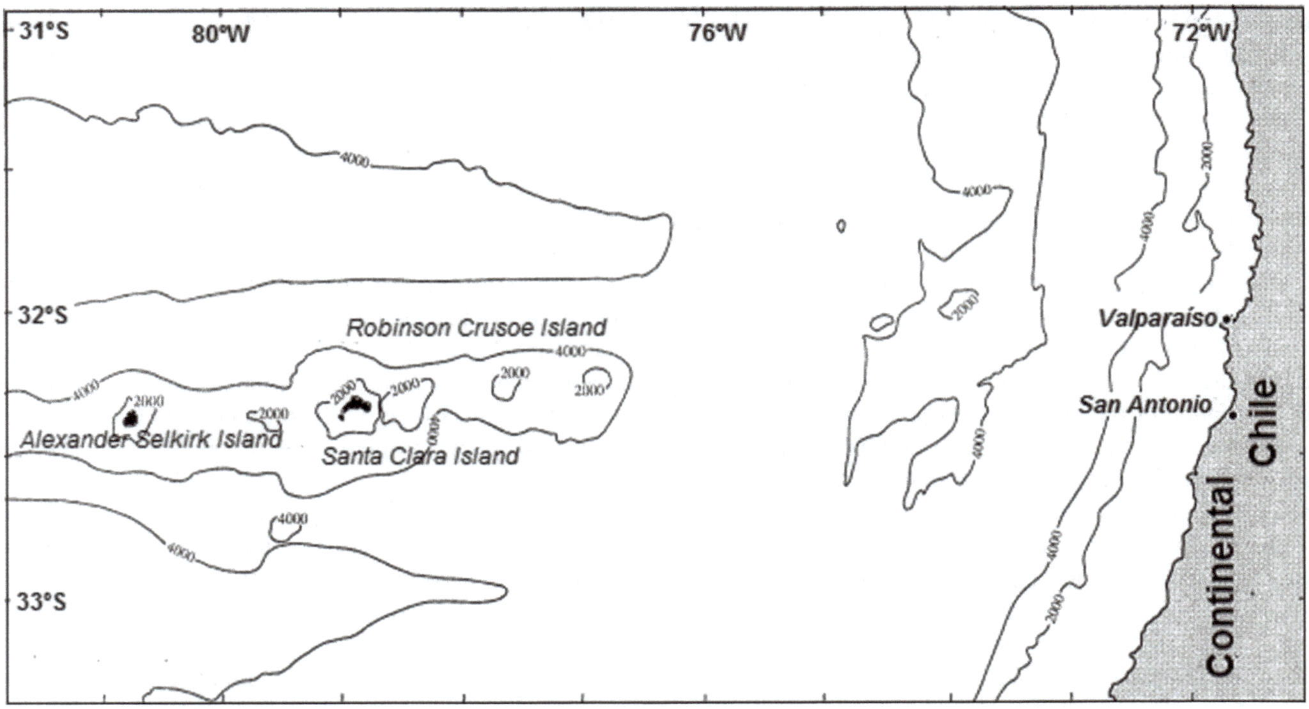

Fig. 2.75 Juan Fernández Archipelago in relation to the Chilean coast

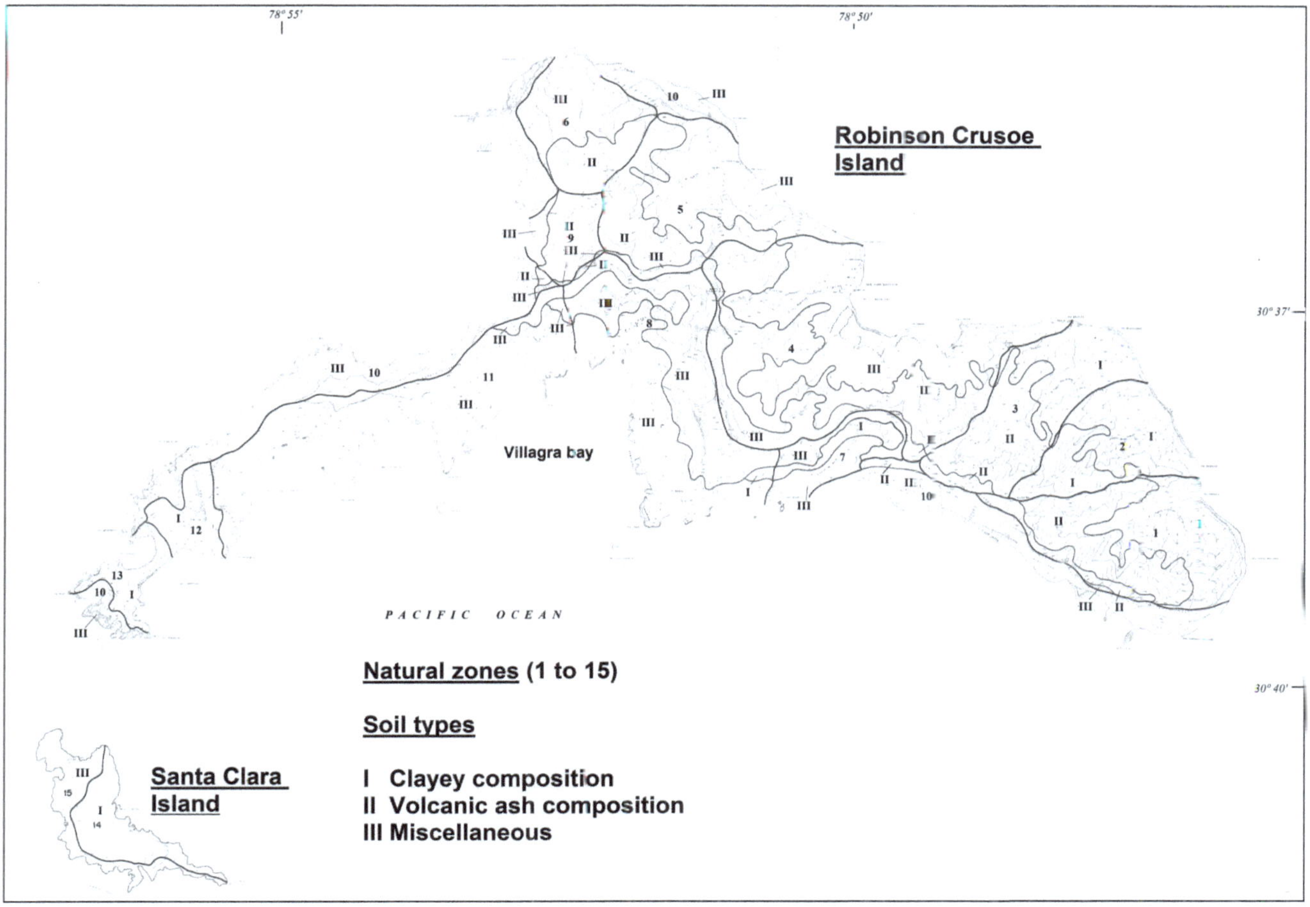

Fig. 2.76 General soil group distribution on Robinson Crusoe and Santa Clara islands. Adapted from IREN (1982)

is high (>40 $cmol_c\ kg^{-1}$). According to IREN (1982) the Vaquerías, Puerto Inglés, Puerto Francés and San Juan Bautista soil series are some examples of the soil groups occurring in these positions.

iii. Soils in depositional positions located in flat to slightly sloping areas, fine textured and with a moderate degree of evolution, displaying in some cases an A–B–C soil horizon distribution. According to Cereceda et al. (1996), these soils may correspond to Fluvents. They occupy 204 ha, approximately 4 % of the total area. Some of these soils in the area close to the airport, in the west zone of the island, show evidence of wind erosion processes.

Finally, the Land Capability of the two islands is illustrated in Fig. 2.78 which gives a general view of the soil and topographical constraints common in this territory.

2.2.4.2 Antarctic Territory

Lagabrielle et al. (2009) provided an integrated review of plate tectonic models of the evolution of the Antarctica-Patagonia connection and robust correlations of seaways and tectonic events along the Scotia and South America plates, and concluded that the opening of Drake Passage (Fig. 2.79) was not steady state since ca. 30 Ma.

Rather, the regions forming the present-day northern limit of this gateway experienced important paleogeographic changes, from deep marine basins to shallow ridges and emerged regions during the late Oligocene and early-mid Miocene. Geological data show that emergence along the North Scotia Ridge and *Tierra del Fuego* was achieved at 23–22 Ma, and has been followed by elimination of the Patagonian Sea in Patagonia. This transition towards more continental sedimentation in southern South America is correlated with more shallow marine conditions in the Austral Basin. This succession of events had a strong influence on the general geometry of Drake Passage, corresponding to a constriction of its northern limit.

A general description of Antarctic soils is included here, extracted from Campbell and Claridge (2009). Coarse fragments (>2 mm) in Antarctic soils are abundant, typically exceeding 50 %. Horizon development is weak, and mostly restricted to colour changes that diminish in intensity with

Fig. 2.77 Soil types on the Robinson Crusoe Island (Cereceda et al. 1996)

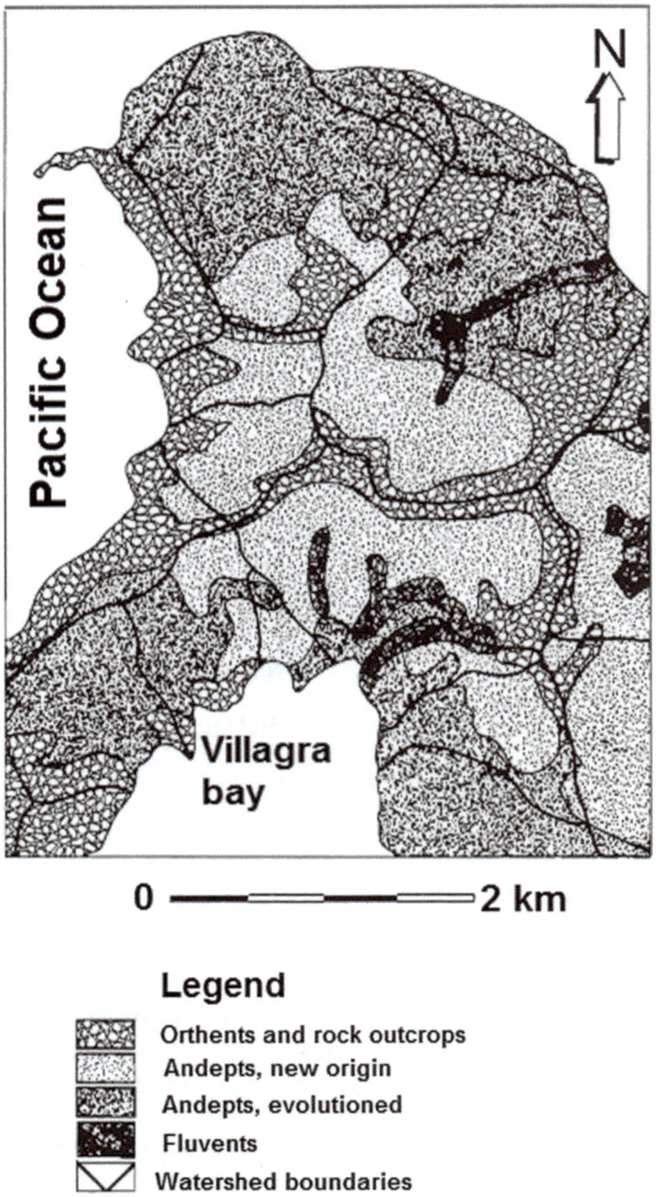

increasing depth, to lithologically related textural changes, or to salt accumulation. The soil surface is usually a stone pavement including loose material derived from fragmentation of surface clasts. On younger surfaces, clasts are mainly angular, coarse and unweathered, while on older surfaces, clast rounding, rock pitting, ventifaction, oxidation and disaggregation may be prominent. Weakly developed structure may be present in upper horizons as a result of freezing when the soil is moist. Where there is an increased proportion of fine material, a thin surface crust may be present. Below the surface, the soil is usually structureless and dusty, except where salt concentrations occur, when the soil material may be firmly cohesive. In older soils, the disaggregation of coarse grained clasts by salt weathering results in rock ghosts that indicate a highly stable soil environment.

The permafrost is generally ice-cemented (Fig. 2.80), but in older and drier soils may be loose. Because of the extreme aridity, the soils accumulate salts derived from precipitation and weathering, the composition and amount of the salts being a function of soil age, composition of the parent material and distance from the coast. Chemical weathering processes are assisted by the salts, which allow unfrozen saline solutions to be present on grain surfaces and cracks in rock particles, even at very low temperatures. Weathering comprises the breakdown of ferromagnesian minerals, releasing Fe and cations to the soil solution. The Fe oxidises and is precipitated on grain surfaces, giving rise to the red colouring of older soils (Fig. 2.81). The cations, especially Ca and Mg, combine with nitric and sulphuric acids arriving in precipitation to make up part of the thick salt

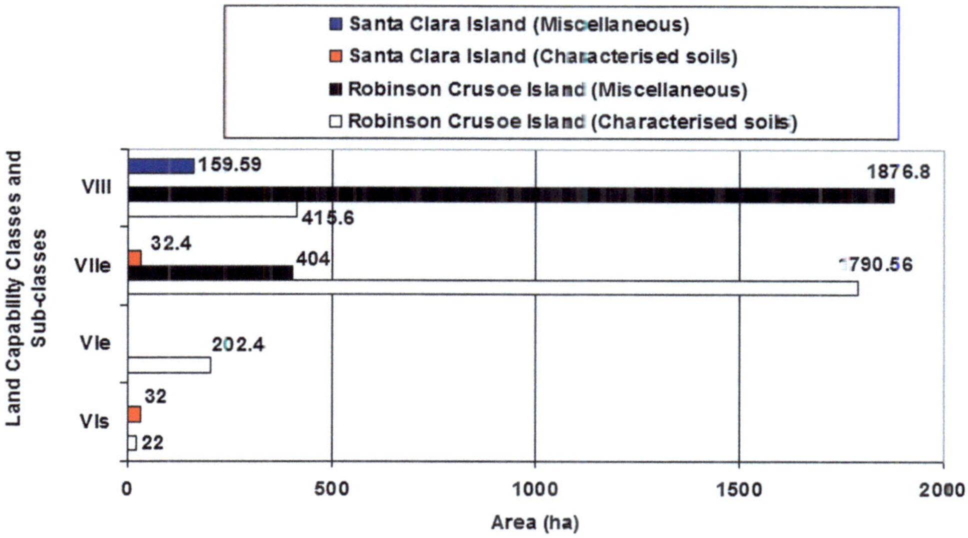

Fig. 2.78 Land Capability of Robinson Crusoe and Santa Clara islands from IREN (1982)

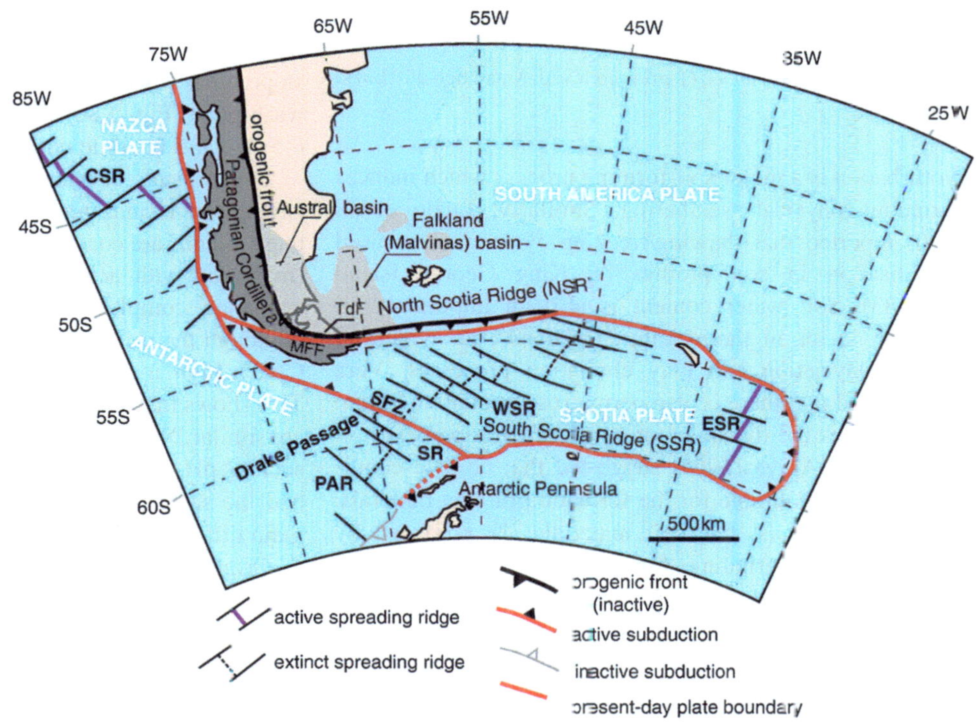

Fig. 2.79 Present-day setting of the Antarctica-South America connection and tectonic environment of Drake Passage (Lagabrielle et al. 2009)

horizons which are found in older soils. The concentrated salt solutions react with silica, also released by weathering, to form secondary clay minerals, and in some cases, zeolites.

Soil surveys carried out during the early 1990s by researchers at the Soil and Engineering Department, University of Chile, in the Antarctic islands are presented is this section, particularly those works carried out in the South Shetland Islands, including Robert (Haberland 1992, Álvarez 1993), Livingston (Henríquez 1994) and King George islands (Luzio et al. 1990). Figure 2.82 indicates the relative geographical position of these islands.

It is important to note that the soil characteristics of the Antarctic islands and the Antarctic continent are different due to dissimilar precipitation and temperature regimes. In particular, climate conditions in the Antarctic islands are more favourable for vegetal growth than those in the Antarctic continent, giving rise to higher soil OM accumulation in the Antarctic islands and having a great impact on soil formation processes.

In the Robert and Livingston islands, Haberland et al. (2008) suggested that the landscape and topography are the main factors affecting soil distribution pattern, whereas

Fig. 2.80 Permafrost under (1 m) a colluvial Antarctic soil (Luzio et al. 2010)

cryoturbation is a major soil-forming process which induces continuous physical weathering. Similarly, Luzio et al. (1990) reported that thawing/freezing cycles have caused haploidisation in soil profiles on King George Island (Fig. 2.83). Soil water content is also an important soil formation factor in these Antarctic islands, as evidenced through gleyisation and clay eluviation processes. Vera (1990) added that there is also permafrost from 40 to 70 cm that influences the dynamics of soil formation processes.

Hervé and Araya (1965) suggested that climate conditions are not favourable for the formation of clay minerals. Thus in these soils, fine soil materials are produced by mineral and rock grinding. Campbell and Claridge (1987) concluded that in the Antarctic there is a lack of significant soil development, with weathering and the action of living organisms having been directly conditioned by climate conditions (Fig. 2.84).

According to soil surveys carried out in the South Shetland Islands, the soils can be broadly divided into mineral and organic (ornithogenic) soils.

Mineral Soils

Figure 2.85 shows a typical poorly developed Antarctic mineral soil. On Robert Island, Haberland (1992) and Álvarez (1993) described 10 profiles, from very thin to moderately deep (17–70 cm), overlying a basaltic rock substratum. In some cases, between 50 and 100 cm, permafrost was described. Deeper soils were found in areas where the permafrost was between 90 and 110 cm. Soils occupy different positions in the landscape with varying slopes, including steep slope areas (30–35 %), drumlins (10–15 %), slightly hilly areas (7–10 %), high terraces (6 %) and marine terraces (2 %).

On Livingston Island, Henríquez (1994) described five profiles and found thin to moderately deep soils, ranging from 37 to 80 cm. These soils occupy high terraces, steep slope areas, marine terraces and hilly areas inside an outwash plain (Fig. 2.86). The soils usually have gentle slopes (–5 %), and in some cases steep slopes (>20 %).

Some results of measured soil properties for Robert and Livingston islands are shown in Appendix. The soils on these islands are coarse textured, generally without structure (massive), with acid to slightly alkaline reaction and an exchange complex dominated by Ca^{2+}, Mg^{2+} and H^{+}. Although the soil colloids are present in low quantities, the CEC is high, mainly due to the presence of zeolite minerals. This shows that soil properties in the South Shetland Islands are similar. Navas et al. (2008) also found that soil properties such as pH, EC, carbonates and soil organic matter content had the same patterns. They added that in these soils, the main differences lie in soil particle size distribution.

Originally, the soils on Robert and Livingston islands were classified as Pergelic Cryorthents and Pergelic Cryochrepts (Haberland 1992; Álvarez 1993; Henríquez 1994), and those on King George Island as Cryochrepts (Luzio et al. 1990), according to Soil Taxonomy and before Gelisols were

Fig. 2.81 Soil profile and landscape at King George Island (Luzio et al. 2010), where coarse texture and structureless soils have been described (single grain or massive)

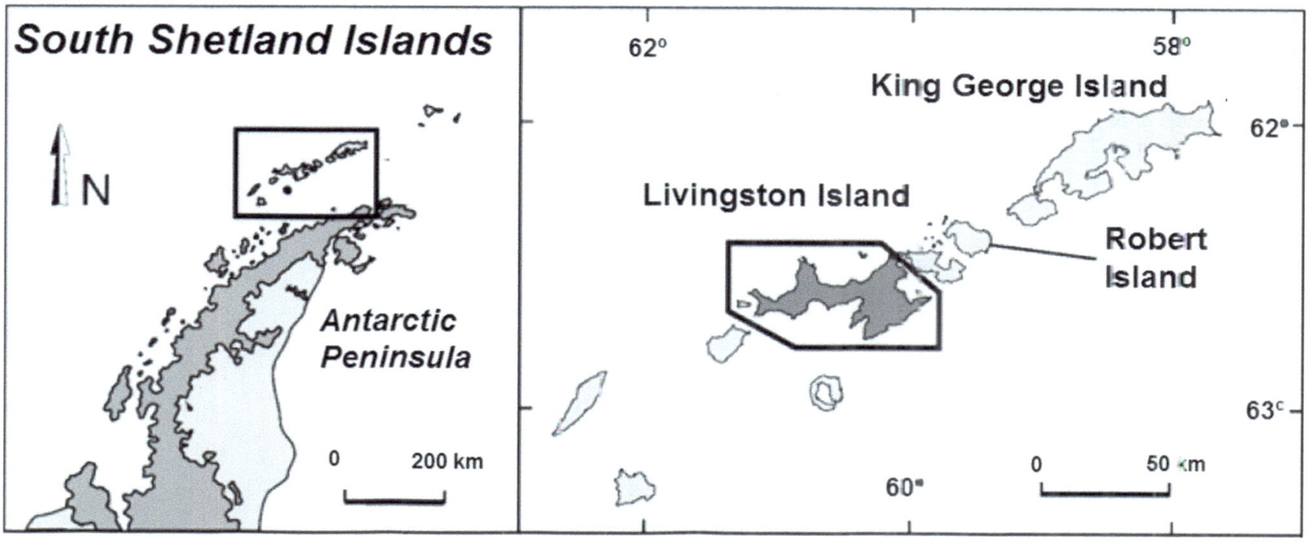

Fig. 2.82 Location of South Shetland Islands and Antarctic Peninsula

Fig. 2.83 Weathering processes (freezing/thawing) occurring at the surface of Antarctic soils (Luzio et al. 2010)

established as a soil order. As regards WRB (IUSS Working Group WRB 2006), Pergelic Cryorthents and Pergelic Cryorthents correlate to Glacic, Cryosols (Dystric) and Glacic, Cambic, Cryosols (Eutric). For these soils, Haberland et al. (2008) suggested that the best approximation according to the Key to Soil Taxonomy (2006) is Typic Haplorthels.

Ornithogenic Soils

These soils are an interesting organic soil group of cold regions, particularly of coastal continental Antarctica, where penguins play an important role in soil nutrient cycles, by during the breeding season bringing huge amounts of organic materials from the sea to the land. These organic materials accumulate to considerable thickness in the form of a friable layer of droppings, feathers, shells, bones and dead penguins. Most soils show cryoturbation features such an organic and organic horizon mixture, displaying particular forms such as wedges and contorted horizons (convoluted). According to Michel et al. (2006), soil organic matter accumulation and associated phosphatisation are marked soil-forming processes in ice-free areas once colonised by penguins. Ugolini and Bockheim (2008) concluded that phosphatisation is the main pedogenic process in these ornithogenic soils, causing acidity and weathering, showing that chemical weathering in maritime Antarctica is more intense than expected.

The ornithogenic soils are distinguished from non-ornithogenic soils by several features such as the presence of

Fig. 2.84 Organism soil formation factor, restricted to weak lichens, mosses and seaweed cover, affecting only few centimetres of upper soil horizons (Luzio et al. 2010)

Fig. 2.85 A typical A–C soil horizon sequence in the Antarctic zone (Luzio et al. 2010)

continuous vegetation cover, lower pH and lower BS. Their chemical properties change rapidly after the abandonment of the rookeries by the penguins. Processes such as CO_2 evolution and N and P release in the sea, formation of oxalic acids in upper horizons and simultaneous concentration of recalcitrant soil compounds (chitin, urates, and phosphate minerals) are indicated by Finkl (2005) and Zhu et al. (2009) in these organic soils.

On King George Island, Michel et al. (2006) describe this organic type of soil, where parent material is composed of moraines covered by weathered volcanic rocks, mainly andesitic basalts and gravels from an upper outwash. Ornithogenic Gelisols are deep and present a coarse active layer with gravel reaching 30 cm in depth and a brown humified histic epipedon. Deeper soils occur along gentle slopes and stable sites abandoned by penguins, where vegetation reaches a maximum height and diversity.

An ornithogenic soil on King George Island is reported by Michel et al. (2006), as a developed, uniform (texture and colour) Lithic Fibristel located on a flat ground moraine, near a present-day penguin rookery.

Fig. 2.86 Antarctic landscape including high terraces and other features within an outwash plain (Luzio et al. 2010)

The upper horizon is formed from fibric remains of mosses, whereas in brownish-coloured deeper horizons organic matter is partly humified. The soil is almost exclusively colonised by mosses due to its poor drainage, although no clear gleying is noticed. Another example is a Lithic Umbriturbel, located on a former lateral moraine covered by debris from an upland nesting bird area, with permafrost at 70–75 cm depth. Topsoil is formed by a thin layer of fibric remains of dead mosses mixed with loose rock fragments, but from 10 to 40 cm it is composed of a stony layer with hydromorphic gleying features, followed by dark humified material mixed with pebbles at 40–60 cm, with fibric material.

Several ornithogenic soils are reported by Simas et al. (2007), also on King George Island (Table 2.7).

Table 2.7 Ornithogenic pedons on King George Island described by Simas et al. (2007)

Pedon	[a]Altitude	Vegetation	ST/WRB	General description
A10	147	Absent	Typic Haploturbel; Turbic Cryosol	Subpolar desert without any ornithogenic influence. Desert pavement overlying oxidised basaltic till. Virtually no horizonation. Single-grain conditions and medium granular structure
A9	87	D > C	Lithic Haplorturbel; Turbi-Leptic Cryosol	Skua nest in subpolar desert with weak ornithogenic influence. Discontinuous vegetation with occurrence of higher plants (*Deschampsia antarctica*) and green algae (*Prasiola crispa*). Formation of a shallow, incipient A horizon. Single-grain condition and medium granular structure
A4	72	D > C	Typic Psammoturbel; Hapli-Turbic Cryosol	Abandoned rookery on the highest and most ancient level of ornithogenic influence. Continuous vegetation cover. Well defined A, B and C horizons, with crumbs in A and medium granular structure in B. Ice-cemented from 45 cm
A3	69	D > M > C	Typic Psammoturbel; Skeleti-Turbic Cryosol	Abandoned well-drained penguin nesting area. Well-defined A horizon below a continuous vegetation cover. Phosphatisation evidenced by whitish discontinuous B horizon starting at 20 cm. Crumbs in A horizon. Single-grains condition and medium granular structure in B horizon. Ice-cemented permafrost at 50 cm
A6	50	D > M > C > L	Andic Umbriturbel; Umbri-Turbic Cryosol	Close to pedon A5 but with a longer time of abandonment as evidenced by the well-developed vegetation cover and presence of A horizon. Ice-cemented permafrost at 60 cm
A1	45	D > L	Andic Umbriturbel; Umbri-Turbic Cryosol	Well-drained uplifted terrace, at the lower part of a slope. Continuous vegetation cover and well defined A horizon. Indirect penguin influence through lateral percolates from former rookeries
A5	45	P	Typic Haploturbel; Andi-Turbic Cryosol	Represents the most recently abandoned rookery, adjacent to the current penguin nesting area. Incipient A horizon, below a well-developed *P. crispa* cover. Ice-cemented permafrost at 65 cm
A2	32	M > D	Psammentic Aquiturbe; Oxyaqui-Turbic Cryosol	Similar to A1 but poorly drained, with predominance of mosses. This site receives a high amount of leacheates from upslope rookeries
A7	23	M > D	Terric Fibristel; Turbi-Histic Cryosol	Moss peat on uplifted marine terrace. Fibric, 50 cm deep, O horizon overlying a mineral horizon. Ice-cemented permafrost at 70 cm from the soil surface
A8	5	D > M > C	Andic Cryofluvent; Gelic Fluvisol	Marine terrace under strong influence of nutrient-rich leacheates from upslope penguin rookeries. Frequent occurrence of phosphatic, white precipitates on rock surfaces. Single grain

M mosses, *L* lichens, *D D. antarctica*, *C C. quietensis*, *P P. crispa*. [a] (m a.s.l.)

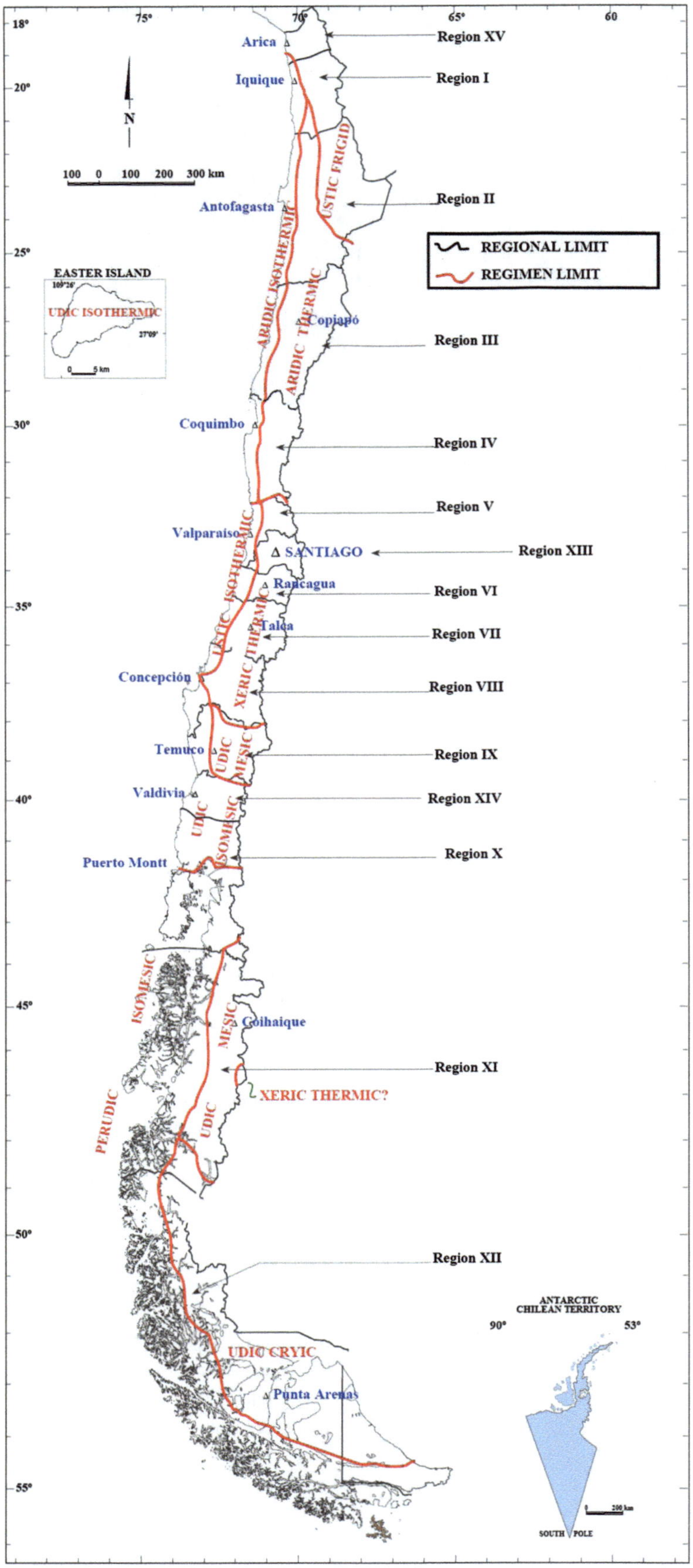

Fig. 2.87 Actual soil temperature and soil moisture regimes for Chile (Luzio et al. 2010)

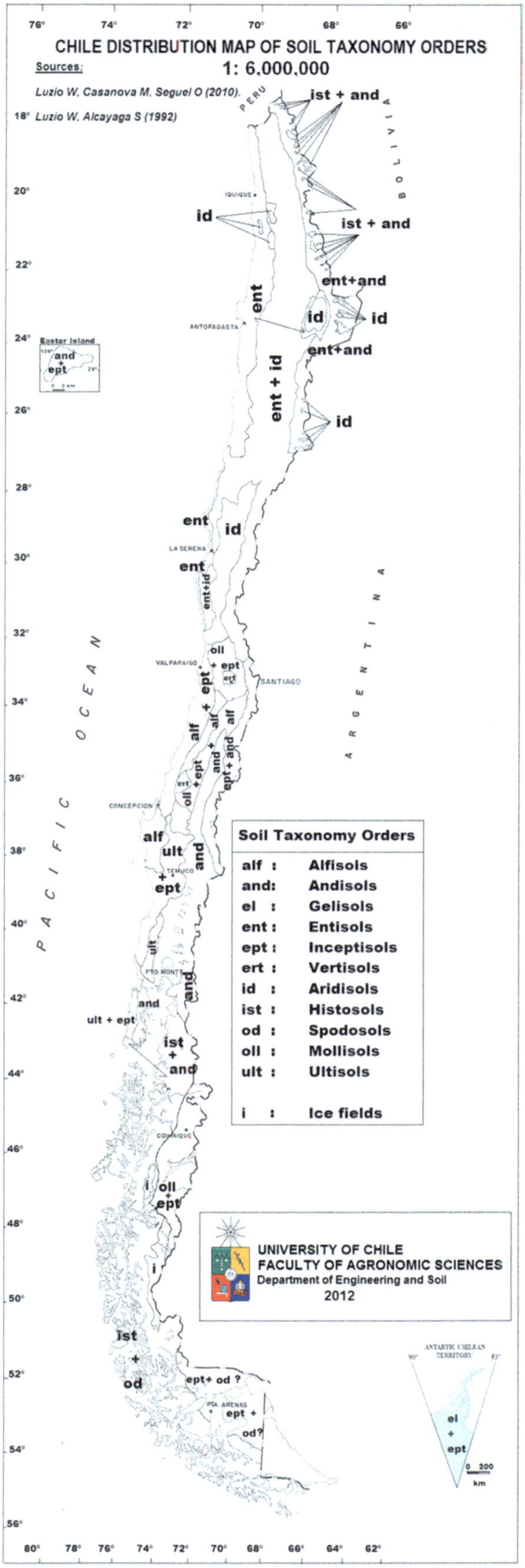

Fig. 2.88 Distribution of current Soil Taxonomy orders in Chile

2.3 A Soil Map of Chile

Although high soil variability is probably present in any particular major soil zone, there are at least two elements that can be considered relatively constant; moisture and temperature regimes in the control section of the soils are studied when soil mapping is the objective (Luzio et al. 2010). Figure 2.87 summarises the soil moisture and temperature regimes for Chile proposed by van Wambeke (1981) and van Wambeke and Luzio (1982) reported by Luzio et al. (2010).

Finally, we present a map showing the distribution of Soil Taxonomy system (USDA) orders observed in Chile (Fig. 2.88). It consists of a broad-based inventory of soils that occur in a repeatable pattern on the landscape, including the probable classification and extent of the soils. The map shows at least 14 different soil mapping units described in the legend at the level of soil orders, represents the state of current knowledge. Each polygon shows the dominant soil order present and identifies the soil order possessing the largest summed component percentage (largest land area) for each map unit. However, this soil distribution map should only be used to compare the general distribution of different soil orders.

Scaled to 1:6,000,000, it is the first informative level for a soil map of Chile, and is also a tool for soil correlation at the continental level. Soil regions are a regionally restricted part of the soil cover characterised by a typical climate, geomorphology and parent material association.

References

Aburto F, Hernández C, Pfeiffer M, Casanova M, Luzio W (2008a) Soil: a work of art of nature. Northern field guide between 30S° and 33S° In: Casanova M, Luzio W (eds) The international conference and field workshops on soil classification. University of Chile, Santiago-Ovalle, Chile

Aburto F, Pfeiffer M, Luzio W, Seguel O (2008b) Norte Chico soil genesis under paleoclimatic change perspective during quaternary and its soil classification implications. In: Casanova M, Luzio W (eds) Proceeding international conference and field workshops on soil classification. Santiago, Chile, 9–18th Nov

Ahumada M, Faúndez L (2009) Descriptive guide of terrestrial azonal hydric vegetation systems of altiplanic ecoregion (SVAHT). Chilean Ministry of Agriculture, Agriculture and Livestock Service. Santiago (in Spanish)

Ahumada M, Araya E, Kühne A, Sepúlveda M (2004) Descriptive guide of prairies and soils from Province of Palena, Region X. Chilean Ministry of Agriculture, Agriculture and Livestock Service. Publication N° 17 (in Spanish)

Alcayaga S, Narbona M (1969) Detailed soil survey of Easter Island. Technical publication N° 3, Development Corporation (CORFO)-Chile, Santiago (in Spanish)

Álvarez P (1993) Soil survey in the Robert Island, South Shetland Islands, Antarctic. Dissertation, University of Chile (in Spanish with English summary)

Amundson R (2003) Soil formation. In: Drever JI (ed) Treatise on geochemistry, vol 5. Surface and ground water weathering and Soils. Elsevier, Amsterdam

Aqueveque C (2008) Analysis of soils developed on coastal old dunes of Central Chile. El Yali, Putú and Chanco systems. Dissertation, University of Chile (in Spanish with English summary)

Araneda M, Avendaño M, Merlo C (2000) Gravimetric model of Santiago basin. Part III Final. IX Chilean Geologic Congress, Puerto Varas, Chile (in Spanish)

Aravena J, Armesto J, Denton G, Fuenzalida H, Garleff K, Heusser C, Pino M, Varela J, Veit H, Villagrán C (1993) The quaternary of Los Lagos Region of southern Chile. Tour-guide of international workshop, Quaternary of Chile, Santiago (in Spanish)

Badano EI, Cavieres LA, Molina-Montenegro MA, Quiroz CL (2005) Slope aspect influences plant association patterns in the Mediterranean matorral of central Chile. J Arid Environ 62:93–108

Baker PE, Buckley F, Holland JG (1974) Petrology and geochemistry of Easter Island. Contrib Mineral Petrol 44:85–100

Baker PE, Gledhill A, Harvey PK, Hawkesworth CJ (1987) Geochemical evolution of the Juan Fernández Islands, SE Pacific. J Geol Soc 144:933–944

Bao H, Jenkinsa K, Khachaturyana M, Chong Díaz G (2004) Different sulfate sources and their post-depositional migration in Atacama soils. Earth Planet Sci Lett 224:577–587

Beinroth FH, Luzio W, Maldonado F, Eswaran H (1985) Proceedings of the sixth international soil classification workshop, Chile and Ecuador. Part II: tour-guide for Chile. Chilean Society of Soil Science, Santiago, Chile

Berger I, Cooke R (1997) The origin and distribution of salts on alluvial fans in the Atacama desert, northern Chile. Earth Surf Proc Land 22:581–600

Bertrand S, Fagel N (2008) Nature, origin, transport and deposition of andosol parent material in south-central Chile (36–42°S). Catena 73:10–22

Besoaín E (1958) Clay mineralogy of some volcanic soils from Chile. Agric Tecnica (Chile) 18(2):110–165 (in Spanish with English summary)

Besoaín E (1968) Imogolite in volcanic soils of Chile. Geoderma 2(2): 151–169

Besoaín E (1985) Genesis and transformation of the clay minerals in volcanic ashes derivated soils of central-south Chile In: Tosso J (ed) Volcanic soils of Chile. Agriculture and Livestock Research Institute (INIA-La Platina), Santiago (in Spanish)

Besoaín E, González S (1978) Mineralogical and genetic relationship in some soils representatives of central-south of Chile. In: Proceeding I national symposium of soil science, Concepción, Chile (in Spanish)

Besoaín E, Peralta M, Massar S (2000) Mineralogy and origin of some volcanic ash soils of continental Chiloé, Chile. Agric Tecnica (Chile) 60(2):127–153 (in Spanish with English summary)

Calderón M, Hervé F, Massone J, Tassinar G, Pankhurst J, Godoy E, Theye T (2007) Petrogenesis of the Puerto Edén Igneous and metamorphic complex, Magallanes, Chile: late Jurassic syn-deformational anatexis of metapelites and granitoid magma genesis. Lithos 93:17–38

Campbell I, Claridge G (1987) Antarctica: soils, weathering processes and environment. Elsevier, Amsterdam

Campbell I, Claridge G (2009) Antarctic permafrost soils In: Margesin R (ed) Permafrost soils. Soil biology, vol 16. Springer, Berlin

Casanova M, Messing I, Joel A (2000) Influence of aspect and slope gradient on hydraulic conductivity measured by tension infiltrometer. Hydrol Process 14(1):155–164

Casanova M, Luzio W, Maldonado R (2007) Correlation between world reference base and soil taxonomy for the soils from the 7th Region del Maule from Chile. J Soil Sci Plant Nutr 7(2):14–21 (Spanish, abstract in English)

Casanova M, Luzio W, Norambuena P, Seguel O, Nájera F (2008) Organic and mineral soils in the Chilean Altiplano In: Casanova M, Luzio W (eds) III Proceeding international conference and workshops on soil classification, Santiago-Chile, 9–18 Oct

Cavieres L, Peñaloza A, Arroyo MK (2000) Altitudinal vegetation belts in the high-Andes of central Chile (33°S). Rev Chil Hist Nat 73(2):331–344

Cepeda J, Novoa J (2006) The high Andean Cordillera of Elqui Valley. In: Cepeda P (ed) Geoecology of desertic Andes. The mountain of Elqui Valley. University of La Serena. La Serena, Chile. (In Spanish, summary in English)

Cereceda P, Brignardello L, Avaca D, Castro C (1996) Environmental constrains to agricultural activities in the Robinson Crusoe Island, Chile. Rev Geogr Norte Gd 23:11–19 (In Spanish with English summary)

CIREN (1996a) Soil survey of the Metropolitan Region of Chile. Natural Resources Information Centre, Publication N° 115. Santiago, Chile (in Spanish)

CIREN (1996b) Soil survey, Region VI of Chile. Natural Resources Information Centre, Publication N° 114. Santiago, Chile (in Spanish)

CIREN (1997a) Soil survey, Region V of Chile. Natural Resources Information Centre, Publication N° 116, 2 vol. Santiago, Chile (in Spanish)

CIREN (1997b) Soil survey, Region VII of Chile Natural Resources Information Centre, Publication N° 121, 2 vol. Santiago, Chile (in Spanish)

CIREN (1999) Soil survey, Region VIII of Chile. Natural Resources Information Centre, Publication N° 117, 2 vol. Santiago, Chile (in Spanish)

CIREN (2002) Soil survey, Region IX of Chile. Natural Resources Information Centre, Publication N° 122. Santiago, Chile (in Spanish)

CIREN (2003) Soil survey, Region X of Chile. Natural Resources Information Centre, Publication N° 123, 2 vol. Santiago, Chile (in Spanish)

CIREN (2005a) Soil survey, Region IV of Chile. Natural Resources Information Centre, Publication 129. Santiago, Chile (in Spanish)

CIREN (2005b) Soil survey, Region XI of Chile. Natural Resources Information Centre, Publication N° 130. Santiago, Chile (in Spanish)

CIREN (2007) Soil survey, Region III of Chile (Copiapó and Huasco valleys). Natural Resources Information Centre, Publication N°135. Santiago, Chile (in Spanish)

Clarke J (2006) Antiquity of aridity in the Chilean Atacama Desert. Geomorphology 73:101–114

Clausen J, Ortega I, Glaude C, Relyea R, Garay G, Guineo O (2006) Classification of wetlands in a Patagonian National Park, Chile. Wetlands 26(1):217–229

CNR (1993) Soil survey of Choapa River basin and their tributaries. National Irrigation Commission, Santiago,Chile (in Spanish)

CNR (1997) Comprehensive study of irrigation and drainage of Magellan, XII Region, vol 2 (in spanish)

Díaz C (1949) Association of soils of the Easter Island. Agric Tecnica (Chile) 9:115–126 (in Spanish)

Díaz C (1958) Soil surveys development in Chile during 1948–1958 decade. Agric Tecnica (Chile) 18(2):59–77 (in Spanish)

Díaz C, Meléndez E (1958) Soil survey of Camarones Valley. Agric Tecnica (Chile) 18(2):78–109 (in Spanish)

Díaz C, Wright C (1965) Soils of the arid zones of Chile. Food and Agriculture Organization of the United Nations, Rome. Soils bulletin N°1. (Appendix)

Díaz C, Avilés C, Roberts R (1959–1960) The great groups of soils in the Magellan Province. Agric Tecnica (Chile) 19/20: 227–308 (in Spanish)

Douglass D (2005) Glacial chronology, soil development and paleoclimate reconstructions for mid-latitude South America, 1 MA to recent. Dissertation, University of Wisconsin–Madison

Douglass D, Bockheim J (2006) Soil-forming rates and processes on Quaternary moraines near Lago Buenos Aires, Argentina. Quat Res 65:293–307

Ewing SA, Sutter B, Owen J, Nishiizumi K, Sharp W, Cliff SS, Perry K, Dietrich W, McKay CP, Amundson R (2006) A threshold in soil formation at Earth's arid-hyperarid transition. Geochim Cosmochim Ac 70:5293–5322

Filipová L, Héld R, Covacevich N (2010) Variability of soil types in wetland meadows in the south of the Chilean Patagonia. Chil J Agric Res 70(2):266–277

Finkl CW (2005) Coastal soils In: Schwartz M (ed) Encyclopedia of coastal science. Encyclopedia of earth sciences series. Springer, The Netherland

Flores-Aqueveque V, Alfaro S, Muñoz R, Rutllant JA, Caquineau S, Le Roux J, Vargas G (2010) Aeolian erosion and sand transport over the Mejillones Pampa in the coastal Atacama Desert of northern Chile. Geomorphology 120:312–325

Gerding V, Thiers O (2002) Characterisation of soils under forest of *Nothofagus betuloides* (Mirb) Blume, in *Tierra del Fuego*, Chile. Rev Ch Hist Nat 75:819–833 (in Spanish)

González G, Cembrano J, Carrizo D, Macci A, Schneider H (2003) The link between forearc tectonics and Pliocene-quaternary deformation of the Coastal Cordillera, northern Chile. J South Am Earth Sci 16:321–342

Gut B (2008) Trees in Patagonia. Birkhäuser Verlag AG, Basel

Haberland J (1992) Soil survey in some tundra communities in the Coppermine Peninsula, Robert Island, South Shetland Islands, Antarctic. Dissertation, University of Chile (in Spanish with English summary)

Haberland J, Luzio W, Vera W (2008) Characterization of some Antarctic soils at the South Shetland Islands, Antarctica In: Casanova M, Luzio W (eds) Proceedings of the international conference & field workshops on soil classification, 9–18 Nov 2008

Haberle SG, Lumley SH (1998) Age and origin of tephras recorded in postglacial lake sediments to the west of the southern Andes, 44°S to 47°S. J Volcanol Geotherm Res 84:239–256

Hardie LA, Eugster HP (1970) The evolution of closed-basin brines. Mineral Soc Am Spec Publ 3:273–290

Henríquez O (1994) Soil survey of soils in the Byers peninsula, Livingstone Island, South Shetland Islands, Antarctic. Dissertation, University of Chile (in Spanish with English summary)

Hervé F, Araya R (1965) Geomorphological and geological study of the Greenwhich, Robert and South Shetland Islands, Antarctica. Dissertation, University of Chile (in Spanish)

Honorato R, Cruz G (1999) Erodability and actual erosion of the soils of Easter Island. Cienc Investig Agrar 26:27–36 (in Spanish with English summary)

Horrocks M, Wozniak JA (2008) Plant microfossil analysis reveals disturbed forest and a mixed-crop, dryland production system at Te Niu, Easter Island. J Archaeol Sci 35:126–142

IREN (1967) Magellan Province. Preliminary inventory of natural resources. Continental zone and *Tierra del Fuego* Island, vol 1. Natural Resources Institute, Chile. Report 21 (in Spanish)

IREN (1976) Inventory of natural resources by perception methods of Landsat satellite. Natural Resources Institute, Chile. Region I (in Spanish)

IREN (1982) Physic resource survey in Juan Fernández Archipelago. Natural Resources Institute, Chile. Report 45 (in Spanish)

IUSS Working Group WRB (2006) World reference base for soil resources. World soil resources report 103. Food and Agriculture Organization of the United Nations, Rome

Juez-Larré J, Kukowski N, Dunai TJ, Hartley AJ, Andriessen PAM (2010) Thermal and exhumation history of the Coastal Cordillera arc of northern Chile revealed by thermochronological dating. Tectonophysics 495:48–66

Kober F, Ivy-Ochs S, Zeilinger G, Schlunegger F, Kubik PW, Baur H, Wieler R (2009) Complex multiple cosmogenic nuclide concentration and histories in the arid Rio Lluta catchment, northern Chile. Earth Surf Process Landf 34:398–412

Lagabrielle Y, Goddéris Y, Donnadieu Y, Malavieille J, Suarez M (2009) The tectonic history of Drake passage and its possible impacts on global climate. Earth Planet Sci Lett 279:197–211

Lagos G, Blanco E (2010) Mining and development in the region of Antofagasta. Resour Policy 35(4):265–275

Le Roux J, Gómez C, Venegas C, Fenner J, Middleton H, Marchant M, Buchbinder B, Frassinetti D, Marquardt C, Gregory-Wodzicki KM, Lavenu A (2005) Neogene-Quaternary coastal and offshore sedimentation in north-central Chile: record of sea level changes and implications for Andean tectonism. J South Am Earth Sci 19:83–98

Le Roux J, Olivares D, Nielsen S, Smith N, Middleton H, Fenner J, Ishman S (2006) Bay sedimentation as controlled by regional crustal behaviour, local tectonics and eustatic sea-level changes: Coquimbo Formation (Miocene–Pliocene), Bay of Tongoy, central Chile. Sediment Geol 184:133–153

Louwagie G, Stevenson CM, Langohr R (2006) The impact of moderate to marginal land suitability on prehistoric agricultural production and models of adaptative strategies for Easter Island (Rapa Nui, Chile). J Anthropol Archaeol 25:290–317

Luzio W, Alcayaga S (1992) Associations map of big soil groups in Chile. Agric Tecnica (Chile) 52(4):347–353 (in Spanish with English summary)

Luzio W, Barros C, Arancibia M, Alcayaga S (1989) The placic horizon in ñadis (Placaquands) and trumaos (Placudands) in the south of Chile. An Edafol Agrobiol (Spain) 48:523–543 (in Spanish with English summary)

Luzio W, Vera W, Carrasco MA (1990) Soils development in cold desert conditions and Antarctic environment of Chile In: Proceeding IV National congress of soil science. University of La Frontera, Temuco, Chile

Luzio W, Casanova M, Vera W, Norambuena P (2002a) Properties and characteristics of hydromorphic soils (*bofedales*) from Chilean altiplano. Chil Soc Soil Sci Bull 18:325–328 (in Spanish)

Luzio W, Norambuena P, Casanova M, Vera W (2002b) Genesis and properties of some soils from the Chilean Altiplano. J Soil Sci Plant Nutr 2(1):35–52 (in Spanish with English summary)

Luzio W, Casanova M, Vera W, Norambuena P (2002c) Hydromorphic soils as forage resource support in the Chilean altiplano. In: Proceeding transcontinental conference, high summit 2002. Latinoamerica (Mendoza), Europa (Milán and Trento), Africa (Nairobi) and Asia (Kathmandú), Mendoza, 6th May

Luzio W, Casanova M, Seguel O (2010) Soils of Chile (Luzio W, ed). University of Chile. Maval Press, Santiago (in Spanish)

Mann D, Edwards J, Chase J, Beck W, Reanier R, Mass M, Finney Loret J (2008) Drought, vegetation change, and human history on Rapa Nui (Isla de Pascua, Easter Island). Quat Res 69:16–28

Marquardt C, Lavenu A, Ortlieb L, Godoy E, Comte D (2004) Coastal neotectonics in Southern Central Andes: uplift and deformation of marine terraces in Northern Chile (27°S). Tectonophysics 394:193–219

Martínez C, Quezada M, Rubio P (2011) Historical changes in the shoreline and littoral processes on a headland bay beach in central Chile. Geomorphology 135:80–96

Meigs P (1966) Geography of coastal deserts. Arid Zone Research XXVIII. UNESCO, Liège (Belgique)

Mella A, Khüne A (1985) Systematic and descriptions of families, associations and series of soils derived from pyroclastic materials in central-south of Chile. In: Tosso J (ed) Volcanic soils of Chile. Agriculture and Livestock Research Institute (INIA La Platina), Santiago (in Spanish)

Michel R, Schaefer C, Dias LE, Simas F, Benites V, Mendonça E (2006) Ornithogenic gelisols (cryosols) from maritime Antartica: pedogenenesis, vegetation and carbon studies. Soil Sci Soc Am J 70:1370–1376

Mikhailov JS (1999) Soils of Rapa Nui (Easter Island). Eurasian Soil Sci 32(11):1177–1180

Mishalski G, Böhlke JK, Thiemens M (2004) Long term atmospheric deposition as the source of nitrate and other salts in the Atacama Desert, Chile: new evidence from mass-independent oxygen isotopic compositions. Geochim Cosmochim Ac 68(20):4023–4038

Moreno H, Varela J (1985) Geology, volcanism and Quaternary pyroclastic deposits from central-south of Chile. In: Tosso J (ed) Volcanic soils of Chile. Agriculture and Livestock Research Institute (INIA La Platina), Santiago (in Spanish)

Muñoz C (2004) Geomorphology of Easter Island. Dissertation, University of Chile (in Spanish with English summary)

Muñoz J, Fernández B, Varas E, Pastén P, Gómez D, Rengifo P, Muñoz J, Atenas M, Jofré JC (2007) Chilean water resources. In: Moreno T (ed) Geology of Chile. Geological Society of London, London

Navas A, López-Martínez J, Casas J, Machín J, Durán J, Serrano J, Cuchi J, Mink S (2008) Soil characteristics on varying lithological substrates in the South Shettland Islands, maritime Antarctica. Geoderma 144:123–139

Norambuena P, Luzio W, Zepeda O, Stern J, Reinoso F (2011) Preliminary survey of some soils from Chilean Altiplano near Iquique. J Soil Sci Plant Nutr 11(2):62–71

Palma T (1993) Physical and chemical characterisation of fierrillo of five ñadis from Region X. Dissertation, University of Chile (in Spanish with English summary)

Paskoff R (1989) Zonality and main geomorphic features of the Chilean coast. Essener Geographische Arbeiten (Paderborn) 18:S237–S267

Pfeiffer M, Mascayano C, Aburto F (2010) Soils of Chilean Patagonia in glacial and periglacial environments. Eurasian Soil Sci 43(13):1430–1438

Pfeiffer M, Le Roux J, Solleiro-Rebolledo E, Kemnitz H, Sedov S, Seguel O (2011) Preservation of beach ridges due to pedogenic calcrete development in the Tongoy paleobay, Chile. Geomorphology 132:234–248

Pino M, Adán L, Seguel O (2004) Geoarchaeology of the area of Calafquén Lake, southwestern flank of Villarrica Volcano. Natl Geol Mining Serv (SERNAGEOMIN-Chile) Bull 61:61–67

Placzek CJ, Matmon A, Granger DE, Quade J, Niedermann S (2010) Evidence for active landscape evolution in the hyperarid Atacama from multiple terrestrial cosmogenic nuclides. Earth Planet Sci Lett 295:12–20

Prellwitz J (2007) A characterization of hyper-arid nitrate soils in the Baquedano Valley of the Atacama Desert, Northern Chile. Dissertation, University, Oxford, OH

Quezada J (2007) The North of Chile Pleistocene coastal uplift: ages coastal terrace ^{21}Ne highest area of Caldera Bay. Rev Geol Chile 34:81–96 (in Spanish with English summary)

Regard V, Saillard M, Martinod J, Audin L, Carretier S, Pedoja K, Riquelme R, Paredes P, Hérail G (2010) Renewed uplift of the Central Andes Forearc revealed by coastal evolution during the Quaternary. Earth Planet Sci Lett 297:199–210

Rici M (2006) Conservation status and ex situ cultivation efforts of endemic flora of the Juan Fernández Archipelago. Biodivers Conserv 15:3111–3130

Risacher F, Alonso H, Salazar C (2003) The origin of brines and salts in Chilean salars: a hydrochemical review. Earth Sci Rev 63: 249–293

Rundell PW, Dillon M, Palma B, Mooney H, Gulmon S, Ehleringer J (1991) The phytogeography and ecology of the coastal Atacama and Peruvian deserts. Aliso 13:1–49

SAG (2002) Soil survey of Chacalluta Valley. Chilean Agriculture Ministry, Agriculture and Livestock Service, Renewable Natural Resources Protection Department, Document N°16 (in Spanish)

Saillard M (2008) Dynamique du soulèvement côtier Pléistocène des Andes centrales: Etude de l'évolution géomorphologique et datations (10Be) de séquences de terrasses marines (Sud Pérou-Nord Chili). Dissertation, Université de Toulouse, France

Saillard M, Hall S, Audin L, Farber DL, Hérail G, Martinod J, Regard V, Finkel RC, Bondoux F (2009) Non-steady long-term uplift rates and Pleistocene marine terrace development along the Andean margin of Chile (31°S) inferred from ^{10}Be dating. Earth Planet Sci Lett 277:50–63

Salazar O, Casanova M, Luzio W (2005) Correlation between World Reference Base and Soil Taxonomy for the soils from the 10th « Los Lagos » Region of Chile. J Soil Sci Plant Nutr 5(2):35–45 (in Spanish with English summary)

Scasso R, Corbella H, Tiberi P (1994) Sedimentological analysis of the tephras from the 12–15 Aug 1991 eruption of Hudson volcano. Bull Volcanol 56:121–132

Schlunegger F, Kober F, Zeilinger G, von Rotz R (2010) Sedimentology-based reconstructions of paleoclimate changes in the Central Andes in response to the uplift of the Andes, Arica region between 19 and 21°S latitude, northern Chile. Int J Earth Sci (Geol Rundsch) 99(suppl 1):S123–S137

Simas F, Schaefer C, Melo V, Albuquerque-Filho M, Michel R, Pereira V, Gomes M, da Costa L (2007) Ornithogenic cryosols from Maritime Antarctica: phosphatization as a soil forming process. Geoderma 138:191–203

Soil Survey Staff (2006) Keys to soil taxonomy, 10th edn. SMSS Technical Monograph 6, New York

Squeo F, Veit H, Arancio G, Gutiérrez JR, Arroyo MTK, Olivares NC (1993) Spatial heterogeneity of high mountain vegetation in the Andean desert zone of Chile. Mt Res Dev 13:203–209

Stolpe N, Zagal E, Sandoval M, Quezada C (2008) Southern field guide. In: Casanova M, Luzio W (eds) The international conference and field workshop on soil classification. Soil: a work of art of the nature. Santiago, Chile

Stuessy TF, Foland KA, Sutter JF, Sanders RW, Silva OM (1984) Botanical and geological significance of K–Ar dates from the Juan Fernández Islands. Science 225:49–51

Takahashi T, Dahlgren R, van Susteren P (1993) Clay mineralogy and chemistry of soils formed in volcanic materials in the xeric moisture regime of northern California. Geoderma 59:131–150

Tan K (2000) Environmental Soil Science, vol 74, 2nd edn. Books in soil, plants and environment. Marcel Dekker Inc., New York

Ugolini FC, Bockheim JG (2008) Antarctic soils and soil formation in a changing environment: a review. Geoderma 144:1–8

Valle G (2012) Chemical characterization of grape berries and wines of cv. Valley. Dissertation, University of Chile (in Spanish with English summary)

van Wambeke A (1981) Soil moisture and temperature regimes of South America. Soil Management Support Service (SMSS) Technical Monograph N°2

van Wambeke A, Luzio W (1982) Soil moisture and soil temperature regimes of soil of Chile. Agric Tecnica (Chile) 42(2):149–159 (in Spanish with English summary)

Vera W (1990) Soil morphology of periglacial areas in the Torres del Paine and the South Shetland islands. In: Proceedings of IV national congress. Chilean Society of Soil Science. Temuco, Chile, 14–16 Nov 1990

Wilson TM, Cole JW, Stewart C, Cronin SJ, Johnston DM (2011) Ash storms: impacts of wind-remobilised volcanic ash on rural communities and agriculture following the 1991 Hudson eruption, southern Patagonia, Chile. Bull Volcanol 73:223–239

Wright C (1965) The volcanic ash soils of Chile. Food and Agriculture Organization of the United Nations Report 2017, Rome

Wright C, Díaz C (1962) Soils and agricultural development of Easter Island (Pascua), Chile. Quarterly report supplement number 1. Chilean Ministry of Agriculture, Santiago, Chile

Wright C, Meléndez E (1961) Soil survey of lower part of Azapa Valley. General Direction of Agrarian Production and Food and Agriculture Organization of the United Nations. Santiago, Chile

Zhu R, Liu Y, Ma E, Sun J, Xu H, Sun L (2009) Greenhouse gas emissions from penguin guanos and ornithogenic soils in coastal Antarctica: effects of freezing-thawing cycles. Atmos Environ 43:2336–2347

3 Management of Soil Properties in Chile

3.1 Chemical Properties

Soil chemical properties are critical in determining nearly all the major soil properties that control plant nutrient availability, the fate of many soil pollutants, microbial reactions in soils, etc. In Chile, climate conditions and parent materials tend to combine to create a pattern of soil chemical properties that varies along the considerable length of the country. Thus, soil chemical properties in the major soil zones, including soil reaction, soil salinity and nutrient availability, are described here. In addition, some management practices for improving crop production are discussed.

3.1.1 Soil Reaction

Soil reaction (pH) in Chilean soils is directly related to the parent material (parental factor), but is also highly influenced by the precipitation regime (climate factor) and vegetation (organism factor). Based on the combined effects of these three soil formation factors, the soil reaction in Chile may be roughly divided into alkaline in the Hyper-arid to Semi-arid zone and acidic in the southern part of the Mediterranean zone. In the Northern Mediterranean zone, a broad range of soil pH can be found as a result of additional influences from the topography. Few studies have been made on the soils in other zones, but the information available to date suggests that the soil reaction in the Rainy and Patagonian zones ranges from acidic to basic, whereas in the Antarctic zone and in the Insular zone it ranges from acidic to neutral.

Figure 3.1 shows soil pH ranges at 0–20 cm depth in 710 pedons in the major soil zones in Chile, based on a number of soil surveys carried out in agricultural areas during the last 40 years by different government agencies (SAG 1974; CIREN 1996a, b, 1997a, b, 1999, 2002, 2003, 2005a, b, 2007; Ahumada et al. 2004). In the Hyper-arid to Semi-arid zone, all the soils described in these studies displayed soil pH_{water} higher than 6.0, and 51 % of the soils had moderately alkaline soil reaction ($pH_{water} \geq 7.9$). In the Mediterranean zone, 52 % of the pedons had moderately acidic (pH_{water} 5.6–6.0) to slightly acidic (pH_{water} 6.1–6.5) soil reaction, with few extreme soil pH soils, for instance only 5 % very strongly acidic and 4 % moderately alkaline soils were included. Considering only two soil surveys carried out in the Rainy and Patagonian zone (CIREN 2005b; SAG 1974) with a limited number of soil descriptions ($n = 23$ pedons) for the extensive area defined, it is possible to find a broad range of soil pH_{water}, but with a dominance of slightly acidic soils (39 %).

The amount of precipitation falling in Chile follows a clear tendency to increase from the Hyper-arid to Semi-arid zone (latitude UTM 7,000 km south; mean annual precipitation (MAP) <150 mm $year^{-1}$) to the southern part of the Mediterranean zone (latitude UTM 5,400 km south; MAP 1,100–2,400 mm $year^{-1}$), whereas it decreases towards the Rainy and Patagonian zone to around 200 mm $year^{-1}$ at latitude UTM 4,800 km south, as shown in Fig. 3.2. Overall, the relationship between MAP obtained from Chilean meteorological stations ($n = 74$) and latitude (UTM) is best fitted to a four-order polynomial equation, with a determination coefficient (R^2) of 0.78. Soil pH follows a similar tendency to precipitation regime, with the relationship between soil pH_{water} at 0–20 cm and latitude (UTM) analysed in 710 pedons from the hyperarid to semiarid zone to the Rainy and Patagonian zone being best fitted to four-order polynomial equation with R^2 of 0.56. Thus as Fig. 3.2 shows, soil pH_{water} at 0–20 cm decreases from alkaline values in the hyperarid to semiarid zone to acidic in the southern part of the Mediterranean zone. Where the MAP decreases again, particularly in the east of the Rainy and Patagonian zone, soil pH increases, in some cases to moderately alkaline values. In this zone, calcium carbonate ($CaCO_3$) observed in surface soil and increasing with depth, originates from dust rather than from pedogenic weathering according to Bockheim and Douglass (2006)

M. Casanova et al., *The Soils of Chile*, World Soils Book Series, DOI: 10.1007/978-94-007-5949-7_3

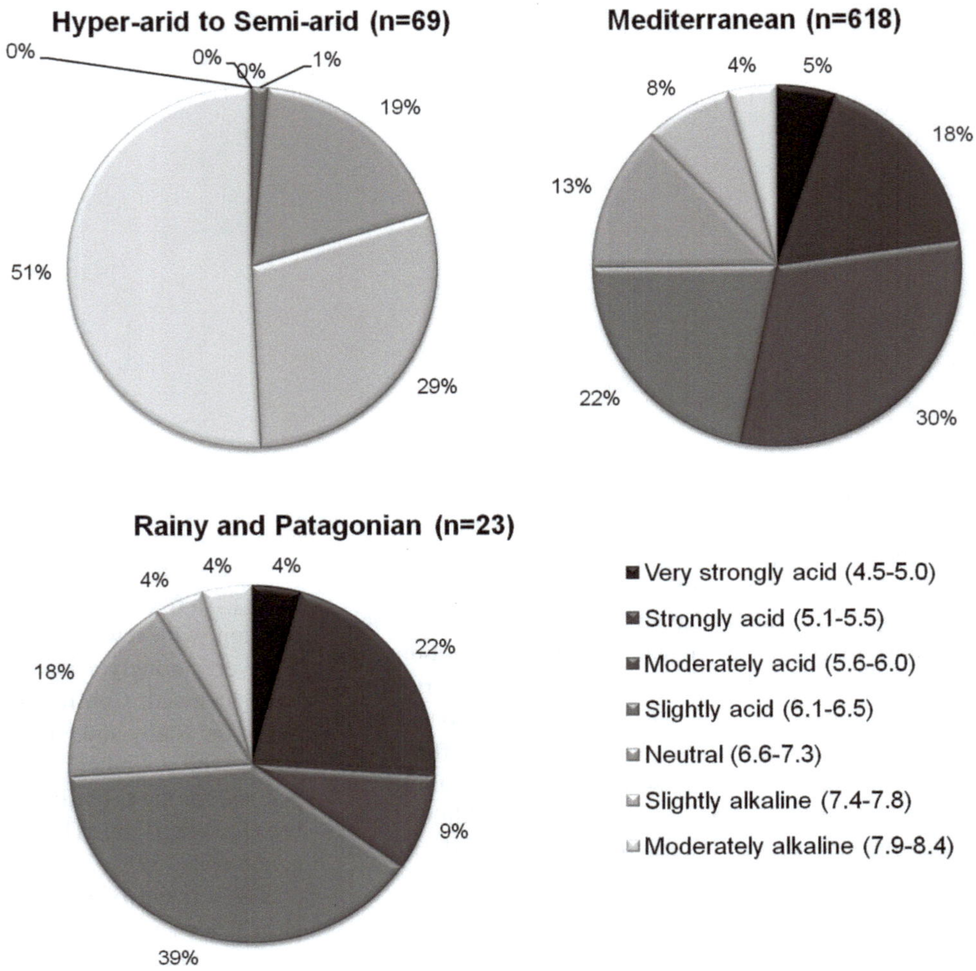

Fig. 3.1 Soil pH_{water} ranges in major soil zones in Chile at 0–20 cm depth; *n* is the amount of pedons analysed

This is because there is minimal evidence of weathering of Ca-bearing minerals in the soil, the amount of $CaCO_3$ in the profile is far in excess of what could be released by weathering, dust collected from the area is enriched in $CaCO_3$ (>4 %), and a source area for the carbonate dust has been identified to the north–west, near Lake General Carrera.

In addition, Fig. 3.2 clearly shows that soil pH reaches its lowest values (around 4.2 at latitude UTM 5,300 km south) in the southern part of the Mediterranean zone, where MAP is over 2,000 mm $year^{-1}$. In contrast, when MAP is lower than 1,000 mm $year^{-1}$, most soils have soil pH above 6.0, while moderately alkaline soils are only found where MAP is lower than 200 mm.

However, in the Hyper-arid to Semi-arid zone, localised areas of acidic soil can be found in the Chilean Altiplano, where pH_{water} of 4.9 (3,000–4,000 m a.s.l.) and 4.2 (4,200–4,500 m a.s.l.) in the surface horizon of Aridisols (Luzio et al. 2002) and Entisols (Norambuena et al. 2011), respectively, have been described.

In the northern part of the Mediterranean zone, localised pockets of acidic soils may also exist, with soil reaction ranging between very strongly acidic (pH 4.5–5.0) and strongly acidic (pH 5.1–5.5). These soils correspond to Inceptisols and Alfisols in terrace or hillslope positions and generally display pH <5.0 and low base saturation (BS <50 %) (Table 3.1).

As shown in Fig. 3.3, in the southern part of the Mediterranean zone most soils have soil pH_{water} below 6.5. In the Central Valley and Coastal Range of the southern part of this zone there is a group of soils, locally named red clayey soils and classified as Ultisols, which display an argillic horizon (B_t) with pH in the range very strongly acidic to strongly acidic and with low Al saturation on the exchange complex (Sadzawka 2006a). In this zone, there are also a number of Andisols, Inceptisols and Alfisols with soil pH_{water} ranging between strongly acidic and moderately acidic.

In soils in the southern part of the Mediterranean zone, the base-forming cations (Ca^{2+}, Mg^{2+}, K^+ and Na^+) have usually left the colloidal complex and show high acidity levels, with the appearance of Al, Mn and H phytotoxicity (Borie and Rubio 1999). The Al phytotoxicity in particular

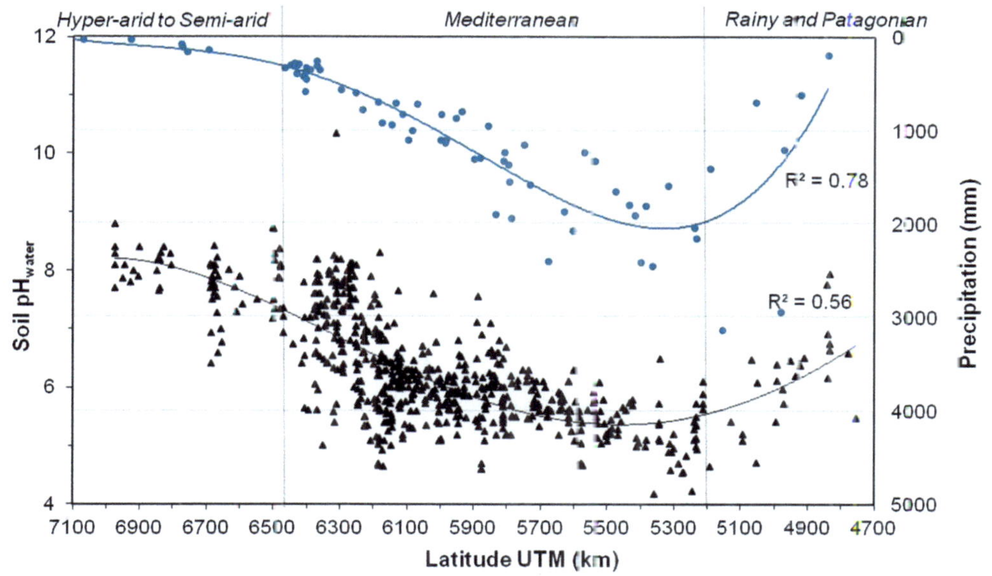

Fig. 3.2 Soil pH$_{water}$ (*triangles*) at 0–20 cm depth between the Hyper-arid to Semi-arid zone and the Rainy and Patagonian zone for 710 pedons. *Circles* are mean annual precipitation from 74 meteorological stations

is an important constraint for plant production in these areas of Southern Chile (Borie and Rubio 1999). For instance, cereals such as wheat and barley growing in very strongly acidic soils with high Al levels display symptoms such as decreased root length (Gallardo et al. 1999, 2005).

Aluminium saturation as a percentage of effective cation exchange capacity (ECEC) has been proposed as a measure of Al phytotoxicity in acidic soils in Chile (Sadzawka 2006a). In soils with Al saturation of ECEC values higher than 5 %, the Chilean Incentives System to Recover Degraded Soils (ISRID) recommends the application of lime (Mora et al. 1999, 2002).

Figure 3.3 shows the negative relationship between measured Al saturation of ECEC and measured soil pH$_{water}$ at 0–20 cm in 169 pedons in the southern part of the Mediterranean zone (CIREN 1999, 2002, 2003). It is clear from the diagram that as soil pH$_{water}$ decreases Al saturation of ECEC increases, fitting a third-order polynomial equation with R^2 of 0.59. In addition, 49 % of analysed samples showed Al saturation of ECEC >5 %, whereas 90 % of soil samples with soil pH$_{water}$ <5.5 had Al saturation of ECEC >5 %.

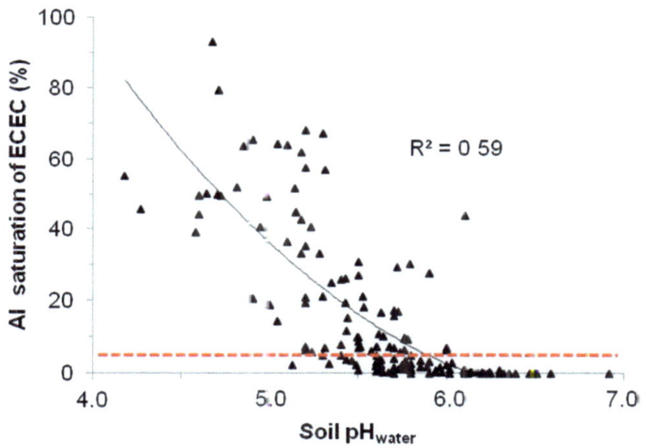

Fig. 3.3 Relationship between Al saturation of the cation exchange capacity (ECEC) and soil pH$_{water}$ at 0–20 cm in soils in the Mediterranean zone ($n = 169$). *Red segmented line* shows Al saturation of ECEC = 5 %

Table 3.1 Classification, location and soil properties at 0–20 cm depth of some acidic soils in the Mediterranean zone (CIREN 1996b, 1997a, b)

Pedon	Latitude UTM (km)	Soil order	Position	pH	BS (%)
Mantagua	6,353	Alfisol	Marine terrace	5.0	35
Carrizal	6,186	Inceptisol	Recent alluvial terrace	4.7	31
Loma Grande	6,148	Inceptisol	Piedmont	5.0	62
Los Lingues	6,174	Alfisol	Remnant alluvial terrace	4.9	58
Pumanque	6,172	Inceptisol	Hillslope	4.7	22
Quinchamalal	6,184	Alfisol	Piedmont	4.7	22
Piuchén	6,184	Alfisol	Hillslope	5.0	–

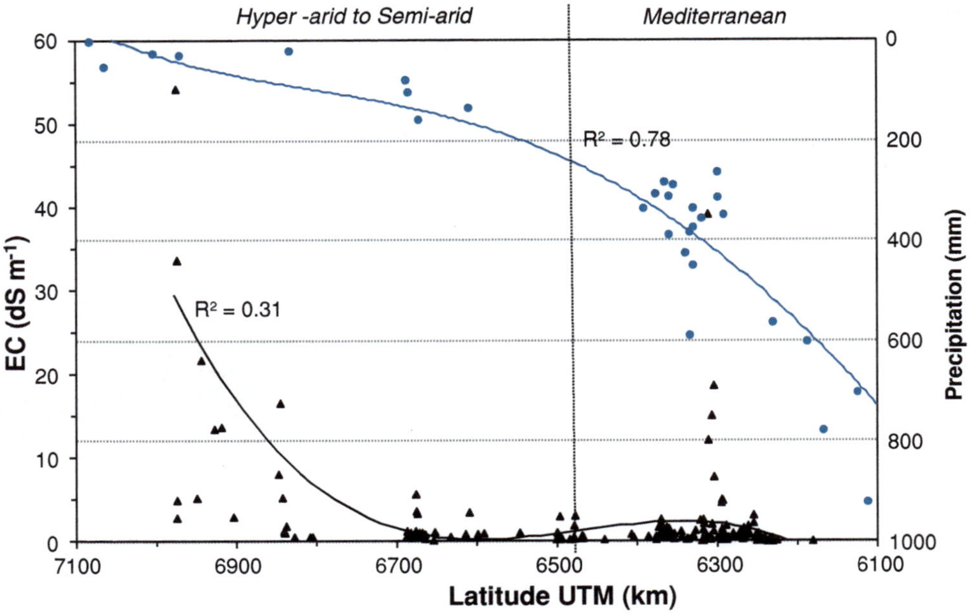

Fig. 3.4 Soil electrical conductivity (EC) at 0–20 cm depth between the Hyper-arid to Semi-arid zone and Mediterranean zone. *Triangles* are EC in pedons ($n = 173$) and *circles* are mean annual precipitation obtained from 30 meteorological stations

3.1.2 Soil Salinity and Sodicity

Soil salinity is well-known to be a serious problem in many parts of the world. Out of the 21 countries worldwide with over 15 % of their land affected by salinity, excluding secondary salinisation caused by poor management of irrigation schemes, three (Paraguay, Argentina and Chile) lie in South America (FAO 2000). Broad statistical estimates (TERRASTAT 2003) indicate that an area of 759,000 km^2 in Chile is affected by salinity and a further 33,000 km^2 by sodicity. Most of this area occurs in Northern Chile, characterised by a succession of North–South aligned ranges and basins occupied by numerous saline lakes and salt crusts, collectively called *salars* (see Chap. 2). Fossil salt crusts are found to the west in the extremely arid Central Valley, while active salars receiving permanent inflows fill many intravolcanic basins to the east in the semiarid Andes (Risacher et al. 2003).

In the Hyper-arid to Semi-arid zone and in the northern part of the Mediterranean zone, salts accumulate naturally due to insufficient rainfall (<500 mm year^{-1}) to remove them from the upper soil layer, as shown in Fig. 3.4.

In the Hyper-arid to Semi-arid zone and the northern part of Mediterranean zone, 69 and 89 % of the total pedons characterised, respectively, are non-saline (EC <2 dS m^{-1}). However, there are several areas that suffer from some degree of salinity (Table 3.2). Most of the soils with extremely saline conditions (EC >8 dS m^{-1}) are located in the Hyper-arid to Semi-arid zone. However, in the northern part of the Mediterranean zone there are some pockets of moderately to strong saline soils that arise from confinement, created by a physical barrier to water flow out of a depression in the landscape.

3.1.3 Nutrient Availability

3.1.3.1 Nitrogen

The first major requirement in agricultural soils is adequate nitrogen (N) for healthy crop growth. In Chile, many N sources are available for supplying N to annual crops and fruit trees. The quantity of N available to plants depends largely on the amounts applied as fertilisers and mineralised from organic N in soils. Although N fertilisation rates have been increasing in Chile in recent decades, in many cases, these have not been proportional to the increase in crop yields. For instance, the current N application rate averaged over all crops is ~120 kg ha^{-1} (World Bank 2008), which is high in comparison to that in other Andean countries (Fig. 3.5). One possible reason for the high N application rates in Chile is that most farmers do not carry out any soil testing or N balance to determine fertiliser/manure applications, or do not select an appropriate N source according to soil characteristics. Overfertilisation is associated with a high risk of N contamination of water bodies (Alfaro et al. 2006; Claret et al. 2011; Salazar and Nájera 2011) or gaseous losses (Casanova and Benavides 2009; Pérez et al. 2010).

During the 1970s, a number of different research groups carried out N analyses in order to determine available N and used this value to calculate recommended N fertilisation rates (Tejeda and Gogan 1970; García and Gandarillas

Table 3.2 Range of soil electrical conductivity (EC) at 0–20 cm depth in Hyper-arid to Semi-arid zone ($n = 68$ pedons) and northern part of Mediterranean zone ($n = 105$ pedons)

Class	EC (dS m^{-1})	Hyper-arid to Semi-arid zone (%)	Northern Mediterranean zone (%)
Non-saline	0–2	69	85
Very slightly saline	2–4	12	8
Slightly saline	4–8	6	4
Moderately saline	8–16	4	3
Strongly saline	>16	9	1

1974). However, it is now generally accepted that N fertilisation rates for crops should be calculated from the N balance before crop sowing (Rodríguez et al. 2001). With this approach, the amount of available N in soil during the growing season is estimated from potential crop yield and quantity and the biochemical quality of crop residues produced and returned to the soil during preceding seasons (Hirzel 2011). Matus and Rodríguez (1994) found that N supplied to soils by mineralisation was close to the soil mineral N content measured at two sites in the Northern Mediterranean zone and two sites in the Southern Mediterranean zone. Table 3.3 shows a simplified estimate of N supply in soils considering net N mineralisation during the growing season in the Mediterranean zone according to soil management and crop residue incorporation, which correlates with previous values reported for this zone (Matus and Rodríguez 1994; Rodríguez et al. 2001).

Lower net N mineralisation rates than in agricultural soils have been reported in studies on evergreen and deciduous forest soils classified as Andisols in the southern part of the Mediterranean zone. For instance, Pérez et al. (2003a) found net N mineralisation values for forest soils in the range 12–30 kg N ha^{-1} $year^{-1}$, Cárcamo et al. (2004) reported values lower than 6 kg N ha^{-1} $year^{-1}$, and recently Staelens et al. (2011) reported a range of 1.7–11.3 kg N ha^{-1} $year^{-1}$. This lower N mineralisation in forest soils compared with agricultural soils is directly related to higher C/N ratio in litterfall reaching the topsoil in forest soils, where immobilisation processes dominate. For instance, Pérez et al. (2003b) studied an evergreen forest in the southern part of the Mediterranean zone and reported litterfall C/N ratio ranging between 41 and 113, whereas Klein et al. (2008) in a *Nothofagus pumilio* forest in the Rainy and Patagonian zone found a litterfall C/N ratio of 58.

In the Mediterranean zone, some studies have been carried out to estimate N addition to soil by rainfall and dry deposition. Although these studies showed values lower than 10 kg N ha^{-1} $year^{-1}$ (Table 3.4), these forms of N addition should be considered in the N budget to reduce application of commercial N fertiliser.

Other alternatives have been studied to calculate optimal N application rate and avoid negative environmental impacts. Some studies have used ^{15}N to assess the recovery of N applied by fertilisation of crops (Pino et al. 2002a) and fruit trees (Pino et al. 2002b), and also to determine biological N fixation in legumes (Campillo et al. 2002, 2005;

Fig. 3.5 Current fertiliser use (cultivated land) in Andean countries including Chile, P–P_2O_5 and K–K_2O (adapted from World Bank 2008)

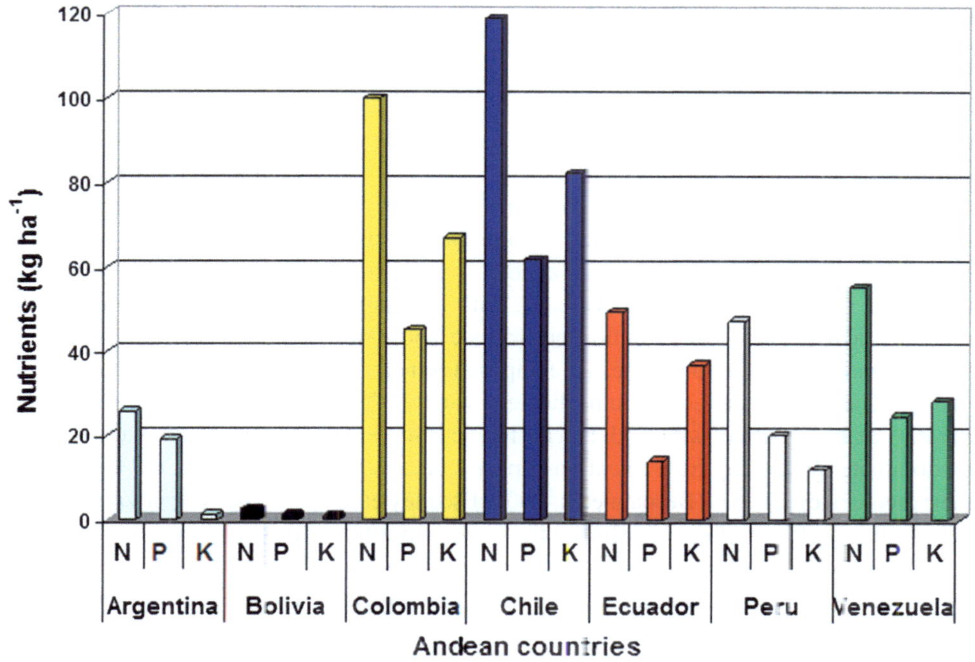

Table 3.3 Estimation of N supply in soils in the Mediterranean zone during the growing season

Soil use management	Nitrogen supply ($kg\ N\ ha^{-1}\ year^{-1}$)
Crop rotation with low yields (>4 year after pasture)	40–60
Crop rotation with high yields (>4 year after pasture)	60–80
Crop rotation (2–4 year after pasture)	80–100
Degraded pasture (1 year after pasture)	100–120
Good pasture (1 year after pasture)	120–150

Table 3.4 Bulk deposition of nitrogen to soils in the Mediterranean zone of Chile

Vegetation	Bulk deposition ($kg\ N\ ha^{-1}\ year^{-1}$)	References
Sclerophyll forest	8.2	Cisternas and Yates (1982)
Evergreen forest	3.3	Godoy et al. (1999)
Pasture	6.9	Oyarzún et al. (2002)
Evergreen forest	3.4	

Pino 2002) and nitrification rates in soils (Huygens et al. 2011). On the other hand, few efforts have been made to test the applicability of computer models for accurate N rate calculations. For example, evaluation of the SimUlation of Nitrogen Dynamics In Arable Land (SUNDIAL) model in Mediterranean zone soils by Zagal et al. (2003) showed that dose rates and fertilisation strategy used by the farmers coincided with those calculated by the model.

3.1.3.2 Phosphorus

Volcanic ash-derived soils cover nearly 50–60 % of arable land in Chile and include Andisols and Ultisols located mainly in the southern part of the Mediterranean zone. These soils are rich in organic matter, possess high specific area, and have pH-dependent CEC. Their surface potential and charge distribution vary substantially with pH and, to a lesser extent, with the concentration of equilibrating solution. Although these soils have many advantages for crop production, one of their main constraints is high phosphate retention, which commonly reaches values over 90 %. When any phosphate is added to the soil, there is immediate competition between plant roots and the soil, and much of the phosphate fertiliser added is retained by soil colloidal surfaces.

Inorganic P retention depends on many physical and chemical properties such as chemical adsorption (Carrasco et al. 1993), but in the Mediterranean zone also soil mineralogy and soil pH. Andisols and Ultisols, with a common mineralogical sequence of evolution but formed under different climate conditions and time, are composed principally of allophane (amorphous aluminium silicate of variable Al to Si ratio), at the surface of which Al- and Fe-humic complexes predominate, presenting high retention capacity because of their greater surface area (Besoaín et al. 2000; Escudey et al. 2001). For instance, Fig. 3.6 shows the strong positive relationship ($R^2 = 0.80$) between phosphate retention and ammonium-oxalate extractable $Al_{ox} + ½Fe_{ox}$ established for Chilean volcanic soils. In most cases, when $Al_{ox} + ½Fe_{ox} > 3$ % phosphate retention in soils is around 100 %.

In Andisols and Ultisols of the Mediterranean zone, phosphate retention is evidenced by low P availability to plants (Montenegro 1989). Thus, the ISRID scheme has focused its efforts on supporting farmers to apply phosphate fertilisers. Plants can present different adaptations to obtain phosphates in high phosphate retention soils. For instance, Sadzawka (1989) examined wheat and lupin growing in a phosphate-deficient Andisol and found that lupin was more efficient in P uptake than wheat because the acidified rhizosphere increased the solubility of P compounds in the soil.

In addition, after the initial rapid allophanic adsorption, any added P is also subjected to reactions resulting in the formation of organic-P forms (Borie and Zunino 1983). Thus, in these soils a large proportion of P is found in organic forms which are not available to plants (Escudey et al. 2001; Borie and Rubio 2003). However, Borie et al. (1983) found that soil fungal populations were particularly active in solubilising these organic phosphates. On the other hand, compared with agricultural soils, forest soils show a higher P cycling rate due to higher fungal activity, root phosphatase activity and organic acid excretion, resulting in higher P bioavailability (Borie and Rubio 2003).

3.1.3.3 Cations

Potassium (K^+), calcium (Ca^{2+}) and magnesium (Mg^{2+}) in a typical soil are present in the solution phase (intensity), but are predominantly adsorbed on the soil's exchange complex (quantity). In Chilean soils, cation availability depends mainly on type and amount of clay minerals, CEC and climate factors such as rainfall amounts (Ruíz and Sadzawka 1986; Gacitúa et al. 2008). In particular, CEC and ECEC have been used to evaluate the capacity of soils to supply plants with K^+, Ca^{2+} and Mg^{2+}. Sadzawka (2006b) related the CEC and the ECEC status to soil order within the Mediterranean zone and found that Histosols and Vertisols presented the highest CEC values, related to the higher amount of exchange sites associated with organo-mineral complexes and clay, respectively (Table 3.5).

Rainfall effects are important for cation availability in some Chilean soils, for example in the Southern

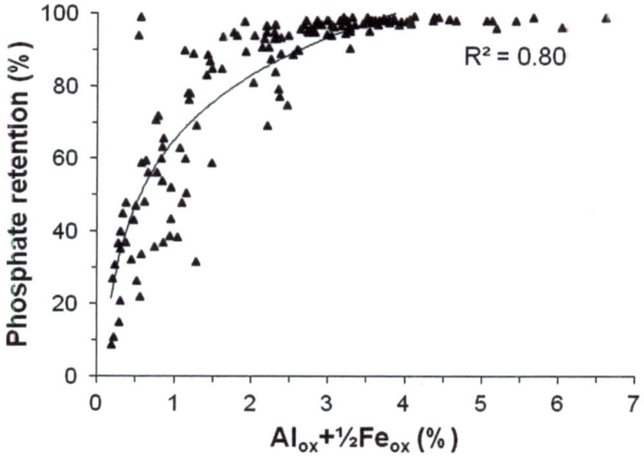

Fig. 3.6 Relationship between phosphate retention and Al_{ox} + ½Fe_{ox} in volcanic ash-derived soils in the Mediterranean zone of Chile ($n = 162$)

Mediterranean zone cation leaching is recognised as a major factor in limiting productivity (Bernier and Alfaro 2006). Most soils in this zone display BS <50 %, and crops usually respond positively to fertiliser application (Montenegro and Rodríguez 1985). Another soil factor that can cause plant nutrient deficiency, mainly of potassium, in the Northern Mediterranean zone is fixation in soils with high 2:1 clay contents (Ruíz and Sadzawka 1986).

3.1.3.4 Sulphur

Sulphur (S) deficiency of crops, which has been reported with increasing frequency over the past two decades on a worldwide scale, is a factor that reduces yield and affects the quality of harvested products (Scherer 2009). In Chile, S applications to soils are mainly recommended for some crops, such as sugarbeet, in the Southern Mediterranean zone, whereas in the hyperarid to semiarid and Northern Mediterranean zones irrigation water contains enough sulphates (SO_4^{2-}) to cover crop requirements. In the Southern Mediterranean zone, adsorption of SO_4^{2-} by soils is an important factor in controlling its mobility and availability to plants. Martínez et al. (1998) found that SO_4^{2-} adsorption decreased with increasing pH of the equilibrium solution and that the ionisation fraction values of organic acids at equilibrium pH were correlated with the amounts of SO_4^{2-} adsorbed.

3.1.3.5 Micronutrients

In Chile, micronutrient availability for plant nutrition is directly related to soil reaction. As in other countries (Rashid and Ryan 2004), in the Hyper-arid to Semi-arid zone and in the Northern Mediterranean zone micronutrients such as iron (Fe), zinc (Zn), manganese (Mn) and copper (Cu) in particular can present deficiencies due to high pH levels.

Table 3.5 Cation exchange capacity (CEC) and effective cation exchange capacity (ECEC) at 0–20 cm in different soil orders in the Mediterranean zone of Chile (from Sadzawka 2006b)

Soil order	CEC		ECEC	
	Range ($cmol_c\ kg^{-1}$)	Average	Range ($cmol_c\ kg^{-1}$)	Average
Entisols	4–36	11	2–13	7
Inceptisols	4–87	16	1–35	7
Alfisols	4–49	19	2–26	10
Mollisols	5–64	22	4–56	17
Ultisols	17–60	23	2–9	5
Andisols	7–76	37	1–18	6
Vertisols	17–75	41	15–67	33
Histosols	71–72	72	–	–

Figure 3.7 shows a number of micronutrients and their relationships to soil pH_{water} at 0–20 cm depth in soils from the hyperarid to semiarid zone and the Northern Mediterranean zone, based on analyses carried out in the Laboratory of Soil and Water Chemistry at the University of Chile using the DTPA extraction method. In these examples, the relationships between micronutrients and soil reaction were best fitted using second-order polynomial equations.

Figure 3.7 shows a clear tendency for declining Fe availability as pH increases. The lower Fe values in soils are most often observed in high pH and calcareous soils located in the Hyper-arid to Semi-arid zone. Similarly, the availability of Zn and Mn decreases with increased soil pH, with most pH-induced Zn and Mn deficiency occurring in the same zones of Chile. For instance, at the Antumapu Experimental Station, which belongs to the Faculty of Agronomy at the University of Chile (33°S), soils pH is around 8.0 and plant species sensitive to low contents of Fe and Mn, such as *Citrus* sp., usually present deficiency symptoms (Fig. 3.8). In calcareous soils of the Hyper-arid to Semi-arid zone, farmers add acid or acid-forming material to dissolve or neutralise $CaCO_3$ and thus decrease pH and increase micronutrient availability (Sierra et al. 2007). However, differing results have been found depending on the pH buffering capacity of the soil.

In contrast, Cu availability does not show a clear correlation with pH (see Fig. 3.7), with high Cu contents even at pH higher than 7.0. This may be related to Cu contamination of irrigated soils due to copper mining activities, which are a potential source of contamination of surface waters with toxic trace elements, as discussed later in the section on Chemical Soil Degradation (Sect. 4.2.2).

Another important micronutrient in Chile is boron (B), with soil nutritional problems in the Hyper-arid to Semi-arid zone being generally related to B toxicity and in the

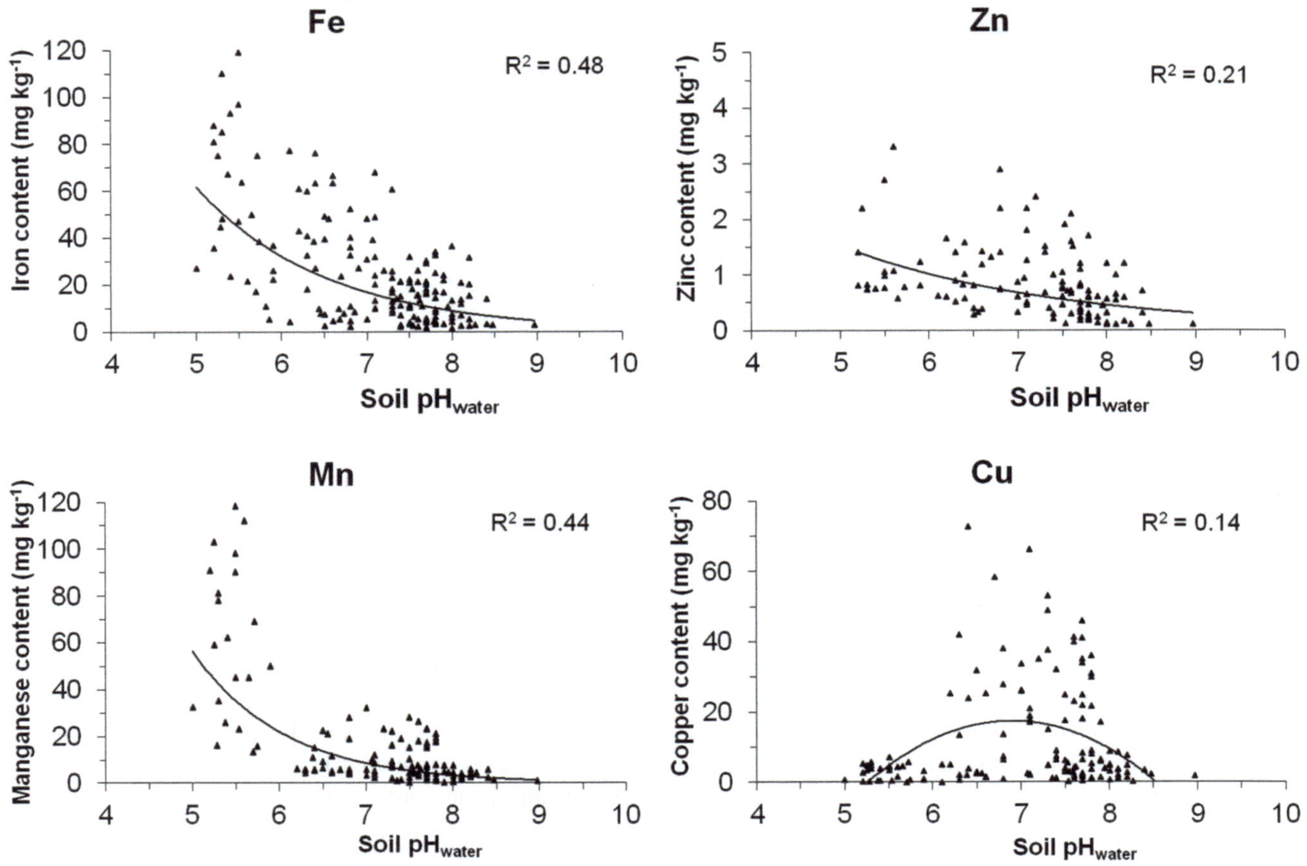

Fig. 3.7 Micronutrient contents (DTPA method) and their relationships with soil pH_{water} at 0–20 cm depth in agricultural soils in the Hyper-arid to Semi-arid zone and Northern Mediterranean zone of Chile. Fe ($n = 150$), Zn ($n = 120$), Mn ($n = 130$) and Cu ($n = 145$)

Mediterranean zone to B deficiency. Figure 3.9 shows B content and its relationship to soil pH_{water} at 0–20 cm depth in soils from the Hyper-arid to Semi-arid zone and the Northern Mediterranean zone. The analyses were carried out in the Laboratory of Soil and Water Chemistry at the University of Chile using a hot water extraction method. Although there is a tendency for B content to decrease as pH increases, particularly over pH 7.5, in the soil pH range 5.0–7.5 a broad range of values can be found. In some cases, the highest B contents occurring in soils are related to high B inputs with irrigation water.

Fig. 3.8 Leaf symptoms of Fe and Mn deficiency in orange trees in the Mediterranean zone of Chile

3.2 Physical Properties

In Chilean soils, soil physical properties are strongly related to OM content and are thus dependent on climate and topography. However, parent material becomes relevant in the case of volcanic ash-derived soils. Thus, the soil physical properties: bulk density, particle size distribution, water retention, structure stability and pore function in soils of the major soil zones are discussed below (see Sect. 2.2).

3.2.1 Bulk Density

Although soil bulk density (Db) is a physical property that is very dependent on soil texture (Brady and Weil 2000), OM input decreases Db because it promotes soil aggregation and abundant coarse pore formation, but also because

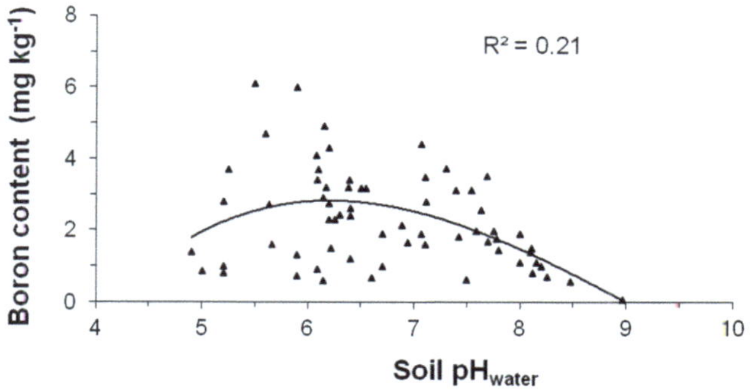

Fig. 3.9 Boron content at 0–20 cm depth and its relationship with soil pH_{water} in agricultural soils from the Hyper-arid to Semi-arid zone and Mediterranean zone ($n = 66$)

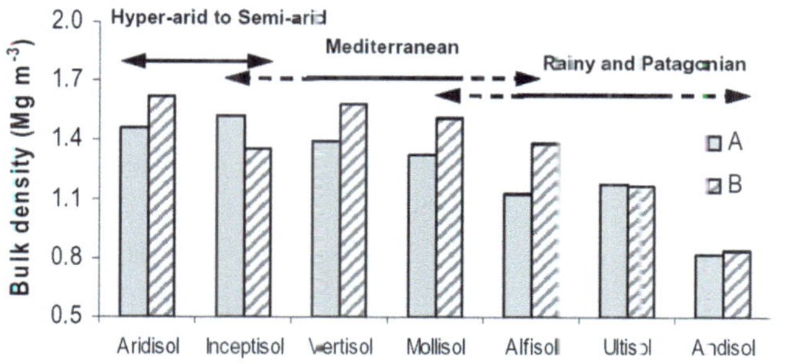

Fig. 3.10 Bulk density values for representative soils (according Soil Survey Staff 2006) along transect (north to south) in Chile. The values of surface (*A horizon*) and subsurface (*B horizon*) are included

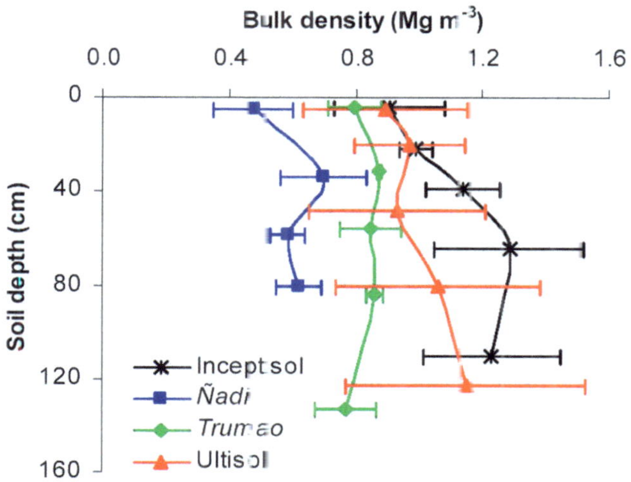

Fig. 3.11 Bulk density values and their variability (±SD) in function of depth for soils influenced by volcanic ashes along Mediterranean, Rainy and Patagonian zones. *Trumao* and *ñadi* are Andisols (see the text)

of the low Db values of the OM itself (Hillel 2004). In this regard, it is possible to define a decreasing Db gradient from north to south in Chile according to dominant soil orders in the major soil zones (Fig. 3.10).

In general, for normal deposited materials, soil Db increases with depth in the soil profile (Hartge 2000). However, this tendency is altered when, for example, a plough pan is created due to excessive machinery traffic (Ellies 1986) or when in natural polygenetic conditions different materials are deposited in alternate events, resulting in variable Db in different soil horizons. In Andisols, the shape of the mineral particles produces very porous soils with high mechanical stability (Ellies 1988), maintaining low Db values with depth, as shown in Fig. 3.11 for representative soils from the Southern Mediterranean zone (data from CIREN 2003).

Inceptisols can be either fresh, coarse volcanic materials or unevolved fine materials as poorly developed soils due to weather conditions or type of parent material, with 10–40 % clay content. In any case, they have low Db values in the first 40 cm because of their high OM content (between 12 and 20 % in topsoil and 4–12 % at 40 cm depth). At the other extreme of soil evolution, Ultisols show high variability of Db values, depending on the parent material (old volcanic ash or metamorphic rocks from the Coastal Range).

Andisols in Chile are locally differentiated into *trumao* and *ñadi*, which are well-drained and poorly drained volcanic soils, respectively. The *ñadis* (in general, a Placaquand) are restricted in depth for a B_{hsm} horizon overlying the glacial deposit, determining the lower depth of these soils (80 cm on average). The poorly drained conditions, with high amounts of water during the year, promote the accumulation of OM in the profile (OM in topsoil >20 %), giving the lowest Db values among all soils. On the other hand, *trumaos* are deeper and slightly denser than *ñadis*, but always with Db values lower than 0.9 Mg m^{-3} according to Soil Survey Staff (2006).

The OM gradient along the major soil zones (see Sect. 2.2) determines a dependency of Db on soil OM content, as is shown in Fig. 3.12 for representative soils under natural or low intensity use conditions.

In conclusion, from north to south there is an increase in OM content due to climate conditions (Padarian 2011) and a decrease in Db values due to the higher amount of OM and the ash-derived mineralogy (Besoaín 1985a). The latter determines the existence of low Db values with relatively low OM contents, moving away from the linear tendency in Fig. 3.12.

Particle density (Dr) varies between normal values in the soils with crystalline mineralogy (2.55–2.75 Mg m^{-3}), while in volcanic ash-derived soils (Andisols and Ultisols from the Rainy and Patagonian zones) the values can be more variable, depending on the OM and iron oxide contents. In Andisols, the values vary between 1.90 and 2.50 Mg m^{-3} (Nissen et al. 2005), owing to the high OM content, but in Ultisols the Dr values are higher (2.50–2.80 Mg m^{-3}), owing to the high iron oxide contents (Besoaín 1985a).

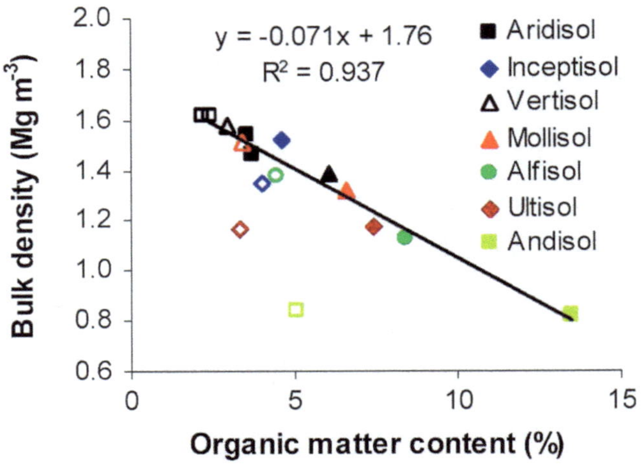

Fig. 3.12 Bulk density values along a soil organic matter gradient for representative soils from different soil orders according to Soil Survey Staff (2006). Full figures are from *A horizon* and empty figures are from *B horizon*. Line adjust excludes the subsurface values of Ultisol and Andisol

Fig. 3.13 Upper 100 cm in a Fluventic Haploxeroll (*left*, Isla Huechún soil series) and in an Entic Haploxeroll (*right*, Codigua soil series), both located in the same alluvial terrace of Mediterranean zone, separated by 20 m. Differences between soils determine a water availability of 16.7 and 6.6 cm, respectively (Ibáñez 2009)

3.2.2 Particle Size Distribution and Water Retention

As can be expected in a country of steep topography with active geological processes, the variability in soil texture is very high, and soils with different textural classes occur within a short distance and in the same physiographical position (Fig. 3.13). This prevents generalisations about the particle size distribution throughout Chile, and it is necessary to interpret the landscape in order to draw conclusions on some possible distributions (Birkeland 1999). In general, the Coastal Range is dominated by fine textured soils (derived from a granitic batholith between the hyperarid to semiarid and Mediterranean zones or from metamorphic rocks in the Mediterranean and Rainy and Patagonian zones). However, owing to erosion processes it is possible to find coarse-textured soils in the same formation and/or in creeks. In the valleys with permanent watercourses (Hyper-arid to Semi-arid zone) and in the Central Valley of the Mediterranean zone, the alluvial terraces are stratified and highly variable in textural classes, but from Chillán city (36°30′S) to the south, medium-textured soils (loam, silty loam, silty) dominate because of the dominance of allophane in volcanic ash-derived soils (Besoaín 1985b).

On the other hand, in the Hyper-arid to Semi-arid and Mediterranean zones the water retention depends on soil texture (Fig. 3.14), while medium to low OM content does not influence this property, giving higher differences between the field capacity and the permanent wilting point (defined by convention as the water retained at pore water pressure of −33 and −1,500 kPa, FC and PWP, respectively) at higher clay contents. Below the adjusted line for PWP (Fig. 3.14), the soil is excessively dry and plants do not have enough energy to absorb water from soil; above FC, there is free drainage of water through coarse pores, ensuring sufficient air-filled porosity for root respiration.

The dispersion around the adjusted lines in Fig. 3.14 depends not only on OM content, but also on clay mineralogy, soil structure, cations, salinity, etc. In general, for these kinds of soils there is a good correlation between water availability and clay content (Fig. 3.15).

Towards the south of Chile, the OM content increases, eliminating the dependency of water retention on clay content. However, the presence of volcanic soils with amorphous minerals creates a problem for correct determination of soil texture (Kimble and Nettleton 1984) because of the high charge of allophane and the risk of destroying particles if ultrasound is used. In this regard, there is a high variation in texture for the Andisols, depending on size of parent material (ashes or lapillis) and time of pedogenesis.

Figure 3.16 shows soil sand distribution in depth observing that Andisols from Chiloé Island present a lower variation and coarser textures than continental Andisols. On the other hand, *ñadis* can show a wide range of textures, depending on local conditions.

Inceptisols from Rainy and Patagonian zone show different textural classes, depending if they are fresh volcanic materials with low evolution or old parent materials eroded or non-evolved as a consequence of aquic conditions (for more details see Sect. 2.2). Ultisols from same zones are fine textured soils, dominating clay loam textures in surface

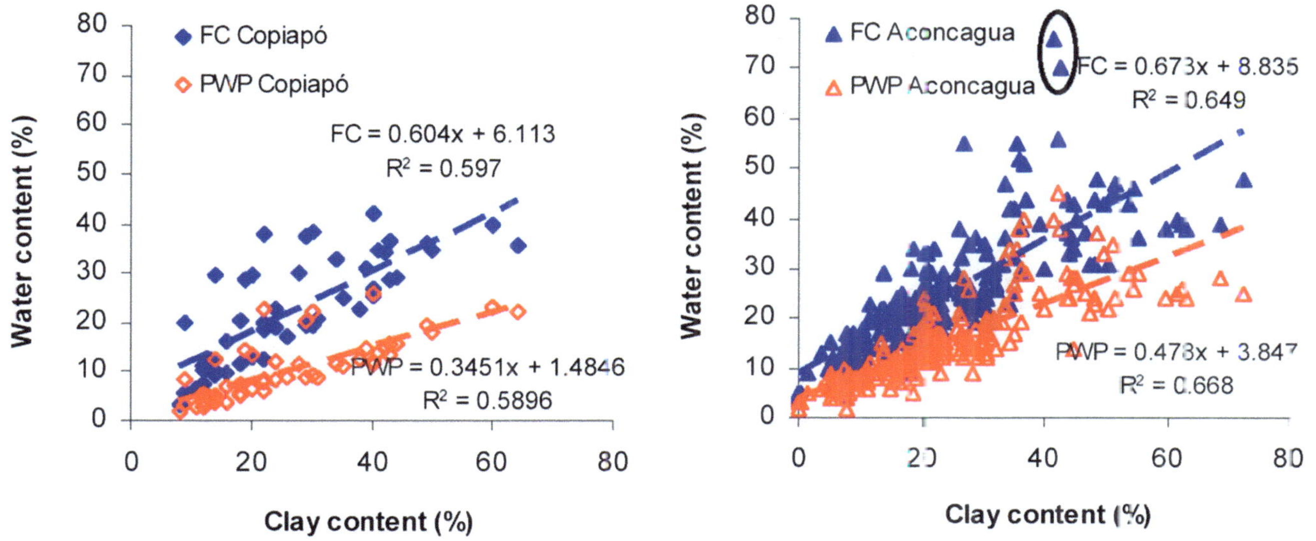

Fig. 3.14 Clay and gravimetric water contents at field capacity (*FC* −33 kPa) and permanent wilting point (*PWP* −1,500 kPa) for Copiapó Valley (Hyper-arid to Semi-arid zone) and Aconcagua Valley (Mediterranean zone). Values into the ellipse of Aconcagua Valley (upper horizons of Histosols) were excluded from the line adjust

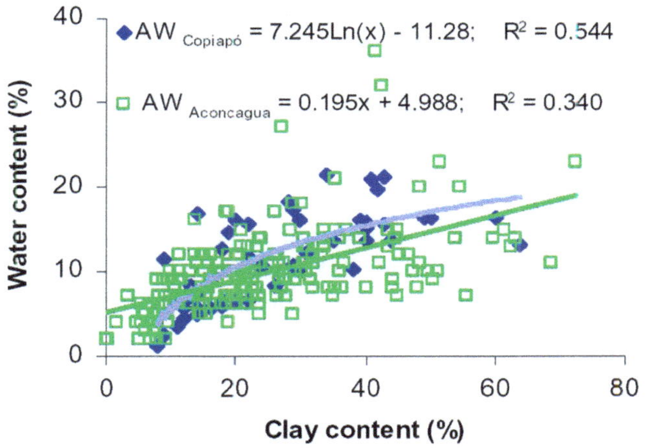

Fig. 3.15 Available water (AW) calculated as the difference in water content at FC and PWP from Fig. 3.14

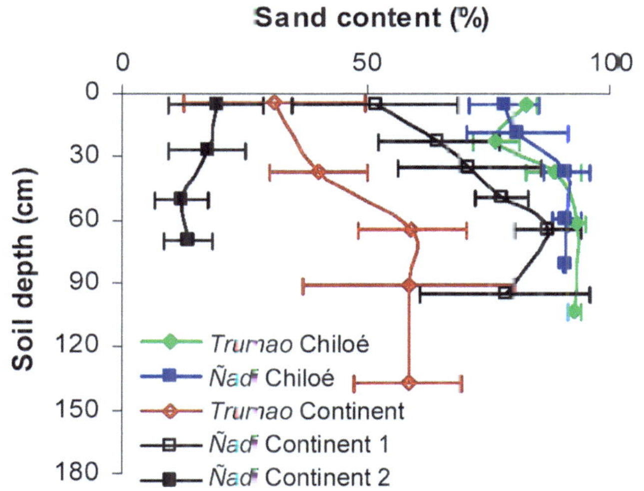

Fig. 3.16 Sand distribution in depth for well-drained Andisols (*trumao*) and poorly drained Andisols (*ñadi*) from continental land and Chiloé Island. *Ñadi* 1 corresponds to typical ash-derived soil and *ñadi* 2 correspond to more organic Andisol (Rainy and Patagonia zone)

and clay textures in depth. Nevertheless, the OM contents and the productive use in both soil orders determine different structural conditions, preventing to find a dependency between clay content and water retention (Fig. 3.17) as in northern regions.

Therefore, regardless of whether there is an effect on other soil properties, there is an increase in water retention from north to south in Chile, mainly owing to increased OM content and the dominance of short-range order minerals (Table 3.6). The only exception is the Andisols in the Rainy and Patagonian zone, which have lower water retention because of their lower pedogenic development.

3.2.3 Structural Stability

The physical and mechanical stability of soils ensures their functioning to maintain adequate productivity. Different approaches exist to evaluate the structural stability of soils, mainly focused on water stability and mechanical strength tests to characterise the behaviour of soils when wetted or submitted to external loads (Nimmo and Perkins 2002; Fredlund and Vanapalli 2002; Hartge and Horn 2009).

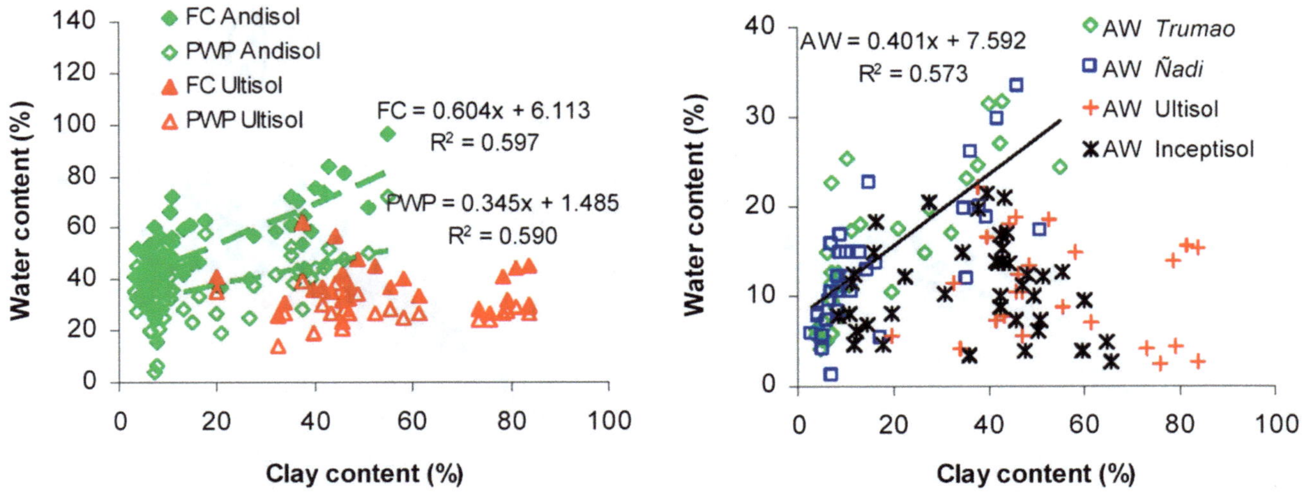

Fig. 3.17 Gravimetric water contents at field capacity and permanent wilting point (*left*) and available water, AW (*right*), for representative soils from Rainy and Patagonian zones. The line adjust of AW considers both kind of Andisols (*trumaos* and *ñadis*)

In relation to stability to water, the SOM content is essential to ensure adequate structure stability because it prevents the fast wetting of aggregates, avoiding air slaking (Chenu et al. 2000). On the other hand, it promotes unions among particles, increasing the contact points (Hartge 2000). According to this, it is obvious that there will be higher stability of Chilean soils influenced by volcanic materials as a consequence of their higher SOM contents, compared with soils from northern zones (Fig. 3.18). However, there is a significant influence of mineralogy and particle shape in volcanic ash-derived soils (Ellies 1988), which explains their high physical and mechanical stability even at low Db values.

The soils in Fig. 3.18 have natural vegetation or low intensity of use, so the results are close to non-altered conditions. The stability of the surface horizon is higher than that of lower horizons as a result of higher SOM content and more intense aggregation processes. On the other hand, there is a strong increase in aggregate stability (decrease in water dispersion) from Region VIII to the south, related to the increase in SOM content and the appearance of Andisols in the Central Valley (Padarian 2011). However, when soils are submitted to intense use, aggregate stability can decrease at critical levels due to SOM losses (Ellies et al. 2000). On the other hand, excessive SOM content can increase hydrophobic behaviour of

Table 3.6 Location, classification and properties for upper horizon of different soils along Chile, according to CIREN (1997a, b, 2003, 2005b, 2007)

Pedon	Region	Soil order	Textural class	Gravimetric water retention (−33 kPa) (%)	Gravimetric water retention (−1,500 kPa) (%)
Amolanas	III	Aridisol	Sandy loam	11.6	5.0
La Capilla	III	Aridisol	Clay loam	25.4	11.8
Encón	V	Inceptisol	Sandy loam	9.0	4.0
Pocuro	V	Mollisol	Loam	23.0	13.0
La Palma	V	Inceptisol	Sandy loam	21.0	9.0
Talca	VII	Alfisol	Loam	20.0	11.0
La Pelada	X	Inceptisol	Sandy loam	30.8	12.6
Los Ulmos	X	Ultisol	Clay	45.4	26.8
Puerto Fonck	X	Andisol	Sandy loam	48.1	29.3
Simpson	XI	Inceptisol	Sandy loam	31.8	19.1
Pollux	XI	Andisol	Sandy loam	31.9	22.0

See sandy loam Inceptisols by comparison

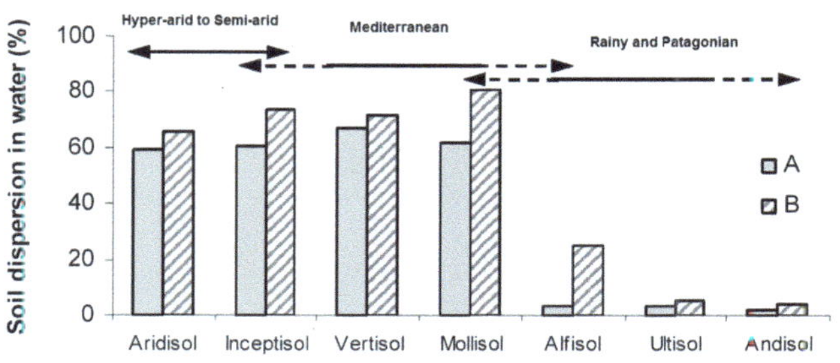

Fig. 3.18 Aggregate stability evaluated as a dry and wet sieving test (Hartge and Horn 2009), where the lowest value (lowest dispersion) denote the higher stability, along transect from north to south in Chile. Values of A and B *horizons* are included

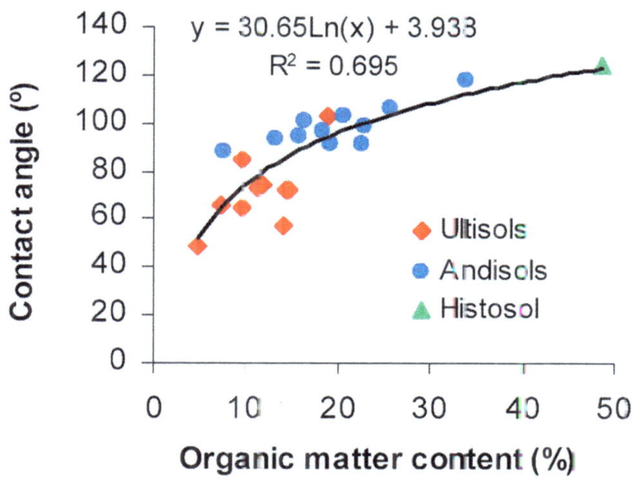

Fig. 3.19 Contact angle of soils from Rainy and Patagonian zone and their dependence on soil organic matter contents (adapted from Orellana 2010)

the soil, resulting in an increased risk of soil erosion (see Sect. 4.2). Figure 3.19 shows the contact angle between water and solid particles, as a hydrophobicity index, for different soils influenced by volcanic ash materials.

Soils with contact angles higher than 90° are hydrophobic (Hallet 2008), a typical condition of Chilean Andisols (Ellies et al. 2005). Other soils in Chile have contact angles lower than 60°, explaining their high dispersibility in water (Fig. 3.18). The common trend for Andisols and Ultisols (Fig. 3.19) denotes a possible common origin for both groups of soils, or at least rejuvenation processes in Ultisols by inputs of new volcanic ash (Besoaín 1985a).

Mechanical stability depends on particle size distribution, soil aggregation, SOM content, bulk density, pore water pressure and other factors related to internal tension, but also on external factors such as type of load (weight of machinery, contact pressure), number of passes and use (Horn 2003). Figure 3.20 shows mechanical strength for different soils from Chile, evaluated with a cone penetrometer in vertical measurements and 24 h after a rain or irrigation.

The nature of Aridisols, with low SOM content and abundant cementing agents, explains their high mechanical strength despite their low stability in water. Intense use can increase the strength of some soils to critical levels, as it is a highly variable soil property. However, even with high intensity of use, Andisols do not reach values of mechanical strength higher than 100 kPa evaluated in wet conditions (Ellies et al. 2000), maintaining a relatively constant behaviour in mechanical properties under wet and dry conditions (Ellies 1988).

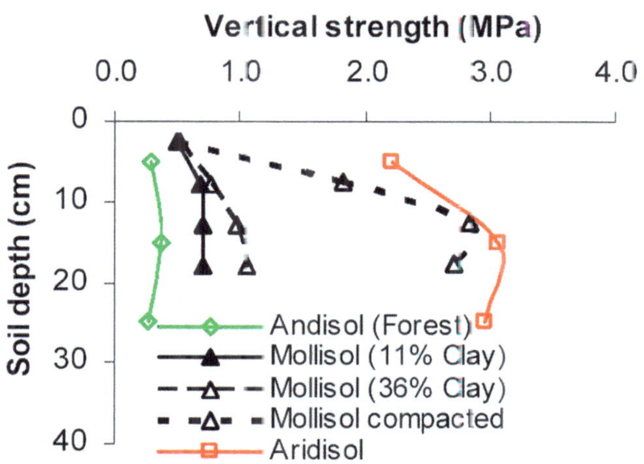

Fig. 3.20 Vertical strength evaluated by cone penetrometer in the ploughed horizon of different soil orders in Chile, 24 h after irrigation. Compacted Mollisol has a clay content of 30 %

3.2.4 Pore Functionality

Basic soil physical properties such as Db and texture allow soil behaviour to be inferred within certain limits, but sometimes it is necessary for a more detailed characterisation to understand specific phenomena, for example gas and water transport, which are dependent on soil structure. In this regard, hydraulic conductivity (K) and pore size distribution are useful for making a correct diagnosis of the structural conditions (Ellies et al. 1997). Many efforts have been made by researchers in Chile to define an appropriate methodology to measure soil K (Pfeiffer et al. 2008; González et al. 2005; Luna 2003). In fact, Fodor et al. (2011) found that the effect of the evaluation method applied for assessing K can be just as significant as the effect of other factors, such as the scale effect, as well as the spatial and temporal variation. Table 3.7 presents values of saturated hydraulic conductivity (K_s) and its variation in Chilean soils.

Table 3.7 Saturated hydraulic conductivity (K_s) mean values for representative Chilean soils

Pedon	Region	Soil classification	$K_s \pm$ SD (m d^{-1})	CV (%)	Clay content (%)
Elisa de Bordos	III	Aridisol	8.51 ± 2.21	18.8	12
La Capilla	III	Typic Haplocambid	3.22 ± 0.92	28.6	31
Isla de Huechún	MR	Fluventic Haploxeroll	2.99 ± 1.80	60.2	22
Santa María	V	Typic Haploxeroll	7.23 ± 1.36	18.8	23
La Lajuela	VI	Ultic Haploxeralf	2.11 ± 0.92	43.6	15
La Obra	VII	Aquultic Haploxeralf	7.10 ± 0.92	13.0	27
Valdivia	X	Typic Hapludand	1.50		25
Malihue	X	Typic Hapludand	7.90		31

La Obra soil from Pfeiffer et al. (2008) and both Hapludand from Ellies et al. (1995). Clay content for Ap horizon is included

The data presented in Table 3.7 were obtained from sites with medium to low intensity of use, and showed low variability according to Jury et al. (1991). As expected, there was no correlation between soil orders or clay content and K_s, a property which is highly variable and dependent on management practices that affect the structure. Nevertheless, Ellies et al. (1997), working in a wide range of volcanic soils, determined a high correlation between coarse porosity (>50 μm) and K_s, showing also that if aggregate stability is high, temporal variation in K_s is low.

Depending on slope gradient and aspect, Casanova et al. (2000, 2003) measured different field non-saturated hydraulic conductivity (K_{ns}) values using a tension disc infiltrometer and established a direct relationship with clay content on Inceptisols from a toposequence in Central Chile. They concluded that with increasing site inclination, K_{ns} increases as result of higher lateral flow, which favours the process of soil slippage. On the other hand, Casanova et al. (2009) reported different K_{ns} values measured in soil monoliths (Table 3.8) and observed that in all soils K_s values were strongly negatively correlated to slope gradient. They attributed this to refraction of water flow, considering the greater frictional or viscous resistance generated by inclination.

Table 3.8 Saturated hydraulic conductivity (K_s) estimated by tension disc infiltrometer in two soil monoliths arranged at different slope gradients

Soil classification	Slope gradient (%)	K_s (mm h^{-1})	Clay content ± SD (%)
Fluventic Haploxerolls	0	1.9	52.5 ± 2.11
	15	1.8	
	20	1.0	
	25	0.8	
Typic Xerochrepts	0	15.0	3.4 ± 0.10
	15	9.5	
	20	6.6	
	25	3.0	

Adapted from Casanova et al. (2009)

Table 3.8 does not include all soils studied by Casanova et al. (2009). Evaluating four soils with different clay contents, they found a direct relationship between clay content and K_s, but it is not possible to generalise this dependence for Chilean soils, as noted for Table 3.7. Recently, Dörner et al. (2010) reported that K_s decreased for Andisols during water infiltration as a function of land use due to particle release, transport and re-sedimentation, which probably affects the pore continuity.

Except for the Andisols, it is not possible to relate K to coarse porosity, because the ability to conduct water also depends on aggregate stability to water. However, the ability to renew pore air is related to soil texture (Ferreyra et al. 2011) and structure (Poblete 2011), the latter particularly when evaluated as changes in coarse porosity as result of soil management (Fig. 3.21).

Again, it is not possible to generalise regarding the behaviour of air flux for Chilean soils, because the results depend on management (which affects the structure, see Fig. 3.21) and soil composition. In general, many soils under agricultural use have some degree of compaction (Ferreyra 2009), but the ability to conduct fluids will depend on the tortuosity, which can be affected by texture and the presence of coarse particles (Seguel et al. 2011).

In the case of soils under the influence of volcanic materials in the Rainy and Patagonian zone, Seguel and Horn (2006) measured values of air flux between 43 and 280 cm h^{-1} for aggregate beds simulating a ploughed horizon in Andisols equilibrated at −6 kPa, while Leiva (2009) measured a wider range of air flux in an Ultisol, with values being closely related to water conductivity.

Finally, Fig. 3.22 shows the dependence of drainage porosity on clay content for different soils in Chile. According to this, Chilean soils could present problems of

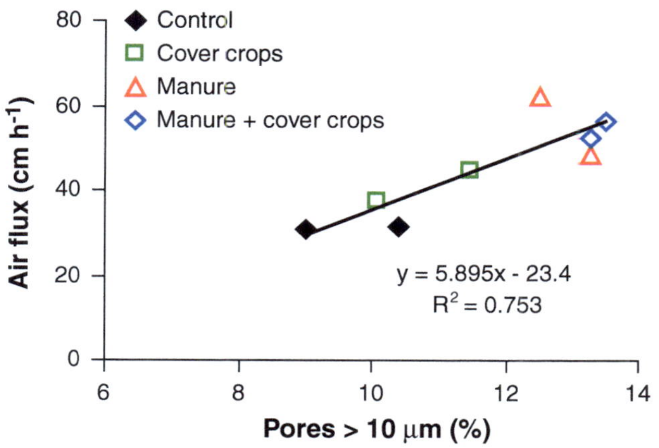

Fig. 3.21 Convection air flux depending on proportion of coarse pores in a compacted Aridisol subjected to different managements during 4 years

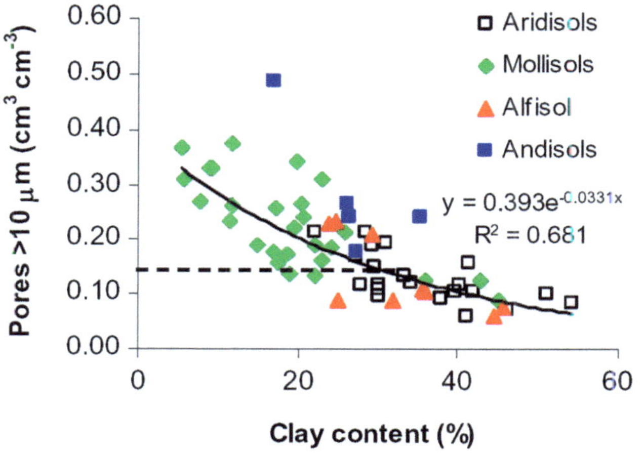

Fig. 3.22 Volume of pores >10 μm in relation with clay content for different soil orders from Chile. Andisols (taken from Ellies and Mac Donald 1984) were excluded from the line adjust. The *horizontal line* demarcates a critical level (0.15 cm^3 cm^{-3}), after Richards (1983)

low air porosity content with about 30 % clay, a very critical level in Aridisols and Alfisols, but not important in Andisols.

3.3 Biological Properties

Low diversity ecosystems are expected to be more vulnerable to global climate change, although they have received less attention than high diversity ecosystems (Wall 2007). As described in Chap. 1, natural barriers (Pacific Ocean, Antarctic, Andes Mountains and Atacama Desert) give an island connotation to the territory of Chile, resulting in moderate biodiversity compared with other South American countries. Badal (2008) concluded that given the smoothness of thermal regimes, the low incidence of extreme temperatures and the generally uniform Chilean weather, the most decisive factor in determining the evolution of the vegetation is the moisture in summer.

Soil biological properties such as soil organic matter (SOM) and soil biodiversity can affect several critical soil functions that support food production and environmental quality. It is well-known that SOM contains large reserves of soil organic carbon (SOC), a major factor controlling global warming. In addition, SOM contains large amounts of plant nutrients (especially N and S) and affects other chemical and physical soil properties, such as cation exchange and water-holding capacity. The activity of the soil biota is also responsible for nutrient transformations in soils and underpins a number of other fundamental chemical and physical soil properties.

These biological properties are revisited in this section, which examines in particular how these soil properties are affected by the soil moisture regime and vegetation in Chile.

3.3.1 Soil Organic Carbon

SOC content is directly related to precipitation regime (climate factor) and vegetation (organism factor). These two soil formation factors mean that soils in Chile may be roughly divided into soils with low SOC stocks (SOC ≤2.5 %) in the hyperarid to semiarid and Northern Mediterranean zones and soils with medium–high (SOC >2.5 %) stocks in the Southern Mediterranean and Rainy and Patagonian zones. For instance, desert and rainy forest landscapes are correlated with low and high SOC levels, respectively, as SOC may be lower than 0.5 % in the Hyper-arid to Semi-arid and in the range 5–20 % in forest soils classified as Andisols in the Rainy and Patagonian zone (Aguilera 2000). Figure 3.23 shows SOC in major soil zones in Chile at 0–20 cm based on a number of soil surveys carried out on 708 pedons in agricultural areas during the last 40 years by different government agencies (SAG 1974; CIREN 1996a, b, 1997a, b, 1999, 2002, 2003, 2005a, b, 2007; Ahumada et al. 2004). In the Hyper-arid to Semi-arid zone all the soils examined in these studies displayed SOC values lower than 5.0 and 96 % of the soils had SOC ≤2.5 %. The Mediterranean zone can be roughly divided into two zones related to the precipitation regime, a Northern or dry Mediterranean zone that has precipitation of 300–1,200 mm $year^{-1}$ and a Southern or wet Mediterranean zone that has precipitation of 1,200–2,500 mm $year^{-1}$. In the Northern Mediterranean zone, 81 % of the pedons analysed had SOC ≤2.5 %, 16 % of pedons had values in the range 2.6–7.5 %, and the highest SOC (20 %) was found in a

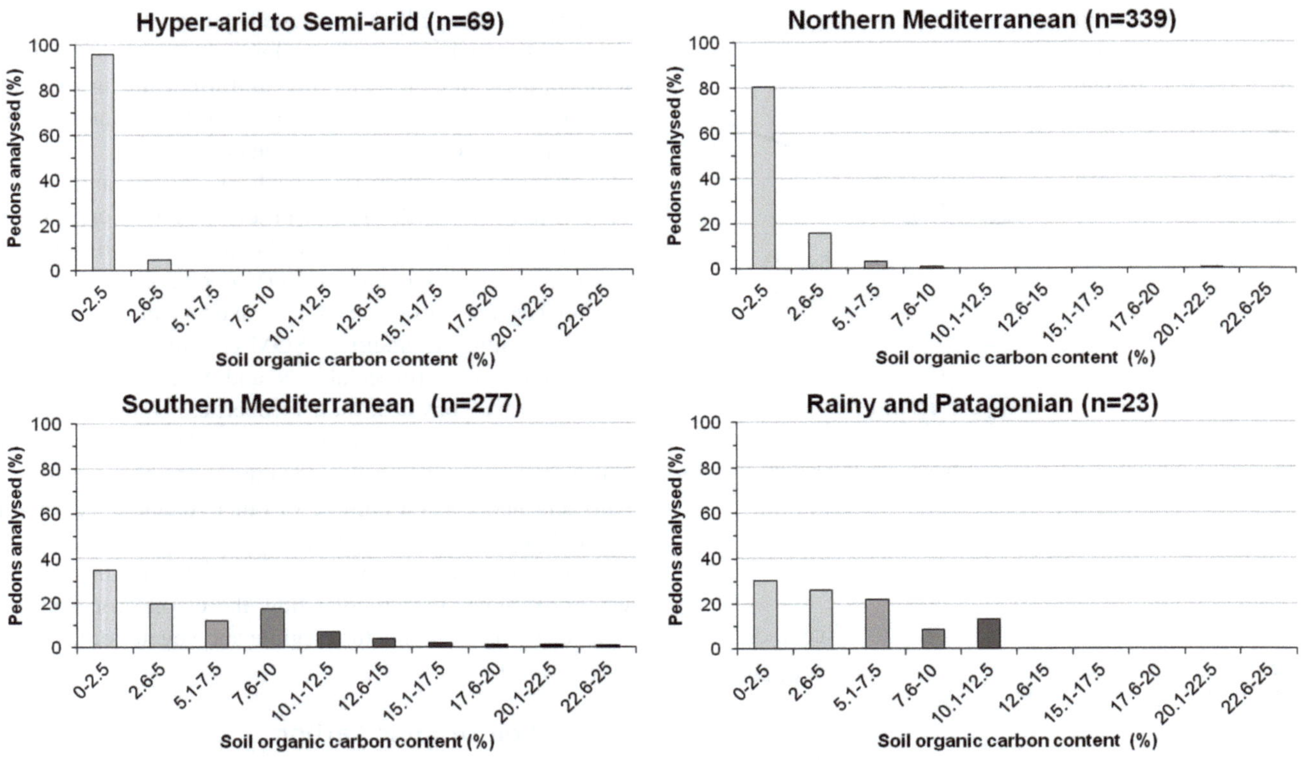

Fig. 3.23 Soil organic carbon ranges in major soil zones in Chile at 0–20 cm depth (n = number of pedons analysed)

Histosol in a lacustrine landform. In contrast, in the Southern Mediterranean zone, soils had high SOC contents, with 45 % of the soils having SOC higher than 5 and 16 % of soils having SOC contents higher than 10 %. Considering only two soil surveys carried out in the Rainy and Patagonian zone (CIREN 2005b; SAG 1974) with a limited number of soil descriptions ($n = 23$ pedons) for the extensive area defined, it is possible to find a broad range of SOC contents, with 56 % of the soils having SOC ≤5 %, 31 % with SOC in the range 5.1–10 % and 13 % of soils with SOC in the range 10.1–12.5 %.

Similarly, as was discussed for soil reaction (Sect. 3.1.1), SOC levels follow the same tendency as precipitation regime. The relationship between SOC stocks at 0–20 cm analysed in 708 pedons between the hyperarid to semiarid zone and the Rainy and Patagonian zone and latitude (UTM) were best fitted to a fourth-order polynomial equation, with R^2 of 0.57. For instance, Fig. 3.24 shows that SOC stocks are directly related to precipitation regime, increasing from levels lower than 2 % in the Hyper-arid to Semi-arid zone to levels higher than 10 % in the southern part of the Mediterranean zone. Furthermore, when the MAP decreases in the Rainy and Patagonian zone, SOC decreases to values around 2 %.

In the *espinal* agroecosystem, the SOC content usually shows a high variability in the upper soil. A possible explanation is that this zone is usually covered by Mediterranean annual prairie, including native *Acacia caven* trees in some patches (see Fig. 3.25). This contributes to microvariation in SOC by creating islands of fertility (Salazar et al. 2011). For example, Olivares et al. (1988) reported greater amounts of SOC and N stocks beneath *A. caven* canopy than in open grassland, which may explain the higher spatial variability in soil properties in the Mediterranean annual prairie. Similarly, Muñoz et al. (2007) in a study in the *espinal* Chilean agroecosystem found that *A. caven* canopy increased SOC stocks in the 0–40 cm layer by 25 % compared with intercanopy.

In soils of the southern part of the Mediterranean zone, Aguilera et al. (1997) noted that Andisols have higher SOC levels than Ultisols, but the fraction distribution in the latter suggests a shift of the more stable fractions to a more labile state. Those authors suggested that the content of humines and humic and fulvic acids may indicate that the OM in Chilean volcanic soil is highly humified.

Some studies have been carried out to determine the effects of agricultural practices on soil carbon fractions. For instance, Zagal et al. (2002a) examined an Andisol from the southern part of the Mediterranean zone and found a clear tendency for light OM fractions to decrease when soil use was intensified.

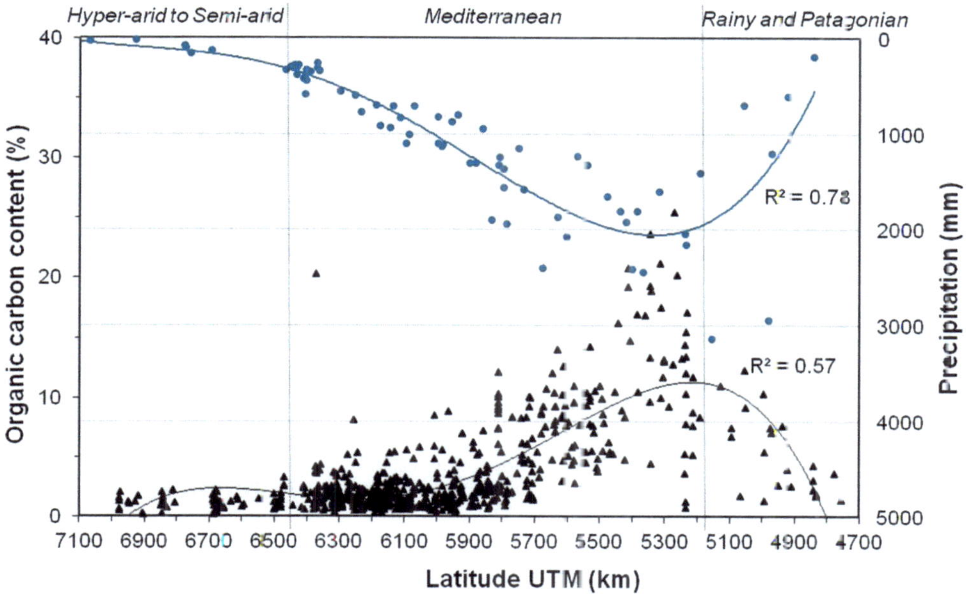

Fig. 3.24 Soil organic carbon (SOC) content at 0–20 cm in pedons between the Hyper-arid to Semi-arid and the Rainy and Patagonian zones. *Triangles* are SOC content in pedons ($n = 708$) and *circles* are mean annual precipitation obtained from 74 Chilean meteorological stations

Fig. 3.25 *Espinal* agroecosystem in the Mediterranean zone of Chile

In addition, the amount of SOM is a key indicator of soil fertility, and has been associated with the capacity of the soil to supply essential elements for plant production. For instance, Matus and Maire (2000) found that in the northern part of the Mediterranean zone, there was a close relationship between soil OC content and nitrogen mineralisation rate. Similarly, Rivas et al. (2007) noted that pristine temperate forest soils in the southern part of the Mediterranean zone receive low N additions by atmospheric deposition, so tree growth depends primarily on the internal recycling of nutrients present in the OM. In addition, Borie et al. (2002) noted that the natural N input processes in these soil systems are carried out by N-fixing microbes such as *Rhizobium* spp. and *Azotobacter* spp., endomycorrhizal activity and microbial synthesis of humic polymers.

3.3.2 Soil Biodiversity

Soil biodiversity reflects the mix of living organisms in the soil, which interact with one another and with plants and small animals, forming a web of biological activity. Biological activity in soils increases or decreases depending on land use and soil management and changes in activity have been evaluated for Chilean soils through different parameters, such as occurrence of species, soil respiration, radioactive tracer techniques, biochemical variables and glomalin content.

Castillo et al. (2010) reported that the diversity of arbuscular mycorrhizal (AM) fungal species is highly variable in agricultural soils in the southern part of the Mediterranean zone, and that the occurrence or absence of AM fungi is likely to depend on the various agronomic inputs that farmers have applied to crops over the years.

Soil respiration, expressed as carbon dioxide (CO_2) emissions, is one measure of biological activity and decomposition which reflects the capacity of soil to support life. Table 3.9 shows some soil respiration rates reported for agricultural and forest soils in Chile. There have also been some soil respiration studies on agricultural soils, for instance Zagal et al. (2002b) examined an Andisol in the Mediterranean zone (36°31′S) and found that the OM content of the different crop rotations correlated significantly with microbial activity.

This CO_2 production considered as carbon (C) mineralisation can be also evaluated using radioactive tracer techniques such as glucose ^{14}C, cellulose ^{14}C and organic residues ^{14}C. Zunino et al. (1982) analysed agricultural soils classified as Andisols in the southern part of the

Table 3.9 Soil respiration rate in different agricultural and forest soils in the Mediterranean zone of Chile

Vegetation	Respiration rate	References
Crop rotation	420–1,046 μg C–CO_2 g^{-1} 30 d^{-1}	Zagal et al. (2002a, 2002b)
Crop rotation	104–353 μg C–CO_2 g^{-1} 10 d^{-1}	Zagal and Córdova (2005)
Espinal forest	2.8–11.8 mg C–CO_2 100 g^{-1} d^{-1}	Carmona et al. (2006)
Conifer forest	55.7 mg C–CO_2 100 g^{-1} d^{-1}	
Nord-patagonia forest	10.4–65.6 mg C–CO_2 100 g^{-1} d^{-1}	
Sclerophyll forest	0.5–0.8 g C–CO_2 m^{-2} h^{-1}	

Table 3.10 Glomalin content in topsoil in different ecosystems in the Mediterranean zone of Chile

Soil order	Vegetation	Glomalin (mg g^{-1})	References
Alfisol	Crop rotation (no-tillage)	1.8–3.6	Borie et al. (2000)
Alfisol	Pasture	1.6–2.1	
Andisol	Mixed forest	44–46	Seguel et al. (2008)
Andisol	Crop rotation	65	Morales et al. (2005)
Andisol	Evergreen forest	88–114	
Ultisol	Crop rotation	9–10	

Mediterranean zone and found that during 4 months, 70–79 % of the C in added glucose and cellulose had evolved as CO_2.

Other studies have determined soil biological properties through biochemical variables. Recently, Lillo et al. (2011) in a study in the high mountain areas (1,140–1,700 m a.s.l.) of the southern part of the Mediterranean zone determined soil biological properties such as fluorescein diacetate hydrolysis, carbon and nitrogen microbial biomass, β-glucosidase, carboximethylcellulase, acid phosphatase, urease and arylsulphatase. They found that these biochemical parameters were sensitive to changes in altitude, with the highest biological activities observed at the lower altitudes. Peirano et al. (1992) in an Andisol in the same zone found a direct relationship between microbial biomass and SOM.

Some studies have evaluated the production of glomalin, a glycoprotein produced by AM fungi, as an indicator of soil biological activity. Borie et al. (2000) studied an Alfisol in the southern part of the Mediterranean zone and found a close relationship among glomalin content, SOC and mycorrhizal hyphal density. Table 3.10 shows soil glomalin content in different ecosystems in Chile.

It is also possible that increasing soil biodiversity by soil inoculation with microorganisms may have some positive effects on crop yield. For instance, Redel et al. (2006) in a study of a strongly acidic Andisol in the southern part of the Mediterranean zone found that when wheat and lupin were inoculated with an exogenous AM fungus, the mineral uptake of Al decreased, which might contribute to reducing toxicity problems in acidic soils.

References

Aguilera SM (2000) Importance of the protection of the soil organic matter. Chilean Soil Sci Soc Bull 14:77–85 (in Spanish)

Aguilera SM, Borie G, Peirano P, Galindo G (1997) Organic matter in volcanic soils in Chile: chemical and biochemical characterization. Commun Soil Sci Plant Anal 28:899–912

Alfaro M, Salazar F, Endress D, Dumont JC, Valdebenito A (2006) Nitrogen leaching losses on a volcanic ash soil as affected by the source of fertilizer. R C Suelo Nutr Veg 6(2):54–63

Ahumada M, Araya E, Kühne A, Sepúlveda M (2004) Guide for soil and pasture description of the Palena Province. Agriculture Ministry-Agriculture and Livestock Service, Report N°17 (in Spanish)

Badal G (2008) Biodiversity of Chile, patrimony and challengers 2nd edn. CONAMA, National Environmental Commission (in Spanish)

Bernier R, Alfaro M (2006) Soil acidity and lime effect. Agriculture and Livestock Research Institute (INIA), INIA Bull 151. Osorno (in Spanish)

Besoaín E (1985a) Mineralogy of volcanic soils from central-south of Chile. In: Tosso J (ed) Volcanic soils from Chile. Institute of Agriculture and Livestock Research–Chilean Ministry of Agriculture, Santiago (in Spanish)

Besoaín E (1985b) Soils. In: Tosso J (ed) Volcanic soils from Chile. Institute of Agriculture and Livestock Research- Chilean Ministry of Agriculture, Santiago, Chile (in Spanish)

Besoaín E, Peralta M, Massaro S (2000) Mineralogy and origin of some volcanic ash soils of continental Chiloé. Chile Agric Tecnica (Chile) 60:127–153 (in Spanish with English summary)

Birkeland P (1999) Soils and geomorphology. Oxford University Press, New York

Bockheim J, Douglass D (2006) Origin and significance of calcium carbonate in soils of southwestern Patagonia. Geoderma 136: 751–762

Borie F, Rubio R (2003) Total and organic phosphorus in Chilean volcanic soils. Gayana Bot 60:69–78

Borie F, Rubio R (1999) Effects of arbuscular mycorrhizae and liming on growth and mineral acquisition of aluminium-tolerant and aluminium sensitive barley cultivars. J Plant Nutr 22:121–137

Borie F, Zunino H (1983) Organic matter-phosphorus associations as a sink in P-fixation processes in allophanic soils of Chile. Soil Biol Biochem 15:599–603

Borie F, Rubio R, Morales A, Castillo C (2000) Relationships between arbuscular mycorrhizal hyphal density and glomalin production

with physical and chemical characteristics of soils under no-tillage. Rev Chil Hist Nat 73:749–756 (in Spanish with English summary)
Borie F, Quinteros J, Aguilera M (1983) Biochemistry of soils derived from volcanic ashes. IV Solubilization of phosphates by soil fungi. Agric Tecnica (Chile) 43(4):371–376 (in Spanish with English summary)
Borie G, Peirano P, Zunino H, Aguilera SM (2002) N-pool in volcanic ash-derived soils in Chile and its changes in deforested soils. Soil Biol Biochem 34:1201–1206
Brady N, Weil R (2000) Elements of the nature and properties of soils. Prentice Hall, Upper Saddle River, New Jersey
Campillo R, Urquiaga S, Undurraga P, Pino I, Boddey RM (2005) Strategies to optimise biological nitrogen fixation in legume/grass pastures in the southern region of Chile. Plant Soil 273:57–67
Campillo R, Urquiaga S, Pino I (2002) Biological nitrogen fixation in white clover through ^{15}N labeling methodologies, on a soil derived from volcanic ashes. J Soil Sci Plant Nutr 2(1):25–34 (in Spanish with English summary)
Cárcamo B, Puentes L, Godoy R, Oyarzún C, Valenzuela E (2004) Soil biological activity in a *Nothofagus obliqua* (Mirb) Oerst forest, south-central de Chile. J Soil Sci Plant Nutr 4(2):14–25 (in Spanish with English summary)
Carmona MR, Aguilera M, Pérez CA, Serey I (2006) Respiratory activity of soil organic horizon of forest ecosystems from centre and south of Chile. Gayana Bot 63:1–12
Carrasco MA, Tapia X, Opazo JD, Benavides C (1993) Kinetics of phosphorus retention in agriculture soils of alluvial and volcanic origin. Agric Tecnica (Chile) 53(4):331–339 (in Spanish with English summary)
Casanova M, Benavides C (2009) Mapping potential losses of N-urea by volatilization in soils of central Chile. J Soil Sc Plant Nutr 9(1):14–25
Casanova M, Salazar O, Seguel O, Noguer V (2009) Hydraulic conductivity on soil monoliths of central Chile. J Soil Sci Plant Nutr 9(3):210–221 (in Spanish with English summary)
Casanova M, Seguel O, Joel A, Messing I, Luzio W, Vera W (2003) Pedotransference functions for hydraulic conductivity in rainfed hillsides. R C Suelo Nutr Veg 3(2):42–48 (in Spanish with English summary)
Casanova M, Messing I, Joel A (2000) Influence of aspect and slope gradient on hydraulic conductivity measured by tension infiltrometer. Hydrol Process 14:155–164
Castillo C, Rubio R, Borie F, Sieverding E (2010) Diversity of arbuscular mycorrhizal fungi in horticultural production systems of southern Chile. J Soil Sci Plant Nutr 10(4):407–413
Chenu C, Le Bissonnais Y, Arrouays D (2000) Organic matter influence on clay wettability and soil aggregate stability. Soil Sci Soc Am J 64: 1479–1486
CIREN (2007) Soil survey, region III of Chile (Copiapó and Huasco valleys). Natural Resources Information Centre, Publication N°135 Santiago (in Spanish)
CIREN (2005a) Soil survey, region IV of Chile. Natural Resources Information Centre, Publication N° 129 Santiago (in Spanish)
CIREN (2005b) Soil survey, region XI of Chile. Natural Resources Information Centre, Publication N° 130 Santiago (in Spanish)
CIREN (2003) Soil survey, region X of Chile. Natural Resources Information Centre, Publication N° 123 Santiago (in Spanish)
CIREN (2002) Soil survey, region IX of Chile. Natural Resources Information Centre, Publication N° 122 Santiago (in Spanish)
CIREN (1999) Soil survey, region VIII of Chile. Natural Resources Information Centre, Publication N° 117 Santiago (in Spanish)
CIREN (1997a) Soil survey, region V of Chile. Natural Resources Information Centre, Publication N° 116 Santiago (in Spanish)
CIREN (1997b) Soil survey of the VII region of Chile. Natural Resources Information Centre, Publication N° 121 Santiago (in Spanish)
CIREN (1996a) Soil survey of the metropolitan region of Chile. Natural Resources Information Centre, Publication N° 115 Santiago (in Spanish)
CIREN (1996b) Soil survey, region VI of Chile. Natural Resources Information Centre, Publication N° 114 Santiago (in Spanish)
Cisternas RE, Yates LR (1982) Nitrogen litterfall and precipitation and its release during litter decomposition in the Chilean piedmont matorral. Plant Soil 67:359–365
Claret M, Urrutia R, Ortega R, Best S, Valderrama N (2011) Quantifying nitrate leaching in irrigated wheat with different nitrogen fertilization strategies in an Alfisol. Chilean J Agric Res 71(1):148–156
Dörner J, Dec D, Peng X, Horn R (2010) Effect of land use change on the dynamic behaviour of structural properties of an Andisol in southern Chile under saturated and unsaturated hydraulic conditions. Geoderma 159:189–197
Ellies A (1988) Mechanical consolidation in volcanic ash soils. In: Drescher J, Horn R, de Boodt M (eds) Impact of water and external forces on soil structure, Catena Supplement 11. Cremlingen-Destedt
Ellies A (1986) Effects of the mechanic consolidation in some Distrandepts and Paleumults of south Chile. Cienc Suelo (Argentina) 2:91–97 (in Spanish)
Ellies A, Mac Donald R (1984) Some physical-mechanical properties of Dystrandepts from south of Chile. Agro Sur 12(2):140–147 (in Spanish)
Ellies A, Ramírez C, Mac Donald R (2005) Organic matter and wetting capacity distribution in aggregates of Chilean soils. Catena 59:69–78
Ellies A, Horn R, Smith R (2000) Effect of management of a volcanic ash soil on structural properties. Int Agrophys 14:377–384
Ellies A, Grez R, Ramírez C (1997) Saturated water conductivity for the soil structure diagnosis. Agro Sur 25(1):51–56 (in Spanish with English summary)
Ellies A, Grez R, Ramírez C (1995) Wetting potential and stability of aggregates in soils with different management. Agric Tecnica (Chile) 55(3–4):220–225 in Spanish with English summary
Escudey M, Galindo G, Förster JE, Briceño M, Díaz P, Chang A (2001) Chemical forms of phosphorus of volcanic ash-derived soils in Chile. Commun Soil Sci Plant Anal 32:601–616
FAO (2000) Land resource potential and constraints at regional and country levels. Land and Water Development Division, Food and Agriculture Organisation of the United Nations, World Soil Resources Report 90, Rome
Ferreyra R (2009) Soil physical properties characterization of soils growth with table grapes in the region V of Chile. Dissertation, University of Chile (in Spanish with English summary)
Ferreyra R, Sellés G, Morales M, Seguel O (2011) Effect of soil air capacity on water relations and vegetative growth of Thompson Seedless grafted on different rootstocks: preliminary results. Acta Hort 889:145–150
Fodor N, Sándor R, Orfanus T, Lichner L, Rajkai K (2011) Evaluation method dependency of measured saturated hydraulic conductivity. Geoderma 165(1):60–68
Fredlund D, Vanapalli S (2002) Shear strength of unsaturated soils. In: Dane J, Topp G (eds) Methods of soil analysis, Part 4. Physical Methods. Soil Science Society of America Book Series 5, Madison
Gacitúa M, Antilén M, Briceño M (2008) K-Ca-Mg binary cation exchange in saline soils from the north of Chile. Aust J Soil Res 46:745–750
Gallardo F, Pino M, Alvear M, Borie F (2005) Effect of aluminum on dry matter production and nitrate reductase activity of two wheat genotypes, growing in nutritive solutions. J Soil Sci Plant Nutr 5:30–36 (in Spanish with English summary)

Gallardo F, Borie F, Alvear M, Baer E (1999) Evaluation of aluminum tolerance of three barley cultivars by two short-term screening methods and field experiments. Soil Sci Plant Nutr 45:713–719

García R, Gandarillas M (1974) Soil analysis calibration in irrigated-wheat: I nitrogen. Agric Tecnica (Chile) 34(3):137–141 (in Spanish with English summary)

Godoy R, Oyarzún C, Bahamondes J (1999) Hydrochemical fluxes in a *Nothofagus pumilio* forest in Puyehue National Park, southern Chile. Rev Chil Hist Nat 72:579–594 (in Spanish with English summary)

González J, Haberland J, Casanova M (2005) Comparison of three methods to determine saturated hydraulic conductivity of soil. Chilean Soil Sci Soc Bull 21:94 (in Spanish)

Hallet P (2008) A brief overview of the causes, impacts and amelioration of soil water repellency. Soil and Water Resour 3(Special Issue 1):S21–S29

Hartge K (2000) The effect of soil deformation on physical soil properties. A discourse of the common background. In: Horn R, van den Akker JJH, Arvidsson J (eds) Subsoil compaction. Distribution, processes and consequences. Advances in Geoecology 32. IUSS, Reiskirchen

Hartge K, Horn R (2009) Die physikalische Untersuchung von Böden. Praxis, Messmethoden, Auswertung. E. Schweizerbart'sche Verlagsbuchhandlung, Stuttgart

Hillel D (2004) Environmental soil physics. Academic Press, San Diego

Hirzel J (2011) Fertilisation of crops in Chile. Agriculture and Livestock Research Institute (INIA-Chillán), Book N° 28 (in Spanish)

Horn R (2003) Stress-strain effects in structured unsaturated soils on coupled mechanical and hydraulic processes. Geoderma 116:77–88

Huygens D, Schouppe J, Roobroeck D, Alvarez M, Balocchi O, Valenzuela E, Pinochet D, Boeckx P (2011) Drying-rewetting effects on N cycling in grassland soils of varying microbial community composition and management intensity in south central Chile. Appl Soil Ecol 48:270–279

Ibáñez D (2009) Effect of three types of soil on the grape quality of the cv *Carménère* in a vineyard located in the Maipo Valley. Dissertation, University of Chile (in Spanish with English summary)

Jury W, Gardner WR, Gardner WH (1991) Soil physics, 5th edn. Wiley, New York

Kimble JM, Nettleton WD (1984) Analytical characterization of Andepts and Andisols. In: Beinroth F, Luzio W, Maldonado F, Eswaran H (eds) Proceedings of the sixth international soil classification workshop. Taxonomy and managements of andisols, Chile and Ecuador, 9–20 Jan 1984. Part 1 Papers, Chilean Society of Soil Science, Santiago

Klein D, Fuentes JP, Schmidt A, Schmidt H, Schulte A (2008) Soil organic C as affected by silvicultural and exploitative interventions in *Nothofagus pumilio* forests of the Chilean Patagonia. Forest Ecol Manag 255:3549–3555

Leiva C (2009) Characterisation of the amount, function and evolution of the porous system of a palehumult when subjected to cycles of structure formation. Dissertation, Austral University of Chile (in Spanish with English summary)

Lillo A, Ramírez H, Reyes F, Ojeda N, Alvear M (2011) Soil biological activities of a temperate forests in altitudinal transect of Conguillío National Park (38°S). Chile Bosque (Chile) 32:46–56 (in Spanish with English summary)

Luna D (2003) Standardisation of field test of hydraulic conductivity for design works of infiltration. Dissertation, Pontifical Catholic University of Chile, Santiago (in Spanish with English summary)

Luzio W, Norambuena P, Casanova M, Vera W (2002) Genesis and properties of some soils from the Chilean Altiplano. J Soil Sci Plant Nutr 2:35–52 (in Spanish with English summary)

Martínez CE, Kleinschmidt AW, Tabatabai MA (1998) Sulfate adsorption by variable charge soils: effect of low-molecular-weight organic acids. Biol Fertil Soils 26:157–163

Matus F, Maire C (2000) Interaction between soil organic matter and texture and the mineralization rates of carbon and nitrogen. Agric Tecnica (Chile) 60(2):112–126

Matus F, Rodríguez J (1994) A simple model for estimating the contribution of nitrogen mineralization to the nitrogen supply of crops from stabilized pool of soil organic matter and recent organic input. Plant Soil 162:259–271

Montenegro A (1989) Preliminary diagnosis of levels of available nitrogen, phosphorus, exchangeable potasium, organic matter and pH in soils of the Araucanía region. Agriculture and Livestock Research Institute (INIA- La Platina) Series N° 14, Chile (in Spanish)

Montenegro A, Rodríguez J (1985) Supply of K in Andisols and Ultisols of the region IX of Chile. Agric Tecnica (Chile) 45:21–27 (in Spanish with English summary)

Mora ML, Cartes P, Demanet R, Cornforth IS (2002) Effects of lime and gypsum on pasture growth and composition on an acid Andisol in Chile, South America. Commun Soil Sci Plant Anal 33(13–14):2069–2081

Mora ML, Schnettler B, Demanet R (1999) Effect of liming and gypsum on soil chemistry, yield and ryegrass grown in an acidic Andisol. Commun Soil Sci Plant Anal 30(9–10):1251–1266

Morales A, Castillo C, Rubio R, Godoy R, Rouanet JL, Borie F (2005) Glomalin levels in soils from two ecosystems of southern Chile. R C Suelo Nutr Veg 5(1):37–45 (in Spanish with English summary)

Muñoz C, Ovalle C, Zagal E (2007) Distribution of soil organic carbon stock in an Alfisol profile in Mediterranean Chilean ecosystems. R C Suelo Nutr Veg 7(1):15–27 (in Spanish with English summary)

Nimmo J, Perkins K (2002) Aggregate stability and size distribution. In: Dane J, Topp G (eds) Methods of soil analysis. Part 4. Physical methods, Soil Science Society of America Book Serie 5, Madison

Nissen J, Quiroz C, Seguel O, Mac Donald R, Ellies A (2005) Variation of matric potential during water movement in Andisols. Agro Sur 33(1):36–47 (in Spanish with English summary)

Norambuena P, Luzio W, Zepeda O, Stern J, Reinoso F (2011) Preliminary survey of some soils from Chilean Altiplano near Iquique. J Soil Sci Plant Nutr 11(2):63–72

Olivares A, Castillo MH, Polzenius G (1988) Modification of microenvironmental characteristics due to presence of *Acacia caven* Mol. II. Influence of herbaceous strata. Av Prod Anim (Chile) 13:41–48 (in Spanish with English summary)

Oyarzún CE, Godoy R, Leiva S (2002) Atmospheric deposition of nitrogen in a transect from the central valley to Andes mountains, south-central Chile. Rev Chil Hist Nat 75:233–243 (in Spanish with English summary)

Orellana I (2010) Comparison of two methods to measure the wetting capacity of soils. Dissertation, Austral University of Chile (in Spanish with English summary)

Padarian J (2011) Geostatistic modeling of soil organic carbon content between Valparaíso and Los Ríos regions (32-40°S). Dissertation, University of Chile, Chile (in Spanish with English summary)

Peirano P, Aguilera M, Borie G, Caiozzi M (1992) Biological activity in volcanic soils in relation to the organic matter dynamic. Agric Tecnica (Chile) 52(4):367–371 (in Spanish with English summary)

Pérez C, Carmona M, Farina JM, Armesto J (2010) Effects of nitrate and labile carbon on denitrification of southern temperate forest soils. Chilean J Agric Res 70(2):251–258

Pérez C, Carmona M, Armesto J (2003a) Non-Symbiotic nitrogen fixation, net nitrogen mineralization and denitrification in evergreen forest of Chiloé Island, Chile: a comparison with other temperate forest. Gayana Bot 60:25–33

Pérez C, Armesto J, Torrealba C, Carmona M (2003b) Litterfall dynamics and nitrogen use efficiency in two evergreen temperate rainforests of southern Chile. Austral Ecol 28:591–600

Pfeiffer M, Haberland J, Kremer C, Seguel O (2008) Comparison of two methods to measure saturated hydraulic conductivity (Ks) as alternative to auger hole method in an Alfisol. J Soil Sc Plant Nutr 8(4):49–56 (in Spanish with English summary)

Pino I (2002) Biological nitrogen fixation in white clover through 15 N labelling methodologies, on a soil derived from volcanic ashes. J Soil Sc Plant Nutr 2:25–34 (in Spanish with English summary)

Pino I, Rouanet JL, Zapata F, Parada AM, Nario A (2002a) Efficiency of recovery of N in plant-soil system on a wheat crop under alternative soil tillage in a Ultisol in the region IX. Agric Tecnica (Chile) 62(2):275–283 (in Spanish with English summary)

Pino I, Nario A, Albornoz MP, Baherle P (2002b) Nitrogen (^{15}N) fertilizer use efficiency in grapes Thompson seedless. J Soil Sc Plant Nutr 2:16–24 (in Spanish with English summary)

Poblete L (2011) Modification of soil physical properties of a fine textured soil from region III of Chile cultivated with grape through the use of organic matter on the row-planting. Dissertation, University of Chile (in Spanish with English summary)

Rashid A, Ryan J (2004) Micronutrient constrains to crop production in soils with Mediterranean type characteristics. J Plant Nutr 27:959–975

Redel Y, Rubio R, Borie F (2006) Effect of addition of crop residues and mycorrhizal fungi on wheat growth and the chemical and biological parameters of an Andisol. Agric Tecnica (Chile) 66:174–184 (in Spanish with English summary)

Richards D (1983) The grape root system. Hortic Rev 5:127–168

Risacher F, Alonso H, Salazar C (2003) The origin of brines and salts in Chilean salars: a hydrochemical review. Earth Sci Rev 63:249–293

Rivas Y, Godoy R, Valenzuela E, Leiva J, Oyarzún C, Alvear M (2007) Biological activity of soils in two *Nothofagus* forest in south-central Chile. Gayana Bot 64:81–92

Rodríguez J, Pinochet D, Matus F (2001) Crop fertilization. LOM Editions, Santiago (in Spanish)

Ruíz R, Sadzawka MA (1986) Potassium fixation by Aconcagua Valley soils. Agric Tecnica (Chile) 46:503–505 (in Spanish with English summary)

Sadzawka MA (2006a) Physical-chemical soil properties in the soils I: reaction (pH), acidity and alkalinity. In: Luzio W, Casanova M (eds) Advances in the knowledge of the Chilean soils. University of Chile-Agriculture and Livestock Service, Maval Press, Santiago (in Spanish)

Sadzawka MA (2006b) Physical-chemical soil properties in the soils II: adsorption and ionic exchange. In: Luzio W, Casanova M (eds) Advances in the knowledge of the Chilean soils. University of Chile and Agriculture and Livestock Service (SAG), Santiago (in Spanish)

Sadzawka MA (1989) Rhizosphere's pH of wheat and lupine. Agric Tecnica (Chile) 49:71–73 (in Spanish with English summary)

SAG (1974) Soil survey in the Chile Chico area. Ministry Agriculture, Agriculture and Livestock Service, Chile (in Spanish)

Salazar O, Nájera F (2011) Water contamination by use of nitrogen fertilizers and manure. Chilean Soil Sci Soc Bull 24:147–152 (in Spanish)

Salazar O, Casanova M, Kätterer T (2011) The impact of agroforestry combined with water harvesting on soil carbon and nitrogen stocks in central Chile evaluated using the ICBM/N model. Agric Ecosyst Environ 140:123–136

Scherer HW (2009) Sulfur in soils. J Plant Nutr Soil Sci 172:326–335

Seguel A, Rubio R, Carrillo R, Espinosa A, Borie F (2008) Levels of glomalin and their relation with soil chemical and biological soil (andisol) characteristics in a relic of native forest of southern Chile. Bosque (Chile) 29:11–22 (in Spanish with English summary)

Seguel O, Horn R (2006) Structure properties and pore dynamics in aggregate beds due to wetting-drying cycles. J Plant Nutr Soil Sci 169:221–232

Seguel O, Baginsky C, Contreras A, Covarrubias J, González C, Sandoval M (2011) Changes in physical properties of a Typic Haplocambid by annual crop culture. J Soil Sc Plant Nutr 11(1):1–15

Sierra C, Lancelloti A, Vidal I (2007) Elemental sulphur as pH and soil fertility amendment for some Chilean soils of Regions III and IV. Agric Tecnica (Chile) 67(2):173–181 (in Spanish with English summary)

Soil Survey Staff (2006) Keys to soil taxonomy. USDA–NRCS

Staelens J, Ameloot N, Almonacid L, Padilla E, Boeckx F, Huygens D, Verheyen K, Oyarzún C, Godoy R (2011) Litterfall, litter decomposition and nitrogen mineralization in old-growth evergreen and secondary deciduous *Nothofagus* forests in south-central Chile. Rev Chil Hist Nat 84:125–141

Tejeda H, Gogan G (1970) Methods to determine available nitrogen in soils with different origin and levels of organic matter. Agric Tecnica (Chile) 30(2):57–64 (in Spanish with English summary)

TERRASTAT (2003) Land resource potential and constraints statistics at country and regional level. http://www.fao.org/ag/agl/agll/terrastat/. Accessed 8 April 2010

Wall D (2007) Global change tipping points: above- and below-ground biotic interactions in a low diversity ecosystem. Phil Trans R Soc B 362:2291–2306

World Bank (2008) World Development Report. Agriculture for Development, Washington

Zagal E, Córdova C (2005) Soil organic matter quality indicators in a cultivated Andisol. Agric Tecnica (Chile) 65(2):186–197 (in Spanish with English summary)

Zagal E, Hirzel J, Vidal I (2003) Evaluation of nitrogen-fertilizer recommendation for annual crops in soils of volcanic origin using a simulation model. Agric Tecnica (Chile) 63(1):94–104 (in Spanish with English summary)

Zagal E, Rodríguez N, Vidal I, Flores B (2002a) Soil light organic matter fractions of a volcanic soil under different agronomic management as an indicator of changes in labile organic matter. Agric Tecnica (Chile) 62(2):284–296 (in Spanish with English summary)

Zagal E, Rodríguez N, Vidal I, Quezada L (2002b) Microbial activity in a volcanic ash soil under different agricultural management. Agric Tecnica (Chile) 62(2):297–309 (in Spanish with English summary)

Zunino H, Borie F, Aguilera S, Martin JP, Haider K (1982) Decomposition of ^{14}C-labeled glucose, plant and microbial products and phenols in volcanic ash-derived soils of Chile. Soil Biol Biochem 14:37–43

4 Human-Induced Soil Degradation in Chile

Land degradation is an old and serious problem for many communities throughout Chile, with adverse impacts on agricultural productivity, food security and rural livelihoods. Numerous connections to other environmental problems such climate change, biodiversity losses, droughts, etc., have been observed, which calls for action-orientated science. It is not the intention of this chapter to review the contested field of definitions of land degradation. Instead, we apply the following wide description of the process: *Land degradation is a long-term loss of ecosystem function and service, caused by disturbances from which the system cannot recover unaided* (UNEP 2007).

Regardless of any ultimate definition, the problem is exacerbated by humans and soil degradation manifests itself in at least two very serious environmental expressions, a quantitative and qualitative reduction in soil quality (Table 4.1).

Only considering the Metropolitan Region, around 14,000 ha of arable soils (Land Capability Classes I–IV) located in the floodplains of Santiago's rivers were lost between 1989 and 2003. In addition, another 5,500 ha of non-arable soils (Land Capability Classes V–VIII) from the Andean piedmont have disappeared as the result of urban land use (Romero and Órdenes 2004). This urban sprawl is illustrated in Fig. 4.1.

As regards land lost to mining, there are 746 abandoned mining dams in Chile which store around 749 million m^3 of tailings and cover an approximate area of 192,107 ha (Ginocchio 2011). Figure 4.2 shows two examples of these dams in extreme zones of Chile, which are normally exposed to seepage, dust, long-term erosion, bio-intrusion etc. Regarding landfills, another important anthropogenic invasion of soil in Chile, at least 300 sites within its territory have been authorised or earmarked for disposal of wastes of different origins (household, industry and construction).

It is also necessary in this context to take account of a major hydroelectric project involving dams in Chilean Patagonia, which will obviously result in quantitative losses of edaphic national patrimony due to flooding of wide trough valleys in a pristine landscape of 5,900 ha in area (artificial lakes on the Baker and Pascua rivers). This project has the potential to satisfy Chile's growing energy needs, but it is imperative try to explore every alternative in order to prevent any long-term damage to zonal natural resources. Pfeiffer et al. (2010) describe Histosols, Entisols and Inceptisols in the zone, but additional area and other important soil orders will be affected along nearly 2,400 km of energy transmission lines (pylon alignment) to northern Chile.

From a general point of view, the mining sector in the centre and north, growing pressure of agriculture in the centre and south, and deforestation in the south of Chile are the main causes of soil degradation caused by man. It is important to note that growth and over-concentration of population and industrial activity around the principal cities are also increasing concerns related to soil degradation processes. Moreover, the scenario becomes more complex and worrying when intense desertification processes are included.

Although generalised two maps presented by FAO (scale 1:10,000,000) for Chile give a general spatial idea of soil constraints and human-induced soil degradation in the country (Fig. 4.3).

Therefore, it is within this context that the present chapter aims to describe the relevant human induced processes in Chile, considering the Barrow (1991) concept of erosive and non-erosive (physical, chemical and organic) soil degradation processes.

4.1 Erosive Soil Degradation

Soil degradation by accelerated erosion is a serious concern, especially in developing countries of the tropics and subtropics. In Latin America and the Caribbean in particular, the extent, severity and economic and environmental impacts of soil degradation are the subject of continuous debate, with uncertain magnitude for the twenty-first century. Moreover, estimates of the regional land area affected

M. Casanova et al., *The Soils of Chile*, World Soils Book Series, DOI: 10.1007/978-94-007-5949-7_4

Table 4.1 Main human-induced soil degradation problems observed in Chile

Quantity reduction		
	Uncontrolled urbanisation	
	Mining wastes deposits and landfills	
	Losses by accelerated erosion	
Quality reduction		
	Physical quality changes	Compaction and crusting
		Subsidence
		Waterlogging
	Chemical quality changes	Excess and depletion of nutrients
		Acidification/alkalinisation
		Salinisation
		Pollution (heavy metals, pesticides, industrial wastes)
	Biological quality changes	Organic matter changes
		Biodiversity changes

are tentative and subjective, with field measurements often technique dependent (Casanova et al. 2010a). Although natural erosion occurs on most landforms in Chile, as soon as land is newly put into production the process increases, triggering accelerated degradation that represents the main focus for most control initiatives.

An abundance of valuable research has been developed in Chile to create initial awareness among land users about the seriousness of the erosion problem in their operations (Rodríguez and Suárez 1946; Elizalde 1970; Peralta 1976; Contreras 1986; Vargas et al. 1998; Gayoso and Alarcon 1999; Lagos 2005; Ruiz 2005). In parallel to proposing soil conservation practices, some engineering solutions are being specified for soil erosion within an environmentally acceptable level of reliability (Pizarro et al. 2009; Lemus and Navarro 2003).

With the conviction that human-induced soil erosion is only the symptom of a deeper problem—incorrect land use and bad management—the main research emphasis has moved to better land management and to identifying how much and where soil is being lost in Chile.

Despite many parallel efforts in soil erosion research in Chile, few initiatives cover other important human-induced erosive processes such as streambank erosion, which is accelerated by deforestation (Fig. 4.4), tillage erosion and harvest erosion (Homer and Casanova 2011). Varnero et al. (2005) report that if grass cover (on soil) in the Metropolitan Region is not maintained, current extraction rates will not only cause a decline in the sustainability of this activity, but also the degradation of soils used for these purposes. According to Arroyo et al. (2005), peat per se comes under the Chilean Mining Code and, like any mineral, is considered to belong to the state, regardless of who the owner of the land may be. Peats are being exploited commercially by private companies (concessions granted) in the extreme south of Chile (Region XI and XII) and transported about 2,500 km to central Chile for use in the fruit, horticulture and mushroom industries.

Even though one could argue that the phenomena of landslides, collapses and so on are natural, it is equally true that human actions (mining, deforestation, fires, road networks, construction etc.) trigger these events to occur with greater severity (Espinosa et al. 1985). Indeed, a clear example is the change in productive land use in the Copiapó Valley from the mid 1970s, with an increase in the area planted with bush vines in the order of 236 % (Castro et al. 2009a, b). This has meant the introduction of significant morphological changes, exceeding in many places the morphodynamic thresholds, and generating impacts on the morphology dynamic. These impacts are significant in terms of increased vulnerability to mass removal occurring in episodes of heavy rains associated with the *El Niño* years, with hazard for the resident population and agricultural workers, and in generating significant loss of infrastructure.

4.1.1 Water Erosion

Conventional and nuclear methods, but also models and remote sensing, have been used in Chile to assess soil water erosion (Casanova et al. 2010b; Zapata et al. 2010).

Erosion pins have been recommended by Stocking and Murnaghan (2001) and used by many Chilean researchers from Coastal Range to Pre-Andean soils (Region VII) to assess sheet erosion and monitor gully head advance (Cuitiño 1999; Cartes et al. 2009). Likewise, Lagos (2006) and Cerda and Jimenez (2003) observed an elevated variability of soil erosion and sedimentation in Region IV and Region V, respectively. Youlton et al. (2010) reported that in Region V, soil erosion increases after building downward ridges (raised beds) along steep hillslopes, in particular during the first winter, but decreases when trees grow, while runoff increases. Thus, after 20 Mg ha^{-1} of soil erosion during the first year, erosion mitigation management (soil cover) reduced erosion by 90 %.

Despite methodological problems connected with measurement of soil erosion (Boix-Fayos et al. 2006, 2007), plots of diverse types and sizes have been utilised in Chile,

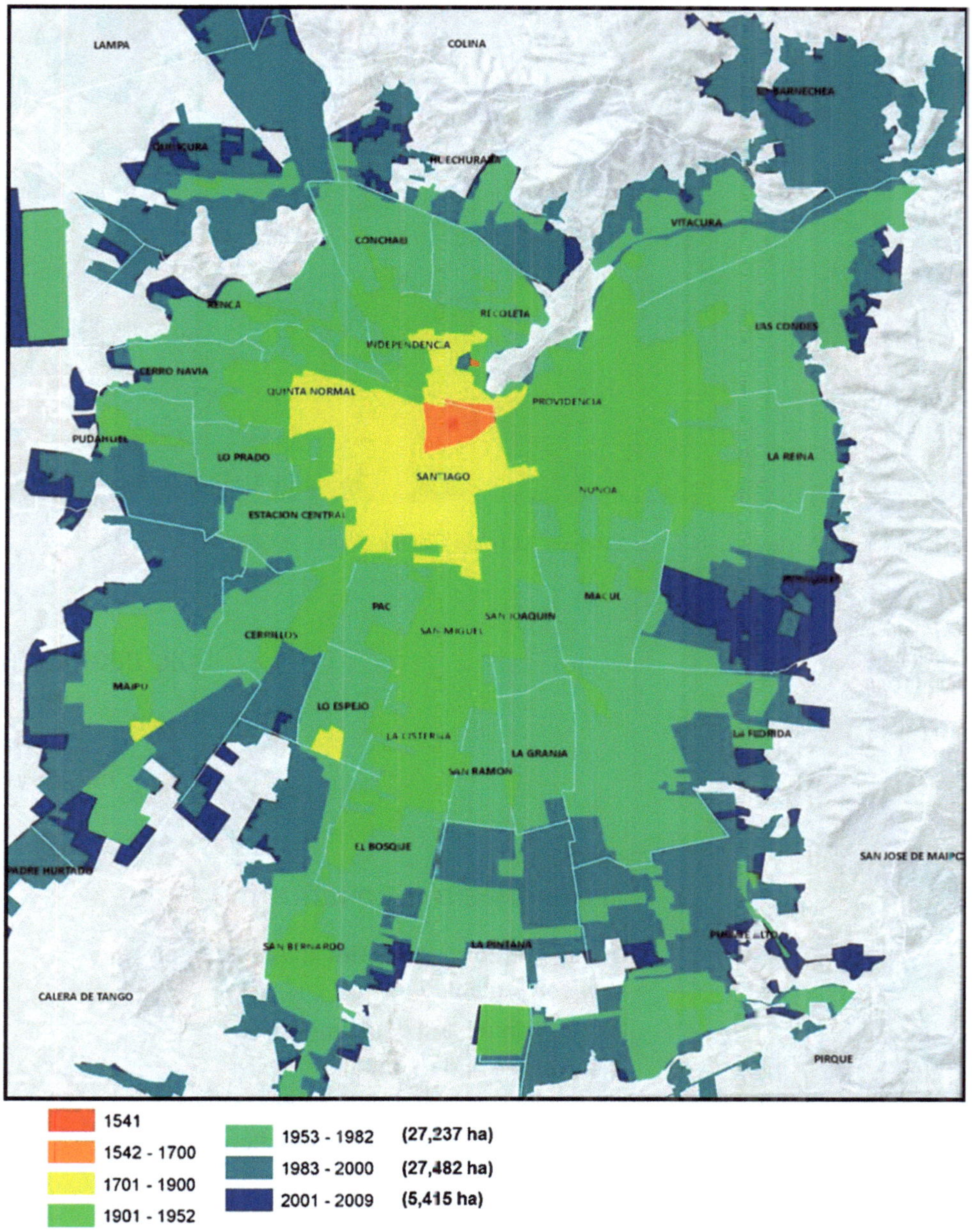

Fig. 4.1 Urban sprawl of Santiago city between 1541 and 2009 (MINVU 2010)

mainly with three objectives: (a) to characterise erosion processes in different zones; (b) to calibrate/verify models, both empirical and physical; and (c) to estimate soil erosion rate for assessing, developing and verifying control methods.

For example, for large and small plots, significant variations in runoff responses to rainfall rates were found by Joel et al. (2002) on hillsides of rainfed central Chile. They concluded that a characteristic minimum rainfall intensity (2 mm h^{-1}) for the site, irrespective of plot size, is required to generate runoff under continued rainfall conditions. This suggests that spatial and temporal variability of runoff generation across hillslides depends mainly upon the heterogeneity of soil hydraulic conductivity and rainfall characteristics. Casanova et al. (2007) studied 18 erosion plots (2 m × 1 m) at the same location (Fig. 4.5) and observed that spatial variability of erosion rate depends on the geomorphological components (slope aspect and gradient) and temporal variability depends on antecedent soil moisture conditions. However, they also confirmed a concept of 'critical slope gradient that represents a soil characteristic inclination threshold on sediment yield.

Fig. 4.2 Mining dams in Region IV (*left* Tambillo) and Region XI (*right* Laguna Verde) of Chile

Fig. 4.3 General environmental constraints (*left*) and human-induced soil degradation (*right*) in Chile (http://www.fao.org, Accessed 15 April 2011)

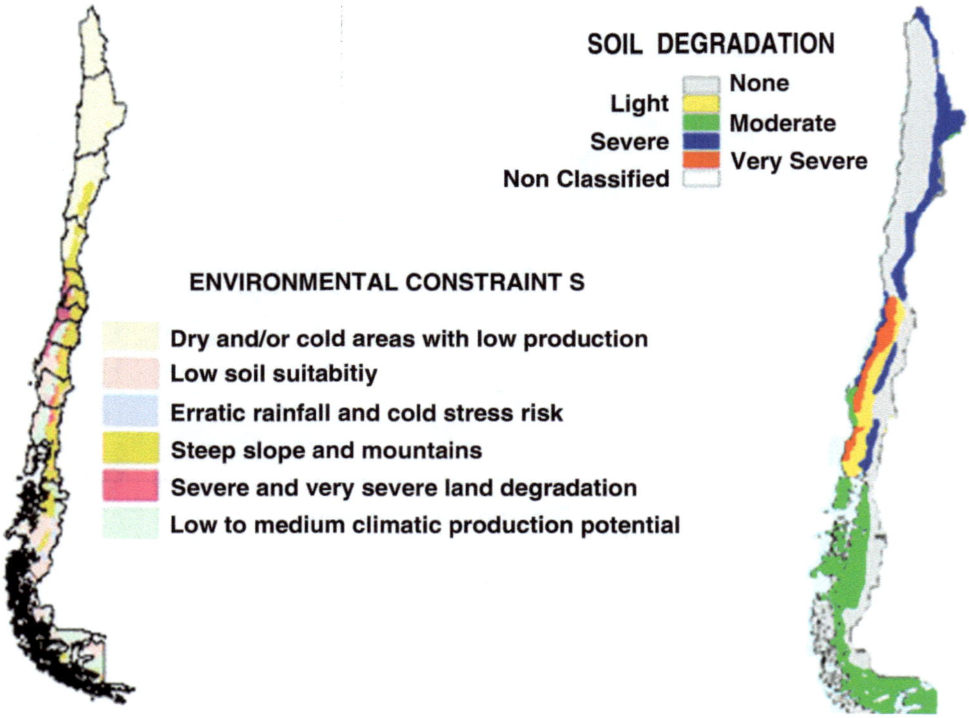

A study evaluating soil erosion as a result of tillage systems was carried out on Andisols (foothills of the Chilean Andes, Region VIII) by Rodríguez et al. (2000). For four soil treatments (conventional tillage, vertical ploughing, direct drilling and permanent pasture on 11 m × 3.6 m plots), annual erosion was found to be 19.3, 5.5, 3.0 and 0.64 Mg ha^{-1}, respectively.

The expansion in recent decades of fruit tree cultivation to steeper hillslopes, considered marginal land for agriculture in the past, is mainly comprised of production of table grapes, avocados and citrus on ridges in the Semi-arid zone of Chile (Fig. 4.6), while in the Mediterranean zone, vineyards for high quality wine production are occupying fragile sloping lands that were formerly protected by native forest.

In Region VI, where vineyards are normally aligned in the direction of the steepest slope (Fig. 4.7), a study is being carried out by University of Chile and the Chilean Commission of Nuclear Energy (Casanova et al. 2009) to assess water erosion control measurements. Mulch and/or organic emulsion were applied to soils (Ultic Haploxeralfs) in 12

Fig. 4.4 Streambank erosion at Region XI (*left* Chile Chico) and Region IX (*right* Laguna del Laja)

Fig. 4.5 Erosion plots at hillside of central Chile (Metropolitan Region)

plots (2 m × 10 m), with preliminary results as shown in Fig. 4.8.

A similar study (Casanova et al. 2011) is being conducted on abandoned and degraded hillsides of Region IV with 9 plots (3 m × 1 m) on the drainage tributary area above a gullied field. Here, emulsions are also being applied to improve soil infiltration rate and reduce runoff (Fig. 4.9).

The universal soil loss equation (USLE) model (Wischmeier and Smith 1978) and its revised version (RUSLE) have been widely used to predict soil erosion in Chile (Brito and Peña 1980; Peña 1983, 1985; Oyarzún 1993; Honorato et al. 2001), with an appropriate database modified to local conditions. Stolpe (2005) compared models (RUSLE-WEPP-EPIC) for Andisols and other models without plots (Honorato and Cruz 1999; Santibáñez et al. 2008; Bonilla et al. 2010) to verify soil losses, using the models to simulate different future scenarios. Figure 4.10 shows the close relationship between estimated and measured soil erosion rates in Regions IV, VI, VII, VIII and IX of Chile.

The quest for techniques as alternatives or complements to the existing methods has directed attention to the use of radionuclides (Zapata 2003). In southern Chile and Patagonia, a comprehensive study with ^{137}Cs and ^{7}Be has been conducted by researchers at Austral University of Chile (Schüller et al. 2000, 2003, 2004a, b, 2006; Sepúlveda et al. 2008), who report a clear advantage of radionuclides over erosion plots and erosion pins. An initial study provided information on selection of reference site and its spatial variability (Schüller et al. 1997), while recent work examined common soil forestry management practices, linear trash barriers (woody harvesting residues) along contour lines and the maintenance of riparian vegetation to act as a sediment filter (Schüller et al. 2010). In agronomic terms, 16 years after implementing zero tillage in southern Chile there was a substantial reduction in soil erosion rates, as measured by ^{137}Cs, of about 87 % (from 1.1 to 1.4 Mg ha^{-1} yr^{-1}). However, such a beneficial effect may be readily lost if the mulch layer of old crop residues is removed or burned. Using ^{7}Be to measure a short-term erosion event occurring just after a dramatic burning event. Schüller et al. (2007) reported substantial soil losses of 12 Mg ha^{-1} over this 27d period of exceptionally wet (400 mm) weather.

On the other hand, a multi-scale approach has been implemented (Mathieu et al. 2007) to produce regional land degradation maps for the Coastal Range of central Chile, which is naturally sensitive to soil erosion, based on remote sensing technology. Radiometric indices were successfully applied to SPOT images to produce land degradation maps, but only broad classes of erosion status were discriminated and the detection of degradation processes was only possible when most of the fertile layer had already been removed.

Recently, the Natural Resources Information Centre (CIREN), a Chilean government-funded institution which

Fig. 4.6 Construction of soil ridges on hillsides with over 100 % slope gradient at Region IV of Chile

Fig. 4.7 Vineyards in sloping land and erosion at toeslope, Region VI of Chile

focuses on natural resources inventory and its corresponding cartography (GIS), released a study on current and potential soil erosion in Chile, which replaced a partial map published previously (IREN-CORFO 1979; Pérez and González 2001). This work, which used qualitative models, geomatic, remote sensing and SIG techniques (Flores et al. 2011), is published at 1:50,000 scale, except for extreme zones and the Andes mountains (1:250,000). A total area of 36.8 million ha (49.1 % of national territory) displays some level of erosion (Table 4.2), which increases from southern to northern Chile, with the north-central zone showing the highest eroded area (Region IV 84 %, Region V 57 % and Region VI 52 %). Human-induced soil erosion is concentrated principally from Region IV to X Region (Fig. 4.11).

4.1.2 Wind Erosion

A few studies on wind erosion in Chilean soils have been reported. The arid climate, flat geomorphology, strong south and south-westerly winds and the characteristics of the superficial sediments in the northern arid zone of Chile justify the study of wind erosion and transport processes in the coastal Atacama Desert (Flores-Aqueveque et al. 2009). Yardangs (wind-abraded ridges of cohesive material) are reported in these dry areas, where deflation is at a maximum, vegetation cover is minimal and sand abrasion is acting over the bare soil surface (Goudie 2008).

The Patagonian steppes occupy the southern tip of the continent from approximately 40 °S, are framed by the Andes to the west and the Atlantic coast to the east and south, and cover more than 800,000 km^2 of Chile and Argentina. Strong winds are a constraint to agricultural development in Patagonia and windbreaks are frequently planted to allow establishment of fruit trees, pasture and horticultural crops, and to protect agricultural crops, livestock and rural houses. Despite being mentioned as one of the main degradation processes in Southern Patagonia (Vött and Endlicher 2001), the conditions and rates of wind erosion in this region have not been studied extensively.

Fig. 4.8 Sediment yield and runoff generated by natural rainfall in plots according to treatments (*T0* control, *TE* emulsion, *TM* mulch, TE + M: emulsion plus mulch) at Region VI, Apalta Valley

Semi-arid steppes of this zone are prone to wind erosion, due to the soil remaining dry for periods during intense spring and summer winds (Gualterio 2006; Dube et al. 2011). Human mismanagement of soils has either initiated or accelerated this degradation process in vast areas of Eastern and Western Patagonia (Gut 2008). Rounded pebbles and gravels associated with glaciofluvial processes are characteristic of the Patagonian steppe soils and are responsible for the formation of extensive desert pavement, where wind erosion has been able to remove finer materials (Paruelo et al. 2007). Another indicator in this zone is related to volcanic activity. After eruption of the Hudson volcano in 1991, around 80,000 km^2 were covered by a layer of ash ranging in thickness from 10 cm to over 100 cm, but today most of this material has been removed from rough landscapes, principally by strong winds.

Most of the coastal dunes in Chile today are the result of soil erosion, with sediments from denuded soils being carried by the rivers to their mouths and then transported by prevailing winds north of the mouth. According to an old inventory by IREN (1966), 74,428 ha are occupied by dunes in Central Chile, between Regions IV and X. Most of the dunes present at that time were found in Region VIII, where they covered 30,709 ha, i.e. 41.3 % of total coastal dunes. An evaluation by Peña-Cortés et al. (2008) of the dynamics of dune systems of the coastal strip of Region IX between 1994 and 2004, defined eight fields of dunes with a total area of 4,597 ha, which represented an overall expansion of 314 ha in the period.

However, a general review of sites covered by dunes today shows that it is possible to observe their presence along all territory (Fig. 4.12). It is clear that this geomorphological feature is not always related to human-induced soil erosion, corresponding mainly to relict dunes, but is an indicator of wind erosivity.

4.2 Non Erosive Soil Degradation (Physical, Chemical and Biological)

The effect of soil formation factors is responsible for the evolution of soils, where the forces of these factors and soil formation processes are always operational in such a manner that under natural conditions a static equilibrium state is never attained. However, the tendencies in soil evolution at any given time are highly sensitive, particularly when there are man–made changes that may generate a new equilibrium towards soil chemical, physical and biological degradation processes. This section reviews the causes of the main soil chemical, physical and biological degradation processes in Chilean soils and some remediation measures are proposed and discussed.

4.2.1 Soil Physical Degradation

From a physical point of view, soil affects plant production through water, heat, air and mechanical conditions, all state variables of a dynamic nature, which affect the flows of matter and energy (Benavides 1992). Physical degradation generates concatenated effects on the soil, which ultimately depend on the soil water condition of the porous system and the stability of the unions between solid particles (Horn 2003).

Fig. 4.9 Erosion plots on hillside in Hyper-arid and Semi-arid zone of Chile (Region IV)

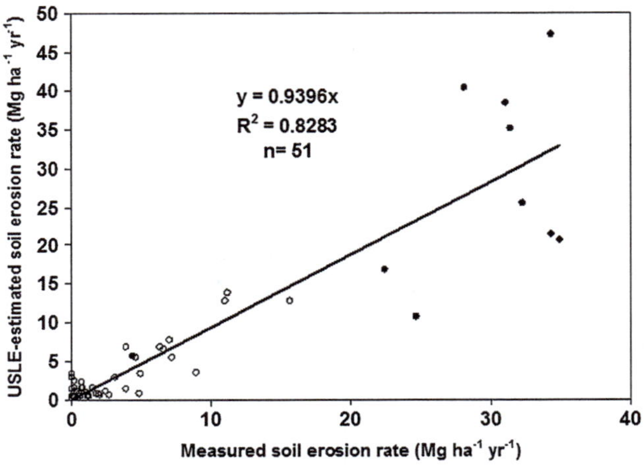

Fig. 4.10 Comparison of estimated and measured soil erosion rate (Stolpe 2005; Honorato et al. 2001). *Black circles* correspond to fallow conditions

Physical degradation may be related to a wide range of problems and processes, such as crusting, reduced permeability, compaction, poor aeration, destruction of the structure and subsidence (Table 4.1). Almost all of these are related to the reduction or alteration of soil porosity. Thus, soil physical degradation is related to direct or indirect human actions that may result in deterioration of properties such as bulk density, structure, aggregate stability, mechanical strength and porosity. All these actions impede the adequate development of roots in the soil, affecting the development of vegetation and rendering the soil more susceptible to degradation by erosion (Casanova et al. 2006).

In Chile, the Ministry of Agriculture, through a special programme for soil recovery (Incentive System for Agro-environmental Sustainability of Agricultural and Livestock Soils, ISRD-S) led by the Agriculture and Livestock Service (SAG), gives incentives and subsidies for the implementation of land reclamation works, including prairie sowing, organic amendment incorporation and implementation of conservation tillage systems (Ruiz 2011). The specific management practices are focused in special areas depending on specific needs, as well as the magnitude of the subsidies (50–90 % of the costs) depending on the kind of farm (size and economic resources).

Analysing the properties of the Chilean soils, it is possible to determine the strong influence of climate and parent material at the regional level in a north–south direction, and the effect of local relief in an east–west direction (Luzio et al. 2010). The climate-relief interaction affects the vegetation that grows on the surface of the land, being less important over time as a soil forming factor.

Thus, it is possible to identify three areas that present different problems related to soil physical degradation. The first includes the Hyper-arid to Semi-arid zone of the country (Sect. 2.2.1) with very low SOM content and a high risk of salinisation; a second zone developed under a Mediterranean climate (Sect. 2.2.2) with intense agricultural activity; and a third one including Rainy and Patagonian zone (Sect. 2.2.3), with soils that have volcanic influence, non-crystalline mineralogy and high levels of OM.

Bearing in mind that the problems are not exclusive to each area and can occur in any area, the limitations of soil physical properties derived from human-induced degradation by pressure on this scarce resource can be grouped into:

Table 4.2 Actual area of soil erosion (in 1,000 ha) in different regions of Chile (Flores et al. 2011)

Region	Non erosion	Light erosion	Moderate erosion	Severe erosion	Very severe erosion	Non apparent erosion	Other uses	Excluded areas	Eroded soil[a]	Studied area
XV	48.8	255.7	171.6	468.7	583.6	0	157.1	0	1,479.5	1,685.5
I	60.8	1.0466	6023	1.1532	838.4	0	524.4	0	3,640.4	4,225.6
II	138.1	1.341.4	3,271.3	3,592.8	2,021.4	0	2,237.3	0	10,226.9	12,602.4
III	178.5	825.2	536.6	2,029.9	629.1	0	3,330.4	36.8	4,020.9	7,566.5
IV	210.0	571.6	1,142.3	1,213.9	492.5	25.5	403.5	0	3,420.4	4,059.5
V	162.0	244.2	324.6	258.2	80.0	162.6	358.4	0	906.9	1,600.0
MR	354.5	93.4	189.5	213.5	186.8	68.0	435.0	0	683.2	1,540.6
VI	331.0	96.3	451.8	197.3	114.5	125.1	322.3	0	860.0	1,638.3
VII	655.9	349.0	416.1	377.8	335.8	453.3	446.2	0	1,478.7	3,034.0
VIII	840.1	393.4	429.1	211.6	148.6	1,443.8	245.4	0	1,182.7	3,712.0
IX	1,131.9	280.5	240.9	243.9	145.9	944.3	182.8	16.3	911.1	3,186.3
XIV	427.0	262.3	197.6	79.6	5.8	688.0	173.5	3.7	545.3	1,837.4
X	750.6	574.6	423.5	138.9	33.3	2,142.4	745.6	24.9	1 170.3	4,833.7
XI	234.7	894.7	743.5	383.1	583.4	4,550.6	3,361.1	0	2.604.6	10,751.0
XII	1,718.8	1,122.9	1,287.8	590.0	761.5	3,088.5	4,247.7	4.3	3.762.1	12,821.4
Total	7,242.6	8,351.9	10,428.3	11,152.4	6,960.4	13,692.1	17,180.7	86.0	36,893.0	75,094.4

[a] Eroded soil includes light, moderate, severe and very severe classes. Unregistered information by remote sensing in zones with vegetal cover >75 % (non apparent erosion)

a. Poor aeration caused by traffic or chemical dispersion by irrigation practices.
b. Compaction and high mechanical strength in medium to fine-textured soils.
c. Soil profile alterations in shallow or coarse-textured soils.
d. Problems associated with high organic matter content.

4.2.1.1 Poor Aeration by Restricted Horizons

Soils from the Hyper-arid to Semi-arid zone have lower average levels of OM (<4 %), making them susceptible to losing their structure owing to low stability. The problem increases when there is a high amount of sodium, which promotes the dispersion of soil. The sodium source may be natural, as is the case of marine terraces such as the Huentelauquén sector (Region IV), or induced by irrigation practices, and reflects the effect of soil stratification on the accumulation of salts (Fig. 4.13).

As a solution to the low OM contents, different organic amendments are used, especially manure and compost, being less used sources based on soluble humic substances. In recent years, cover crops have also been used to improve the soil physical properties in orchards (Baginsky et al. 2010; Cortés 2011). The advantage of adding OM is that it increases the stability of the structure, promotes a stable and functional porosity and allows air and water flow to the roots in an adequate amount and time. In coarse-textured soils OM also increases water retention, avoiding water stress (see Sect. 4.2.1.3).

As regards pore function, Fig. 4.14 shows the effect of two practices in an alluvial, deep, clay loam and stratified Typic Haplocambid (CIREN 2007) in Region III (Copiapó Valley). During 3 years, a manure application at doses of 40 Mg ha^{-1} and a crop rotation (maize/broad bean/barley) was implemented. The control was subjected to surface tillage with chisel plough, thus maintaining a high coarse porosity to 10 cm depth (Poblete 2011).

Even though the manure is applied on the surface, there is an effect on water dynamics in the soil profile in the form of higher infiltration rates (Fig. 4.14a). Nevertheless, this higher amount of water may result in a cooler soil, decreasing the rate of root development (Baginsky et al. 2010). In coarse-textured soils, growing crops (specifically grasses) can improve water infiltration (Seguel et al. 2011).

The high coarse porosity of the control as a result of tillage and the high porosity of the manure explain the high capacity for air flow observed at the surface (Fig. 4.14b). However, the flow decreases sharply at 20 cm depth because of a plough pan in the control and pore discontinuity in the manure treatment (Keller 2011). On the other hand, the crop rotation favours a continuous pore system at depth, maintaining the air flow capacity to renew oxygen to the roots. In fine-textured soils in particular, excessive traffic and poor irrigation practices affect the oxygen

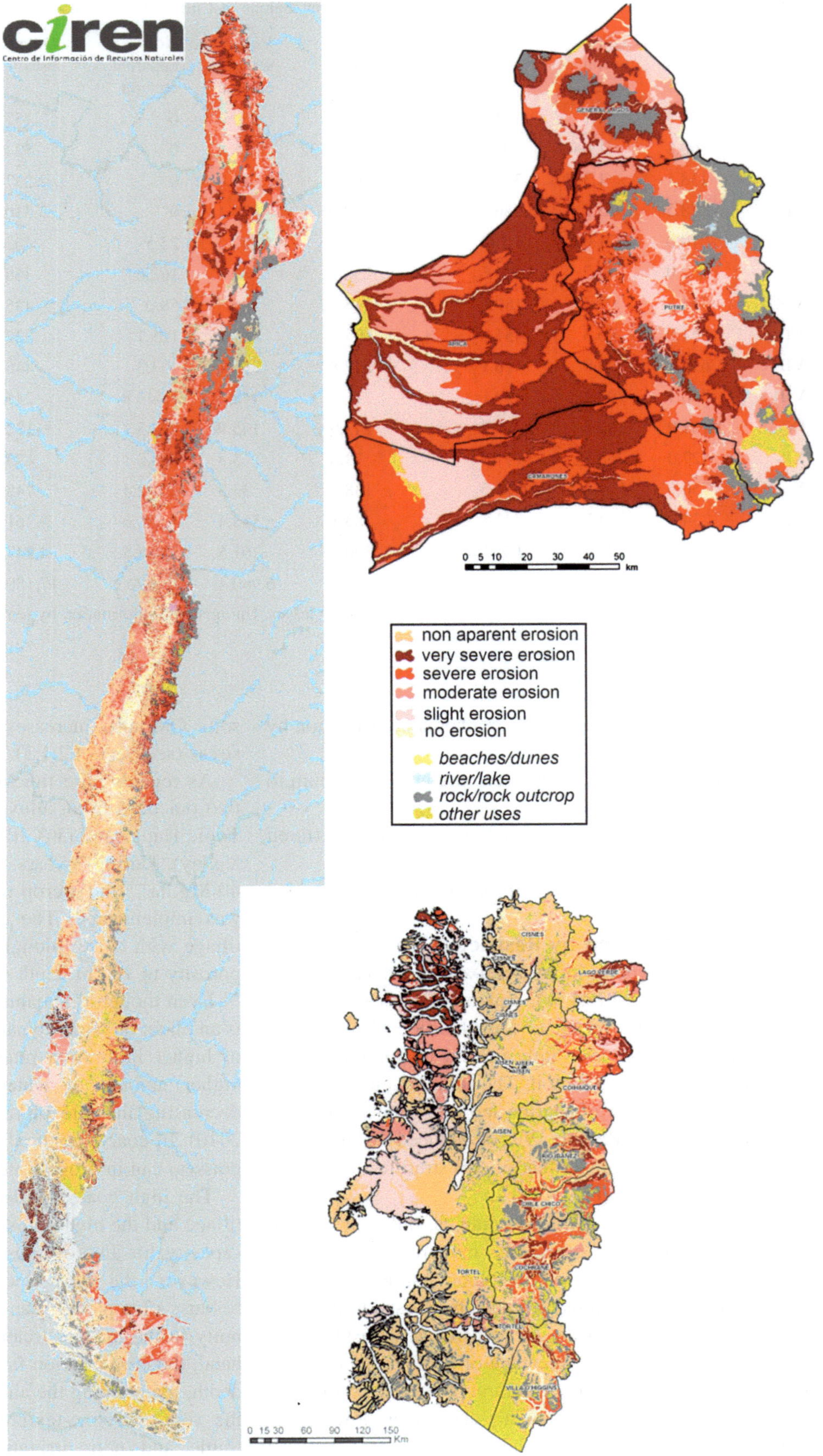

Fig. 4.11 Soil erosion map of Chile including a zoom of Region XV (*above*) and Region XI (*below*)

Fig. 4.12 Locations in Chile with dunes present. Examples: Region IV (*above*) and Region VII (*below*)

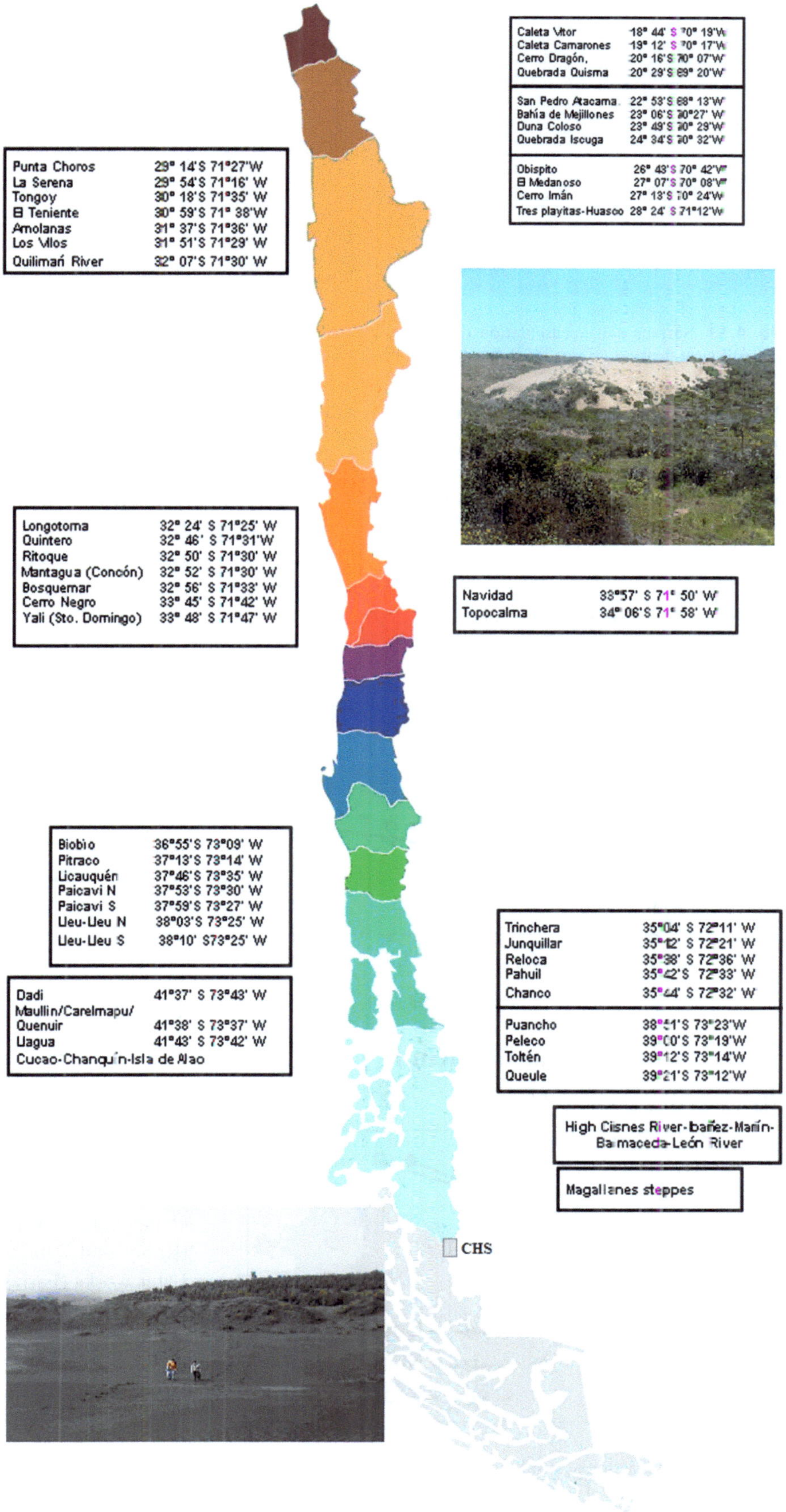

Fig. 4.13 Surface salt accumulation (*left*) caused by drip irrigation and soil stratification (*right*) increases the risk of salinisation

diffusion rate in the soil (waterlogging) resulting in low productivity (Ferreyra et al. 2011). The Chilean government and the private sector, coordinated by the National Commission for Irrigation (CNR), have made significant

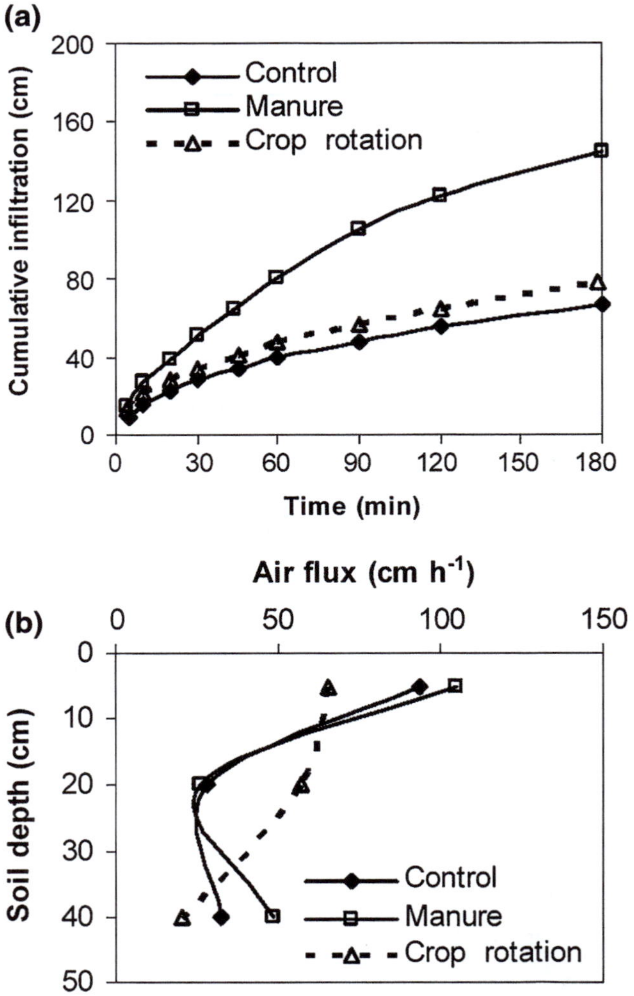

Fig. 4.14 Soil function as result of soil management. **a** Infiltration of water by a ring method test. **b** Air flux as a function of soil depth, performed by a laboratory test with samples equilibrated at −33 kPa

investment efforts for improving water management and irrigation infrastructure and resolve poor soil drainage. In fact, Law 18,450 (Development of Private Investment in Irrigation and Drainage Works) dating from 1985 not only allows an important irrigated area in Chile to be increased, but also improves water use efficiency, which is relatively low in the country. The programme is preferentially focused on regions where irrigation is most necessary for agriculture, with the public sector subsidising up to 75 % of the construction, refurbishment and equipment costs for minor irrigation or drainage works.

Both salinity and sodicity cause physical problems, decreasing the osmotic potential of water and dispersing the soil, respectively, affecting water uptake and the renewal of oxygen to the roots (Hillel 2004). For instance, a problem promoted by the use of manure is the salt intake, as shown in Fig. 4.14. At the end of the season, the manure treatment resulted in a surface EC of 6.6 dS m^{-1} and exchangeable Na value of 3.9 $cmol_c$ kg^{-1}, compared with 0.9 dS m^{-1} and 0.7 $cmol_c$ kg^{-1}, respectively, when a crop rotation was used. Meanwhile, the control without any management reached EC values of 11.5 dS m^{-1} as result of capillary rise, showing that soil physical properties may increase the effects of soil chemical degradation processes such as soil salinisation.

4.2.1.2 Compaction and High Mechanical Impedance

The first consequence of land use change from native vegetation to a mechanised productive use is a reduction in soil OM content. This decrease causes a loss of structural stability, resulting in compaction processes. In agricultural soils, the main factors responsible for compaction are excessive traffic, the use of farm equipment that exceeds the bearing capacity of soil and tillage at unsuitable soil water content (Ellies 1988; Cuevas and Ellies 2001). These result in a change in the proportion of pores with water and air (mainly loss of coarse pores) and an increase in mechanical resistance to root development (Ellies et al. 2000; Ellies 1999; Seguel et al. 2009). All the intense production areas in Chile are subject to this problem.

Early studies by Ellies (1986, 1988, 1999) evaluated changes in physical properties of Andisols and Ultisols (Rainy and Patagonian zone) with different use intensity. Later works (Farías 2009; Seguel et al. 2009; Fuentes 2010; Poblete 2011) included Alfisols, Mollisols and Aridisols. Table 4.3 presents some important results related to changes in soil physical properties as result of compaction.

The direct consequence of tillage is to incorporate the superficial organic mulch into the Ap horizon, making it more accessible to the action of soil microorganisms. As Table 4.3 shows, in general there is a reduction in OM as a consequence of soil use. This results in an increase in soil

Table 4.3 Soil physical properties as result of compaction processes by intense use

Pedon	Land use	Soil depth (cm)	Organic matter content (%)	Bulk density (Mg m^{-3})	Coarse pores (>10 μm) (%)	Mechanical stress (kPa)
Valdivia (Andisol)	Forest	0–20	19.3	nd	34.4	176[a]
	Grassland		17.1	0.68	26.6	215[a]
Cudico (Ultisol)	Forest		8.9	nd	30.4	125[a]
	Grassland		7.3	1.03	18.8	160[a]
La Lajuela (Alfisol)	Forest	0–10	10.5	1.13	23.0	115
		10–30	5.7	1.38	20.8	151
	Vineyard	0–10	4.8	1.53	23.5	349
		10–30	4.6	1.59	11.0	283
Santiago (Mollisol)	Grassland	0–10	3.6	1.29	17.4	114
	Rotation	0–10	3.7	1.53	16.0	197
La Capilla (Aridisol)	Grapes	0–10	2.8	1.46	19.2	nd
		10–30	nd	1.62	11.7	nd
	Grapes + OM	0–10	6.2	1.02	17.1	nd
		10–30	nd	1.47	14.9	nd

Valdivia and Cudico soils: Ellies (1986, 1988) and Ellies et al. (1995a), La Lajuela soil: Seguel et al. (2009), Santiago soil: Fuentes (2010), La Capilla soil: Poblete (2011) with organic management
nd non determined
[a] Bearing capacity in wet condition, the others are strength of air-dry aggregates

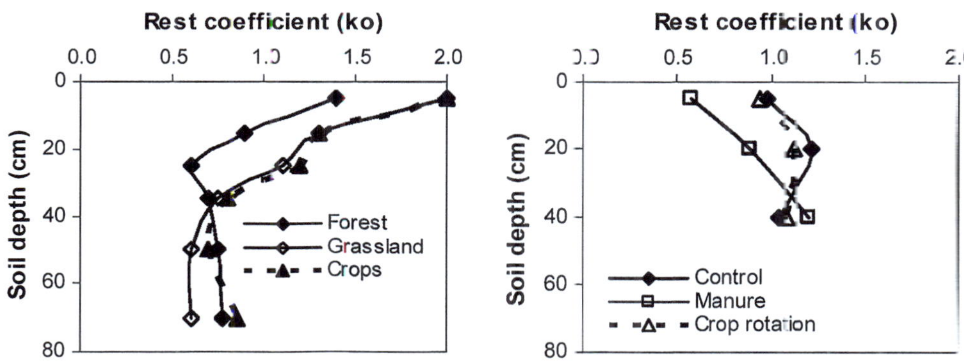

Fig. 4.15 Stress at rest coefficient (k_0) determined by penetrometer for an Andisol (*left* Contreras 2006) and an Aridisol (*right* Poblete 2011) with different management

bulk density (Db) values, with coarse porosity losses. According to Fuentes et al. (2011), it is possible to discriminate soils which by nature are more sensitive to others, i.e. soils with high clay content (Ultisols and Alfisols), which show great changes over time and space. In Andisols, the Db values are still lower than 0.9 Mg m^{-3}, even under conditions of high degradation (Ellies et al. 1996).

Under agricultural use without organic amendments, while the superficial tillage reduces soil densification, it is common to find a plough pan at 10–30 cm depth, as in the case of Alfisols and Aridisols, with low amounts of coarse pores. While a first effect of coarse porosity loss is reduced ability of soil to conduct air and water (Fig. 4.14), increased mechanical strength is also observed, hindering crop root growth (Table 4.3, Fig. 4.15).

Values of stress at rest coefficient (k_0 = horizontal tension/vertical tension) over 1.0 denote excessive mechanical strength (Bachmann et al. 2006), because of the rearrangement of soil particles or aggregates into a denser soil matrix. The surface of Andisols can be strengthened by critical stress levels, and with use there is a transmission of stress to depth, but the shape and the rigid nature of volcanic-derived minerals maintain high levels of coarse porosity (Table 4.3), enabling rooting (Ellies 1999; Ellies et al. 2000). The results for Aridisols in Fig. 4.15 (right) are from the same trial as Fig. 4.14, and only with high amounts of organic amendments does k_0 decrease below a critical level (<1). Considering the properties of Aridisols, with poor structure development and low OM content, natural sites (with native vegetation) probably develop high values of k_0.

Fig. 4.16 *Left* raised bed 1.2 m high built with San Julián soil (Typic Haplotorrert) and planted with 2-years-old mandarin trees; *right* Holocene sand dunes (Typic Usticpsamment) with pine forest

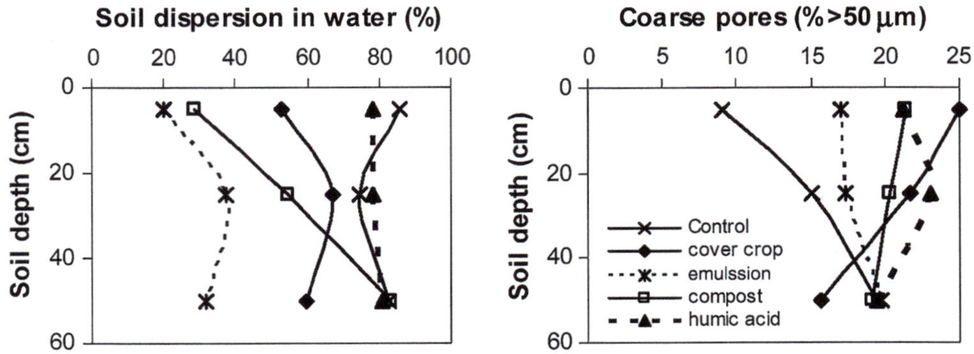

Fig. 4.17 Soil properties as a function of depth in a raised bed of San Julián soil series (Typic Haplotorrert). *Left*: macroaggregates dispersion in water (the lower value, the better stability); *right*: proportion of coarse pores. Adapted from Cortés (2011)

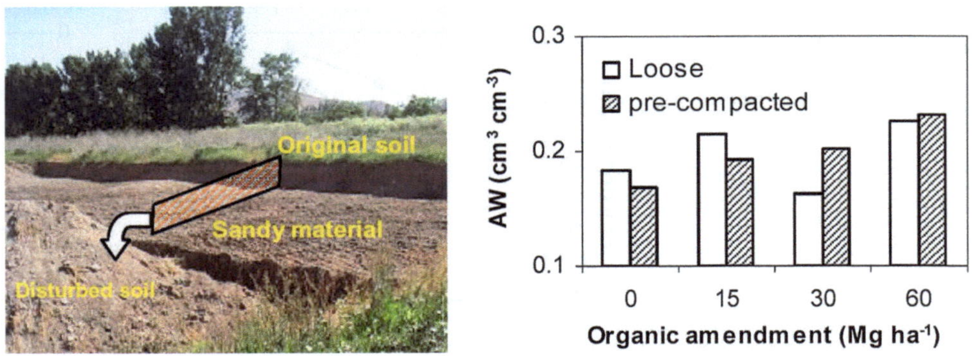

Fig. 4.18 Subsoil sand extraction from Rinconada Lo Vial soil series (Typic Xerochrept). *Left* disturbed/loosened soil deposited in the trench left by previous sand extraction; *right* available water (AW) as a function of organic amendment dose (OM). Adapted from Rodríguez (2011)

4.2.1.3 Soil Profile Modification

Many soils in Chile lack adequate depth due to poor structure or water retention due to coarse texture, both of which are influenced by continuous geological processes (volcanic eruptions, alluvial fans) alternating deposits or renewing the parent material. In the case of depth limitations, soils in the north of Chile (Hyper-arid to Semi-arid zone) restricted by cemented or massive horizons are common but the *ñadi,* an important soil type in the Rainy and Patagonian zone, is limited by a B_{hs} horizon at 50 cm depth on average. In shallow soils of the Hyper-arid to Semi-arid and Mediterranean zones, a raised bed 1.5 m high (locally called *camellón*) has been implemented during the past decade (Fig. 4.16) and causes mixing of the pedogenesis of the entire profile (Cortés 2011). On the other hand, artificial drainage of *ñadis* promotes settlement and a loss of soil function (Ellies 2001). Obviously, the rainy conditions provide greater potential in the latter case.

Coarse-textured soils are randomly distributed throughout the country, being associated with Holocene alluvial deposits, close to volcanoes and in dunes along the Coastal Range (Fig. 4.16, right). Typical coarse-textured soils derived from volcanic materials with alluvial transport are those grouped in the Arenales soil series from the Rainy and Patagonian zone (see Sect. 2.2.3), which are suitable for forest but have marginal agricultural use in flat areas. Even though these are naturally limited soils, in both cases (shallow and coarse-textured soils) they are being brought

Table 4.4 Soil properties (0–10 cm) related to water repellence in different Chilean soils

Soil	Land use (yr)	OM (%)	Wetting angle (°)	Water dispersion (%)
Valdivia (Andisol)	Forest (0)	24.3	105	2
	Grass (55)	16.0	79	4
	Crops (123)	11.7	70	7
Huiti (Andisol)	Forest (0)	33.8	118	nd
Cudico (Ultisol)	Forest (0)	12.6	83	3
	Crops (95)	6.4	48	75
La Lajuela (Alfisol)	Forest (0)	10.5	nd	3
	Vineyard (8)	4.8	nd	78

Adapted from Valdivia and Cudico soils: Ellies et al. (1995a, b), Huiti: Orellana (2010), La Lajuela soil: Seguel et al. (2009)
nd non determined

into agricultural use as a result of the need for new production area, and the impact on soil properties will affect the ecology of the area and promotes degradation unless correct management is implemented.

When a shallow soil is modified by building a raised bed, mixing of A_p and B horizons results in a loose material, which consolidates because of its low physical and mechanical stability, resulting in a reduction in coarse porosity. As shown in Fig. 4.17, it is necessary to use amendments to prevent excessive consolidation.

A soil disturbed by raised bed construction will have low physical and mechanical stability, losing coarse porosity and diminishing the capacity for air and water flux. The soil in Fig. 4.17 has clay content ranging between 44 and 52 %. Different soil amendments (commercial organic products such as humus and humic acids) improve the soil properties but to different extents depending on the property analysed, and some may need to be combined to restore degraded soil.

In the case of poorly drained soils, when an artificial drainage system is installed, natural soil settlement takes place, caused by OM and coarse porosity losses. In Frutillar soil series, a typical *Ñadi* (Typic Placaquand) from the Rainy and Patagonian zone, 50 years of use after drainage caused soil settlement from 90 to 50 cm deep and loss of OM from 36 to 10 %, resulting in a decline in water availability from 400 to 120 mm (Ellies 2001).

When coarse-textured soils have no agricultural use, they generally serve a valuable ecological role. However, the limited agricultural area in Chile creates pressure on the soil resource, causing sectors that should not be in productive use, such as dunes, to be cultivated and affecting ecological niches. One example is the sand dunes near La Serena city (~30°S), which have been levelled, amended with organic materials and subjected to potato production with drip irrigation. While there is no risk of settlement in sandy materials, the excessive organic amendment could promote water repellence (Ellies 1978) and preferential flow, increasing the risk of pollution.

A special case is the extraction of sand from subsoil for construction purposes, which alters the superficial soil (Fig. 4.18). In this case, the soil is totally disturbed and deposited in a ditch beside the trench.

The soil from Fig. 4.18 was classified as II LCC, with slight restrictions by 80 cm soil depth. Subsoil sand extraction results in a totally loose material, without bearing capacity and with low fertility index, diluted along the profile. Restoration of soil physical properties requires organic amendment use, complemented with a pre-compaction work. Efforts have been made to restore these soils (Macaya and Gallardo 2007), but still there is a lack of effective legislation to protect the soil resource.

4.2.1.4 Organic Matter Content and Physical Soil Degradation

Soil OM ensures physical–mechanical stability and therefore the functionality of the pore system, but high amounts of OM could result in higher water repellence, creating a risk of erosion or generating preferential water flow into the

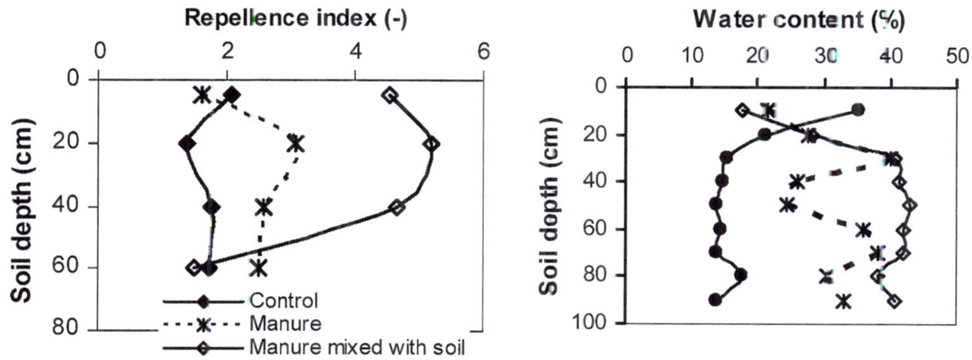

Fig. 4.19 Negative effect of high doses of organic amendment (3 years with 40 Mg ha^{-1}) in a Typic Haplocambid. *Left* soil repellence index (>4, excessive repellence). *Right* irregular available water (AW) distribution into the soil profile. Adapted from Keller (2011)

soil profile. Table 4.4 shows some soil properties related to hydrophobicity and aggregate stability in water.

The high OM in soils derived from volcanic ash ensures good aggregate water stability (low dispersion), even after 120 years of farming, and subsequent OM mineralisation. Nevertheless, there are potential problems with the hydrophobicity conferred by the OM (assessed as wetting angle) causing water repellence, depending on the type of OM and its interaction with mineral particles. This phenomenon is especially critical in Andisols from the Region VIII (Peña 1992), where dry summers promote the highest expression of water repellence (Orellana 2010) promoting runoff and erosion. Even in unsaturated conditions, the hydrophobicity may promote lower hydraulic conductivity at sites that naturally have higher OM contents (Nissen et al. 2006). However, with an adequate crop rotation a sufficient OM content can maintain hydraulic conductivity values that exceed rain intensity (Sandoval et al. 2007).

The difference between the Andisols shown in Table 4.4 is that Valdivia soil is a Hapludand (well-drained Andisol or *trumao*) and Huiti soil is a Duraquand (poorly drained Andisol or *ñadi*) (CIREN 2003). When these Duraquands are drained, their high shrinkage capacity and OM mineralisation cause irreversible contraction and settlement (Dörner et al. 2009), affecting pore functionality. In fact, an initial hydraulic conductivity close to 9 m d^{-1} has been measured in such soils, but after a second rotation of pine and 55 years of use it may decline to less than 4 m d^{-1} (Ellies et al. 1995b).

On the other hand non-volcanic soils (Table 4.4), with lower OM contents and lower water repellence risks than Andisols, show reduced aggregate stability under agricultural use, and under strong rain events are easily dispersed and sealed, favouring runoff processes. Ultisols and Alfisols, located in the Coastal Range, are particularly sensitive and erosion rates will depend on soil management, crop rotation and tillage system (Ellies 2000; Traub 2010). Finally, as shown in Fig. 4.19, it is also possible to generate problems with excessive applications of organic amendments or by particular situations of organic inputs (Sagardía et al. 2008). These problems are related to water flow continuity, water repellence and soil temperature (Keller 2011; Baginsky et al. 2010).

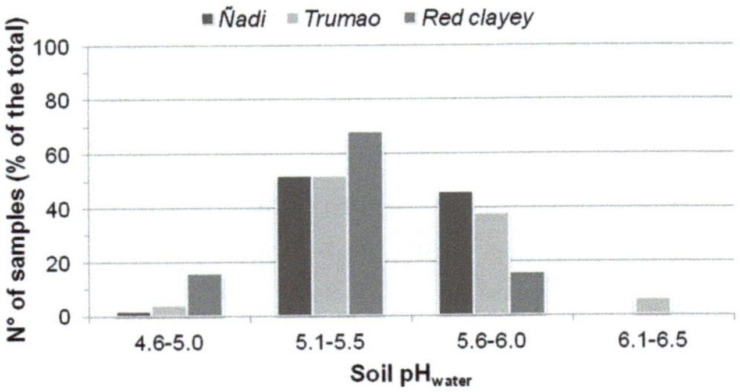

Fig. 4.20 Frequency of pH_{water} in soils at the Region X of Chile (Bernier and Alfaro 2006)

In NW Patagonia, Candan and Broquen (2009) reported highly water-stable aggregates under different types of vegetation (>80 % of >0.25 mm aggregates) in local Andisols (Udivitrands and Haplovitrand). Furthermore, the highest OM contents are found in the smallest-sized aggregates and there is a significant correlation between stable soil aggregate formation and OM, Al activity, base content and soil reaction, suggesting that these factors can potentially be used as edaphic indicators of aggregate stability in volcanic ash soils.

4.2.2 Soil Chemical Degradation

The main chemical soil degradation processes related to soils in Chile are:

a. Soil acidification.
b. Soil salinisation.
c. Soil contamination.

Table 4.5 Frequency of soils in the Mediterranean and Rainy and Patagonian zones with soil pH_{water} below 5.8 at 0–20 cm (Sadzawka 2006)

Region	Latitude	Soils with pH_{water} < 5.8 (%)
Metropolitan	(32–36°S)	1
V		14
VI		37
VII		35
VIII	(36–43°S)	37
IX		54
X		88

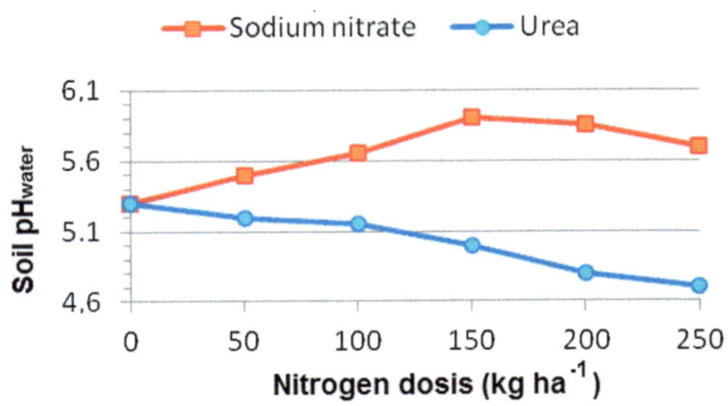

Fig. 4.21 Effect of nitrate sodium and urea in the soil pH in a strongly acid soil in the Rainy and Patagonian zone of Chile (Bernier and Alfaro 2006)

Fig. 4.22 Maize production in the Mediterranean zone with high NH_4^+-forming fertiliser application

It is important to note that the vulnerability of soils to these chemical degradation processes is mainly dependent on the initial state of the soil and its intrinsic soil properties, which are discussed in this section.

4.2.2.1 Soil Acidification

In Chile, accelerated soil acidification is tending to increase in some areas, and in the short term may cause a serious soil degradation problem, reducing agricultural production. As soil pH decreases, Al^{3+} saturation increases and plant growth and yield may be impaired, depending on their susceptibility to Al^{3+} toxicity.

Soil acidification is well-known as a natural process, but in many zones in Chile that contain a high proportion of soils with soil $pH_{water} < 5.8$ (Sadzawka 2006), which are considered moderately acid (see Table 4.5), agricultural practices and pollution from industrial, mining and other human activities may accelerate the acidification process. Soils in the Rainy and Patagonian zone are particularly susceptible, with 37–88 % of all soils having $pH_{water} < 5.8$ and with sufficient rainfall to leach out most of the base-forming cations (Ca^{2+}, Mg^{2+}, K^+ and Na^+), leaving the colloidal complex dominated by H^+ and Al^{3+}.

Areas with the highest percentage of acidic soils are located in Region X, dominated by Andisols and Ultisols (Fig. 4.20). These soils present a high risk of acidification and usually show a high Al saturation of the ECEC (effective cation exchange capacity) where ISRID (Incentives System to Recovery Degraded Soils) has subsidised farmers to apply lime. Several studies have confirmed that addition of lime to neutralise the acidity and reduce the toxic effect of Al^{3+} increases yield, forage quality, botanical composition and nitrogen symbiosis by legumes in pastures (Campillo et al. 2005; Bernier and Alfaro 2006). On the other hand, studies such as that by Inostroza-Blancheteau et al. (2008) suggest that liming is costly, laborious and not very effective and propose as an alternative a search for genetic variability in the genome of cropping grasses and/or their wild relatives to resist Al^{3+}.

One of the possible factors that may intensify soil acidification in the Rainy and Patagonian zone is continued application of NH_4^+-containing or NH_4^+-forming fertilisers that produce H^+ during nitrification (Bernier and Alfaro 2006; Campillo and Sadzawka 2006). In addition, following the application of NH_4^+, the adsorbed cations are subjected to replacement by NH_4^+ and can thus be potentially leached from soils (Sadzawka 2006). For example, a study by Bernier and Alfaro (2006) in a strongly acidic soil (pH 5.3) compared the effects of nitrate sodium and urea, where the maximum doses of N (250 kg ha^{-1}) as urea clearly generated a significant acidification effect on soil (Fig. 4.21). Similarly, Campillo and Rodríguez (1984) in a study in two Andisols found that application of urea may acidify the soil to levels where Al and Mn are toxic to crop development.

There is also a risk of soil acidification in the Mediterranean zone, in soils under maize production with high NH_4^+-forming fertiliser applications, mainly as urea, where in some cases N application can reach 500 kg N ha^{-1} (Fig. 4.22).

Other possible causes include continued application of organic wastes to soils, forming strong inorganic acids which lead to increased soil acidity. On the other hand, more local sources of soil acidification are acid deposition of sulphur dioxide (SO_2) and nitrogen oxides (NO_x) in areas close to copper smelters in the Mediterranean zone, where the impact on soil pH is more evident in the topsoil (Ginocchio et al. 2004; León and Carrasco 2011).

4.2.2.2 Soil Salinisation

In the Hyper-arid to Semi-arid zone, potential problems in irrigated soils have been identified due to high evaporation rates and low annual rainfall leaving accumulated salts in upper soil horizons. In this zone, the agricultural activities are mainly concentrated at the bottom of a number of valleys, where a river usually provides irrigation water throughout the growing season.

Table 4.6 Annual accumulation of salts at 0–50 cm by irrigation (8,000 m^3) in the Copiapó and Huasco valleys, Region III

Valley	Source irrigation	EC ($dS\ m^{-1}$)	Salt accumulation ($Mg\ ha^{-1}$)
		3.4	18
Copiapó	Tube well	2.2	11
	Ditch	5.8	30
Huasco	Ditch	4.3	22

Adapted from Benavides (2011)

In this zone two productivity areas can be identified: (i) valleys with intermittent rivers (the Lluta, Azapa and Loa valleys) with some horticultural crops, orchard fruits and lucerne production (ii) valleys with permanent watercourses (the Copiapó, Huasco, Elqui, Limarí and Choapa valleys) with some horticultural crops and intensive production of orchard fruits, such as avocado, citrus, table grapes and olives (see Chap. 2).

Torres and Acevedo (2008) reported that in the Lluta and Azapa valleys the salinity is increasing, causing a marked salt accumulation effect which is seriously affecting crop production in these areas. They noted that the electrical conductivity of the irrigation water (EC_{iw}) is high, with values of >2 and >1 dS m^{-1} in the Lluta and Azapa valleys, respectively, which may contribute to the problem of salinity of these soils. In the Lluta Valley in particular, the salinity problems are more important in poorly drained soils located in lowlands, where EC_{iw} can reach values of 6 dS m^{-1}.

In the Lluta Valley, the concentration of boron (B) in irrigation waters ranges between 9 and 29 mg L^{-1}, which may cause toxicity in plants in irrigated soils.

Similarly, in the valleys with permanent watercourses (between 27 and 32°S), salt build-up may be caused by irrigation with poor quality water. For instance, Hugo (2008) noted that salinisation has been observed in many sectors of irrigated soils in these valleys. Since fruit-bearing species such as table grapes do not tolerate high concentrations of salt or sodium, this can delay the start of harvest and seed germination, decreasing the overall productivity.

In a study in the Copiapó Valley, Sierra et al. (2001) reported that the main rivers used for irrigation demonstrated a medium level of salinity (>0.76 dS m^{-1}), but in the valley 76 % of the soils showed EC > 4 dS m^{-1} so the use of irrigation water poses a risk of soil salinisation. As an example, Table 4.6 shows salt build-up by irrigation with saline water in the Copiapó and Huasco Valleys calculated by Benavides (2011), where accumulation of soluble salts at 0–50 cm in the soil can reach 30 Mg ha^{-1} yr^{-1}.

In the Elqui Valley, where the main B input to the soil comes from irrigation water, table grapes have shown visual symptoms of B toxicity, which are correlated with high B contents in tissue analysis ranging between 135 and 376 ppm (Valenzuela and Narváez 1983). In addition, correction of soil B deficiency is complex due to the narrow range between sufficient and toxic levels, and there is a high risk of generating B toxicity in sensitive plants to excess B. For instance, Lavín (1988) in an experiment in the Mediterranean zone, evaluated the same B application by fertigation to 21 different fruit species, and found that fig, kaki, mulberry, pistachio raspberry and walnut showed some degree of B toxicity.

Another possible cause of soil salinisation in these valleys may be the use of fertilisers with a high saline index. Oyarzún et al. (2008) noted that with a larger contribution of fertilisers, the salinity will increase in the root area unless the irrigation is applied at washing rate to keep the original saline balance.

It is well-known that plants exhibit a wide range of tolerance to salinity and specific ions. One possible solution for these areas has been the use of salt-tolerant plant species or the selection of varieties more resistant to high salt levels in soil. Table 4.7 summarises some studies of salt-tolerant species carried out in the Hyper-arid to Semi-arid zone of Chile.

Native species such as quinoa (*Chenopodium quinoa*) have been cultivated for centuries in the Hyper-arid to Semi-arid zone in soils with high salinity, even with EC values close to 10 dS m^{-1} (Delatorre et al. 1995). In addition, Delatorre and Pinto (2009) have taken advantage of quinoa by selecting plant material in the nursery according to relative salt tolerance, where they found that *Amarilla* selection was the most tolerant to salts.

In another study, Ferreyra et al. (1997) in a field experiment near Calama town (Region II) evaluated the effects of irrigation with saline water (EC_{iw} 8.2 dS m^{-1} and B content of 17 mg L^{-1}) on the growth and yield of 42 crop species, including local varieties. They found that plant growth and yield were higher than expected from published information, which they attributed to the milder climate in Chile compared with that in Riverside-California, where much of the salt and B tolerance data have been obtained. They added that the productivity of the local variety of sweet corn (*Zea mays* L.) suggests that it is more salt-tolerant, which has arisen as a consequence of seed selection practised since the time irrigation began in the Region, which predates the sixteenth century. In the Lluta Valley too, a variety of sweetcorn (*Zea mays* L., *amylacea*) have arisen as a consequence of seed selection, suggesting that it is extremely tolerant to salinity and high B levels (Bastías et al. 2004). Other species studied include jojoba (*Simmondsia chinensis*), some clones of which have been found to be particularly resistant to salinity (Botti et al. 1998a), which could be a profitable alternative for the Hyper-arid to Semi-arid zone. Botti et al. (1998b) found that the most salt-resistant clone

Table 4.7 Summary of some studies (Sotomayor et al. 1994; Benavides 2011; Botti et al. 1998a; Martínez et al. 2009) in salt-tolerant species carried out in the in Hyper-arid to Semi-arid and Mediterranean zones

Region or valley	Specie	Yield	Soil			Water		
			pH	EC (dS m^{-1})	B (mg kg^{-1})	pH	EC (dS m^{-1})	B (mg L^{-1})
Lluta Valley	Olive	36–73 (kg/tree)	7.3–7.7	2–13	31–55	7.7–8.4	2.2–3.4	8–23
CopiapóValley	Olive	12–37 (kg/tree)	7.7	5.2	4	–	3.4	–
Regions I and Region IV	Jojoba	81–1,131 (g seed/plant)	–	2–38	–	–	1–7	–
	Quinoa	1–7 (Mg ha^{-1})	7.5–8.1	1–4	–	–	–	–

had some differences in morphological and anatomical parameters compared with those grown under non-saline conditions, such as the lowest stomata and trichome density and the largest stomatal size.

Other studies (Ferreyra et al. 1997; Botti 2000; Silva et al. 2010) have focused on finding salt-tolerant fruit trees, bushes and CAM plants by evaluating new varieties or rootstocks in trees to be used in the Hyper-arid to Semi-arid zone of Chile. These species include olive (*Olea europaea* L.), pomegranate (*Punica granatum* L.), fig (*Ficus carica* L.), pistachio (*Pistacia vera* L.), caper bush (*Capparis spinosa* L.), aloe vera (*Aloe barbadensis* M.) and pickly pear (*Opuntia ficus-indica* L.). For example, Sotomayor et al. (1994) studied the natural adaptation of some olives in the Lluta Valley, where some trees can survive under extreme saline conditions. Other studies are looking at the introduction of more sensitive species to salinity. For instance, Castro et al. (2009a, b) evaluated the tolerance to saline irrigation waters for different rootstocks in avocado, because it is one of the most sensitive species to salinity (Fig. 4.23) and has been intensively planted in the Mediterranean zone of Chile. They found that the *Nabal* rootstock was the most tolerant to salts, by retaining the highest chloride concentration in the roots and greatly limiting the concentration found in the leaves.

On the other hand, more advanced irrigation techniques have been introduced in the Hyper-arid to Semi-arid zone (drip and sprinkler systems), which can control rhizosphere salinity by keeping soil water content high and free of salts near to the emitters, while the salts are accumulated in the boundary between the moist and dry zone. In the Copiapó Valley, Osorio and Césped (2000) found that in irrigated table grapes the soils showed EC > 5 dS m^{-1}, with the highest salinity values at the mid-point between two drip lines and in first 40 cm of the soil profile. The contents of Na, Cl and B were also high in relation to the tolerance range for table grapes. They tested several combinations of irrigation systems and found that single drip line and microjet sprinkler irrigation systems provided better control of the salinity in the root zone.

Fig. 4.23 Leaf symptoms of chloride excess in avocado in the Mediterranean zone of Chile

4.2.2.3 Soil Contamination

Although the principal criteria used to estimate trace element threats are bioaccumulation, toxicity and persistence (Kabata-Pendias 2011), trace elements such as cadmium (Cd), cobalt (Co), chromium (Cr), copper (Cu), nickel (Ni), antimony (Sb) and zinc (Zn) can reach concentrations in soils that are toxic to plants and microorganisms (McBride 1994), but lead (Pb) and mercury (Hg) should be of concern as serious human health risks. In Chile, a number of studies have found that soil heavy metals pollution occurs mainly near Cu mines, Cu smelters and mine tailings, particularly in the Hyper-arid to Semi-arid and Mediterranean zones (González et al. 1984, 2008; González and Bergqvist 1986; González and Ite 1992; Ginocchio 2000; De Gregori et al. 2003; Ginocchio et al. 2004; Montenegro et al. 2009; Neaman et al. 2009). It is important to note that although the wastes from Cu mining activities in Chile can broadly vary in physico–chemical properties and total metal contents, they usually have high total Cu levels (Ginocchio 2011).

Even agricultural soils have been found to be polluted by heavy metals several kilometres from the source, where the

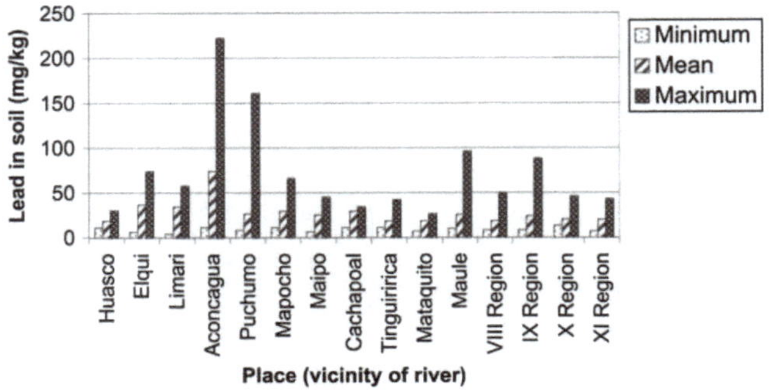

Fig. 4.24 Lead content in soils in the vicinity of various Chilean rivers (Tchernitchin et al. 2006)

magnitude of heavy metal pollution of the soils near these sources depends on the type of chemical element and the possibility of its dispersion by wind or irrigation water, affecting in this way extensive areas of agricultural soils (Romo-Kröger et al. 1994; González 2000; De Gregori et al. 2003). Pizarro et al. (2010) noted that rivers of central-northern Chile are exposed to heavy metal pollution from different sources, such as mining activities, natural orogeny processes, volcanic activity and geology. They suggested that mining pollution is the main process contributing to the increasing annual trend in As and Cu. Tchernitchin et al. (2006) evaluated the Pb content in soils in the vicinity of several Chilean rivers and suggested that in some cases the Pb originated from natural sources, while in others, it was anthropogenic and coincided with high Cu concentrations (Fig. 4.24). Similarly, González (2000) noted that in three valleys of the Mediterranean zone, high Cu concentrations in soils were associated with the soil concentrations of other metals such as Pb, As, Cd and Zn in the Puchuncaví (32°40′S) and Maipo (33°37′S) valleys, and with Pb, Cd and Zn in the Aconcagua (32°55′) Valley.

On the other hand, Biester et al. (2002) found an increase in Hg accumulation rates in moorlands in the Rainy and Patagonian zone within the past 100 years, and suggest that this is at least partly attributable to global dispersion of Hg derived from anthropogenic sources in the Northern Hemisphere.

Table 4.8 shows a list of studies examining heavy metal soil contamination, particularly by Cu, due to the great importance and impact of the copper mining activities carried out from the extreme north to the Mediterranean zone of Chile.

These studies clearly show that the total Cu in soils near copper mining activities centres far exceeds the normal range found in agricultural soils (McBride 1994; Kabata-Pendias 2011; Hooda 2010). These observed elevated values in soils are a consequence of Cu mining activities, and therefore pose a range of environmental and health risks. Copper is retained in soils essentially indefinite, because it is not degradable and consequently these Cu contaminated-soils pose a long-term risk of increased plant uptake and leaching, with potentially adverse implications for the wider environment, including human health. In some cases, the concentrations of other trace elements such as Zn, Pb, Cd, As, Sb and Hg were higher than the range of means for polluted soils worldwide, which also entails serious risks for human health (Higueras et al. 2004; Tchernitchin et al. 2006; Hugo 2008).

In addition, De Gregori et al. (2003) found a clear decrease in Cu, As and Sb concentrations in soils with increasing distance from a Cu smelter in Region V (Fig. 4.25). They highlighted the transport process downwind from the smelter or within the combined influence of the smelter and the tailings dams.

In a monitoring experiment in the same area, González and Ite (1992) found that in 9 years period the Cu content in some soils had increased by a factor of 6.5. Similarly, Hugo (2008) reported that at close proximity to Cu mining complexes, the Cu concentration in the soil exceeded the natural content 100 fold, easily surpassing the maximum tolerance limit for plants.

Kelm et al. (2009) added that once Cu is deposited in soils due to transport by wind, there is potential for acidity surges and Cu mobilisation in topsoils after rainfall. To avoid this transport, liming application may decrease Cu concentrations and Cu^{2+} activity in the soil solution (Muena et al. 2010). It is important to note that to predict the phytotoxicity and environmental risk of Cu the chemistry/mineralogy of mine materials, soil chemical conditions and plant physiological status should be also considered (Badilla-Ohlbaum et al. 2001; Ginocchio et al. 2002, 2009). These studies suggest that hot spots of heavy metal-contaminated soils are located around Cu mining activities, which are increasing in scale over time. Therefore, ameliorative measures should be carried out to stabilise the heavy metal in contaminated soils and to avoid its spread in the environment.

Another possible source of heavy metals to soils is fertiliser application. Molina et al. (2009) analysed the trace element composition of 22 fertilisers currently used in Chile and found that phosphorus (P) fertilisers had the highest trace element concentrations. They also noted that the long-term use of P fertilisers may increase the levels of As, Cd and other trace elements in agricultural soils.

On the other hand, few studies have been done in Chile to evaluate the impacts of persistent toxic chemicals in Mediterranean zone soils. Barra et al. (2005) evaluated the distribution of 15 polycyclic aromatic hydrocarbons

Table 4.8 Summary of some studies of soil contamination by total copper (Cu), zinc (Zn), chromium (Cr), lead (Pb), cadmium (Cd), arsenic (As) and antimony (Sb) in agricultural zones near copper mining areas in the Hyper-arid to Semi-arid and Mediterranean zones of Chile

Region	Cu	Zn	Cr	Pb (mg kg^{-1})	Cd	As	Sb	References
V	30–100[a]	10-200	–	56–225	1–5	–	–	González et al. (1984)
V	242–7,921	60–512	–	39–717	2–12	18–724	–	González and Ite (1992)
VI	162–751	138–153	–	47–50	0.2–0.4	–	–	Badilla-Ohlbaum et al. (2001)
I-II-V	11–530	–	–	–	–	3–202	0.4–11	De Gregori et al. (2003)
V	45–680	125–174	–	21–105	0–1	–	–	Ginocchio et al. (2004)
V	92–872	–	–	–	–	–	–	González et al. (2008)
V	310–640	–	–	–	–	–	–	Neaman et al. (2009)
IV	–	–	5–45	2–129	1–7	–	–	Montenegro et al. (2009)
V	60–800	85–220	–	29–103	–	–	–	Muena et al. (2010)

[a] Normal ranges of total heavy metal contents in agricultural soils [adapted from McBride (1994), Kabata-Pendias (2011) and Hooda (2010)]: As = 2.2–165 mg kg^{-1} ; Cu = 2–109 mg kg^{-1} ; Sb = 0.19–1.77 μg kg^{-1} ; Cd = 0.06–1.1 mg kg^{-1} ; Zn = 17–125 mg kg^{-1} ; Pb = 11–145 mg kg^{-1} ; Cr = 7–221 mg kg^{-1}

(PAHs), seven polychlorinated biphenyls (PCBs), and three organochlorine pesticides in topsoils in a watershed. They found that PCB levels in soil samples were very low and the level of chlorinated pesticides was generally low and reflected the historical use of pesticides. However, the PAH levels found were more related to local sources of contamination near the sampling areas, such as forest fires and the presence of boilers fed with wood residue pellets, where the reported data could be of some concern. In another study, Flores et al. (2009) studied simazine adsorption behaviour in two agricultural soils and suggested that it is mainly governed by simazine-organic matter interactions and simazine-clay interactions. In addition, Alister et al. (2005) reported that water quantity has a significant effect on the quantity of simazine moving downward through a soil in the Mediterranean zone. Nario et al. (2009), in a study on the risk of environmental contamination due to the application of pesticides in the same zone, highlighted that runoff is the main process for pesticide transport in the landscape.

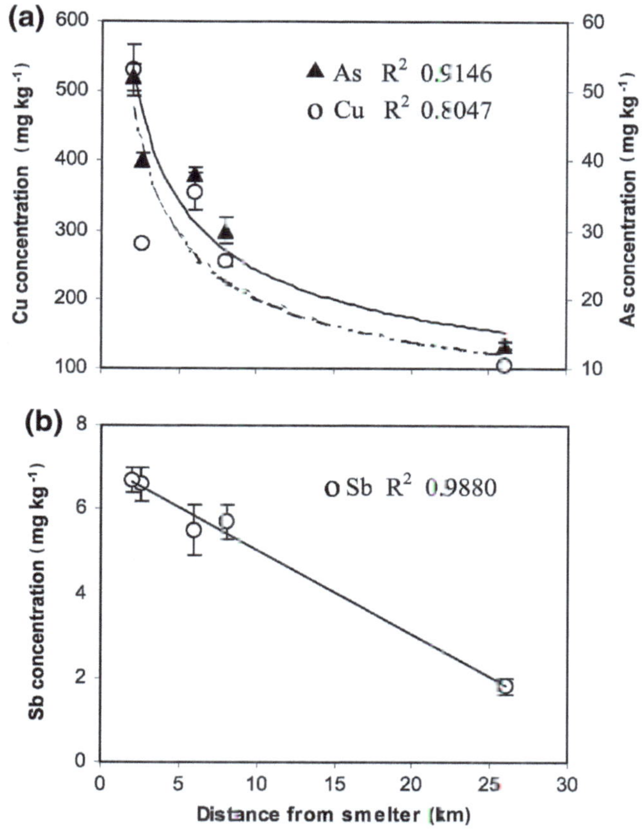

Fig. 4.25 Relationships between the element concentration in soil from Puchuncaví Valley, and the distance from Ventanas industrial complex, Region V of Chile; **a** copper (O) and arsenic (▲) and **b** antimony (O) (De Gregori et al. 2003)

4.2.3 Soil Biological Degradation

Soil organic matter (SOM) and biodiversity can decline due to biological soil degradation processes, leading to a reduction in soil functions such as decomposition and recycling of chemical and organic materials, as well as control of water and gas flows. This soil degradation problem is particularly important in Chile, for instance Ellies (2000) estimated that many soils have lost from 20 to 50 % of their SOM stocks in surface soil since cultivation began at the end of the nineteenth century.

In the Mediterranean zone, Ovalle et al. (1990) noted that possible causes of this continued decrease in soil organic carbon (SOC) stocks include inappropriate agricultural methods, excessive woodcutting and systematic overgrazing. This zone has a common natural agroecosystem, locally named *espinal*, where the dominant vegetation is *Acacia caven* trees with a spring grass cover rich in

Fig. 4.26 Degraded *espinal* in the Mediterranean zone of Chile (Region V)

annual plants (named *Mediterranean annual prairie*) (Silva and Lozano 1986; Ovalle and Squella 1988). In particular, unsustainable management of the vegetation in the *espinal* has progressively lowered the SOM content and soil fertility, as evidenced by decreasing forage production and lower coverage of trees (Stolpe et al. 2008), as shown in Fig. 4.26.

For instance, Muñoz et al. (2007a) found that at 0–40 cm depth in the soil the degree of coverage of *A. caven*, ranging from native forest to much degraded *espinal*, directly affected the C stocks, decreasing these when the tree coverage diminished and soil use intensity increased, as shown in Fig. 4.27.

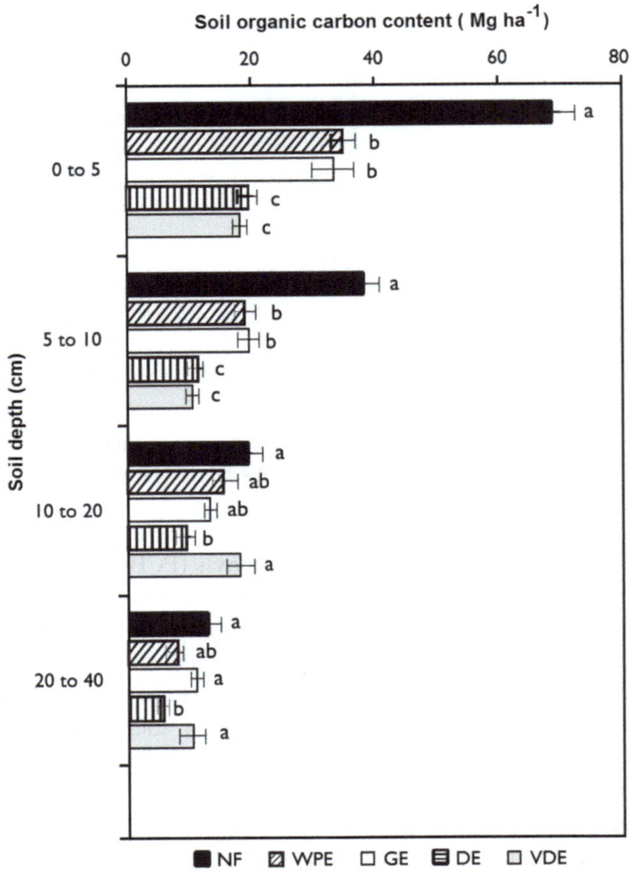

Fig. 4.27 Soil organic carbon stocks under canopy of *Acacia caven* in the *espinal* ecosystems in the Mediterranean zone soils. Ecosystem studied ($n = 8$): native forest is NF, well-preserved *espinal* is WPE, good *espinal* is GE, degraded *espinal* is DE and much degraded *espinal* is VDE. (Muñoz et al. 2007a)

In addition, Muñoz et al. (2008) in a study of the C distribution, structure and functional properties in different SOC fractions in *espinal* ecosystems with different land surface coverage, found that the aromaticity of SOC in the intermediate fraction was lower in well-preserved *espinal* (33 %) than in degraded *espinal* (50 %). They noted that these data reflect the effect of greater soil use intensity under degraded *espinal* due to the inverse relationship between *A. caven* land coverage and soil use intensity.

In the Rainy and Patagonian zone, there is also some evidence that organic matter (OM) in Chilean volcanic soils is being degraded after human intrusion. Heredia et al. (2007) found that soils under native forest showed higher OM level than agricultural soils. They found that in agricultural soils the soluble C or labile C is increased and contributes greatly to the solubilisation, mobility and availability of plant nutrients, but also enhances losses of organic C by lixiviation. Similarly, in this zone Ramírez et al. (2003) developed a study that compared the floristic composition and SOM content in native forest and anthropic prairie in an Andisol and an Ultisol. They found that compared with the forest, the OM at 0–20 cm in the prairie decreased from 15.5 to 13.6 % and from 11.5 to 5.1 % in the Andisol and Ultisol, respectively, with the floristic composition following the same tendency. In a study in Andisols in this zone, Undurraga et al. (2009) concluded that measurements of dissolved organic C (DOC) and N (DON) reflect non-intensive and intensive soil management and can be used as good biological indicators of changes in SOM, as more sensitive parameters than soil total C and N. In the same zone, Alvear et al. (2005) reported that tillage/residue management induced significant changes in soil biological activities, such as microbial biomass and soil enzyme activities. For instance, they found that no tillage increased C and N of microbial biomass in comparison with conventional tillage.

On the other hand, loss of biodiversity is also considered a soil biological degradation symptom, because soil organisms are a key components in a number of crucial processes, such as OM decomposition, nutrient cycling, nutrient retention, C sequestration, N fixation and pollutant

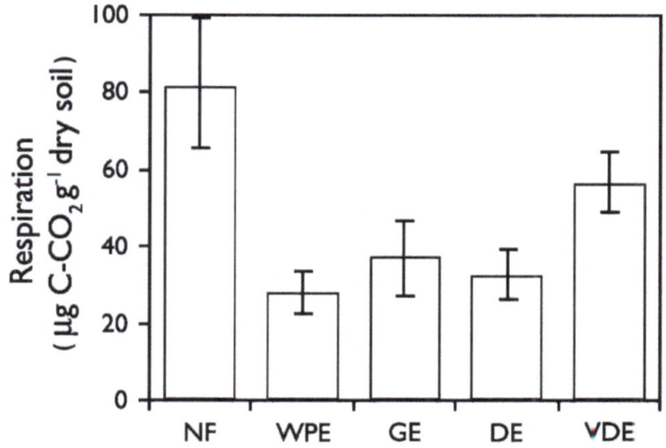

Fig. 4.28 Microbial respiration (µg C–CO_2 g^{-1} dry) in soil under canopy at 0–5 cm in ecosystems studied ($n = 8$): native forest is NF, well-preserved *espinal* is WPE, good *espinal* is GE, degraded *espinal* is DE and very degraded *espinal* is VDE (from Muñoz et al. 2007b)

degradation, among others. Fuentes and Varnero (2011) noted that in the Mediterranean zone the alteration of *espinal* ecosystems not only affects the C pools in the soils, but also the microbial communities and their enzymatic and biochemical activities. Therefore, they recommend that to evaluate the soil biological degradation the focus should not only be on quantification of the SOC stocks, but also on the role of microbial communities in the soils. On the other hand, microbial activity in soils has been related to soil respiration rate. For instance, Muñoz et al. (2007b) in the Mediterranean zone found that the respiration of microbial communities was affected by ecosystem degradation at 0–10 cm soil depth under the tree canopy, with soil respiration in native *espinal* ecosystems greater than in other degraded ecosystems (Fig. 4.28).

Therefore, ameliorative measures to recover OM and biodiversity are needed in soils affected by degradation processes. During recent decades, some efforts have been made by government to afforest degraded soils in the Hyper-arid to Semi-arid zone, including some trees (*Prosopis tamarugo*) and bushes (*Atriplex nummularia* and *Atriplex repanda*) (Olivares and Gastó 1981; Ormazábal 1991).

Aronson et al. (2002) noted that agroforestry systems are a promising approach to rehabilitating damaged agroecosystems, sustaining profitable agricultural production, restoring soil fertility on degraded lands and biological conservation. Similarly, Aronson et al. (1993) highlighted the beneficial effects of including nitrogen-fixing legume to optimise the amount of atmospheric N fixed in a degraded arid or semi-arid land ecosystem, to promote the development of associated plants, and to improve soils and enhance possibilities for the spontaneous or assisted return of native plants and animals. Ovalle et al. (1999) added that water resources in the Mediterranean zone, especially precipitation and runoff, must be better managed. Considering the latter point, Salazar et al. (2006) reported that the use of runoff water harvesting combined with agroforestry systems in the central zone of Chile can be beneficial for increasing SOC and N stocks, which indicates that these land management practices can be used for restoration of degraded soils and potential SOC sequestration in degraded *espinal* ecosystems. Figure 4.29 shows a field experiment combining agroforestry with water harvesting at the Germán Greve Silva Experimental Station, University of Chile, which started in 1996 in the Mediterranean zone (Salazar and Casanova 2011). It can be seen that after 14 years, the system has been able to survive under extreme scarcity of water.

In the same study, Salazar et al. (2011) reported that after 12 years, the treatments with water harvesting had higher OM contents (Fig. 4.30) than the control, which comprised degraded natural pasture. This suggests that this practice had positive effects in improving the water content in the soil, stimulating biomass production of trees (*Acacia saligna*) and the decomposition of litter and favouring the process of root turnover. They used *A. saligna* as N-fixing trees, and also found that agroforestry combined with water harvesting showed the highest accumulation of N content as a result of increased litter and root turnover (see Fig. 4.30).

It is important to note that some management practices that aim to restore degraded soils may have unexpected negative effects in the short term. For instance, Pérez-Quezada et al. (2011), in a field experiment in the Hyper-arid to Semi-arid zone, evaluated the conversion of ecosystems from natural conditions with arid shrub land (moderately disturbed by grazing) to afforested conditions (2-year-old plantation of *A. saligna*). They found that stocks of ecosystem C decreased from 32.4 in the natural places to 19.8 Mg ha^{-1} at the afforested sites, mainly due to loss of C from the soil C pool at 0–50 cm depth during site preparation for afforestation (Fig. 4.31).

In one of the few studies carried out in the Rainy and Patagonian zone, Klein et al. (2008) analysed SOC and the organic soil horizons in five stands at different stages of development (Fig. 4.32): intact native forest (NI); a 3-year-old shelterwood stand (S3); an 8-year-old shelterwood stand (S8); a 14-year-old stand that was initially treated with shelterwood and subsequently final cut (10 year after the first intervention) (S14), and a 25-year-old stand subjected to a exploitative intervention (E25). Short-term effects of shelterwood cuts on SOC were only detected in the upper 10 cm of the mineral upper horizons, where comparison of the SOC per unit area between the NI and S3 stands indicated a significant net C loss of 5 Mg ha^{-1}, whereas long-term effects of shelterwood cutting on SOC could not be verified for this type of forest.

Fig. 4.29 Agroforestry under water harvesting field experiment located in the Mediterranean zone. Initial conditions in 1996 (*left*) and 2010 situation (*right*)

It is important to note that the development of computer simulation models has also provided tools to predict the effects of management practices to recover OM and biodiversity in soils affected by biological degradation processes. However, in Chile few soil carbon model applications have been carried out. In one study, Stolpe et al. (2008), using the *Century* model, predicted that less intensive management could gradually increase soil organic C (3.19 to 3.86 Mg ha^{-1} in 100 years) in degraded *espinal*, thereby improving the general quality of the soil and agroecosystem. In another modelling approach, Salazar et al. (2011) used the *ICBM/N* model and predicted that in a degraded *Mediterranean annual prairie* system, there was a slight trend for decreasing OM over time, which may indicate a negative OM balance, whereas a degraded *Mediterranean annual prairie* system with introduced *A. saligna* trees increased the OM stocks.

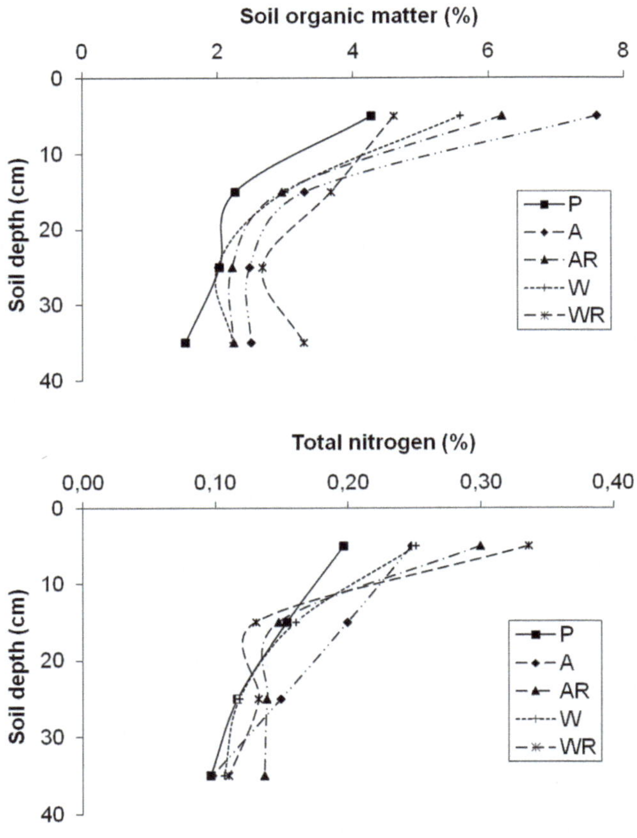

Fig. 4.30 Soil organic matter content and total nitrogen content in different soil layers (year 2008) for the treatments control (P), agroforestry (A), agroforestry with water harvesting (AR), woody perennial (W) and woody perennial with water harvesting (WR) from Villarroel (2012)

4.3 Desertification

Ecosystems in semi-arid and arid regions around the world appear to be undergoing various processes of degradation commonly described as desertification, probably caused by land misuse, soil mismanagement and a harsh climate. Desertification is a stage of extreme degradation that includes environmental and socio-economic aspects. It involves soil degradation as an inseparable term of deficient sustainability of ecosystem, which covers a wide variety of interactive phenomena, both natural and anthropogenic. In this sense, basic investigation has to precede monitoring, because it is not possible to know what to monitor if basic processes are unknown and, the impacts on inhabitants are ignored.

Much of the agricultural production in Chile derives from areas subject to the effects of desertification: the irrigated valleys in the northern Hyper-arid to Semi-arid zone of the country down to Region IV, and south of Santiago down to Region VII (Fig. 4.33). The areas considered vulnerable amount to about 45 % (340,000 km^2) of the national land surface, affecting 1.5 million inhabitants (Beekman 2007).

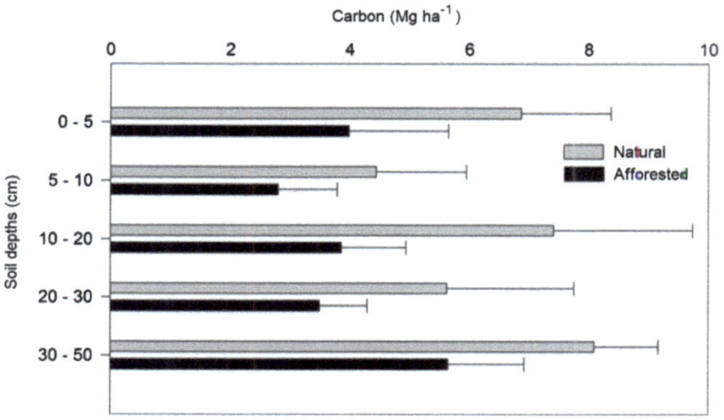

Fig. 4.31 Soil carbon content at five soil depth intervals in natural and afforested conditions of an arid Mediterranean shrubland in Chile (Pérez-Quezada et al. 2011)

In ecosystems subject to desertification, the vegetation cover is highly variable, with large patches of bare ground where solar radiation reaches the surface without interference. Therefore, plant cover is an important factor for combating desertification, because it protects the soil surface from raindrop impact, enhances infiltration and retards runoff (also reducing spatial variability in both), resulting in arresting land degradation. Optimal production, as dictated by the farmer's particular needs, comprises a well-planned combination of productive system components, so agroforestry emerges as the best option to establish diverse, stable and productive systems. Using either native or exotic species is a potent and efficient tool for biological recovery, allowing higher productivity than pristine vegetation under the same ecological conditions. The presence of trees modifies site microclimate in terms of temperature, relative humidity and wind speed, among other factors. With respect to spatial niche differentiation in agroforestry systems, the general expectation is that deep-rooted trees will not compete with, but rather complement, shallow-rooted annual crops (Daudin and Sierra 2008). Consequently, yield advantages for components in agroforestry systems within semi-arid environments occur only when profile soil water status remains high throughout the growing season, or when the trees are managed with a reduced transpiration surfaces (McIntyre et al. 1997). However, in the case of arid and semi-arid rainfed farming, there is insufficient rainfall to maintain such plant cover.

4.3.1 Coquimbo Region and Patagonia, Two Emblematic Cases of Desertification in Extremes Zones of Chile

Desertification is particularly important in Region IV, which is recognised at planetary level and characterised by a steeply landscape and a semi-arid climate. Rural development in this region takes place under constant tension between use of the natural resources and economic activity. Favourable agro-ecological and climate conditions in the valleys have stimulated the rapid expansion of fruit production, notably table grapes and grapes for processing into liquor (*pisco*), both with outward orientation. However, the dynamism of the fruit sector has not translated into an improvement of quality of life for most people in the region. Moreover, the complex and diverse land tenure (Agricultural Communities) in rainfed conditions contribute to increase this tension. On the whole, this situation may be further exacerbated by the interactions among adverse climate factors, with recurrent drought episodes, few torrential and destructive rain events, and a historical reduction in the amount of rainfall (Fig. 4.34). These features mean that this region is extremely prone to runoff and soil erosion, especially when the soil is unprotected by vegetation. Natural regeneration of plants in many arid zones occurs only for every 5 or 7 years (or less frequently), when two successive years of favourable conditions occur (Hahm and Wester 2004).

Land degradation in this zone has been attributed to increased population pressure on natural resources and to

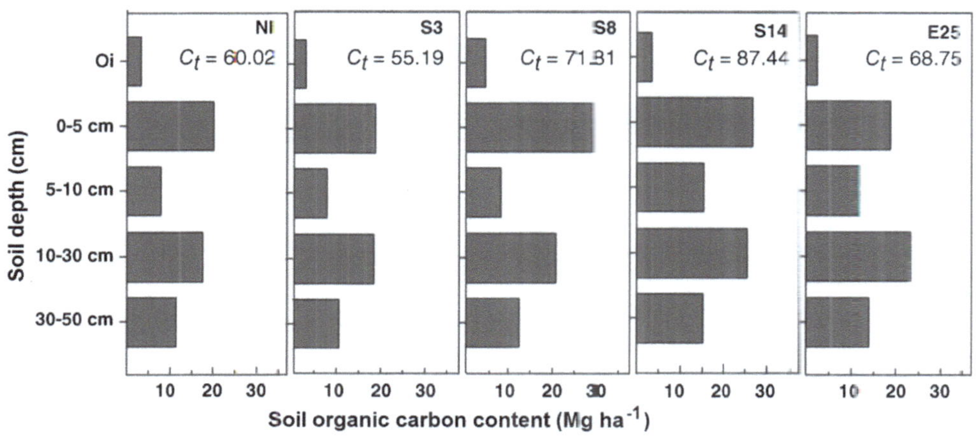

Fig. 4.32 Soil organic carbon (C) in the different forest stands. The number in each graph indicates the sum of total C for the entire soil profile. Sampling number was 40 for the O_i layers and 20 for the mineral soil layers (Klein et al. 2008)

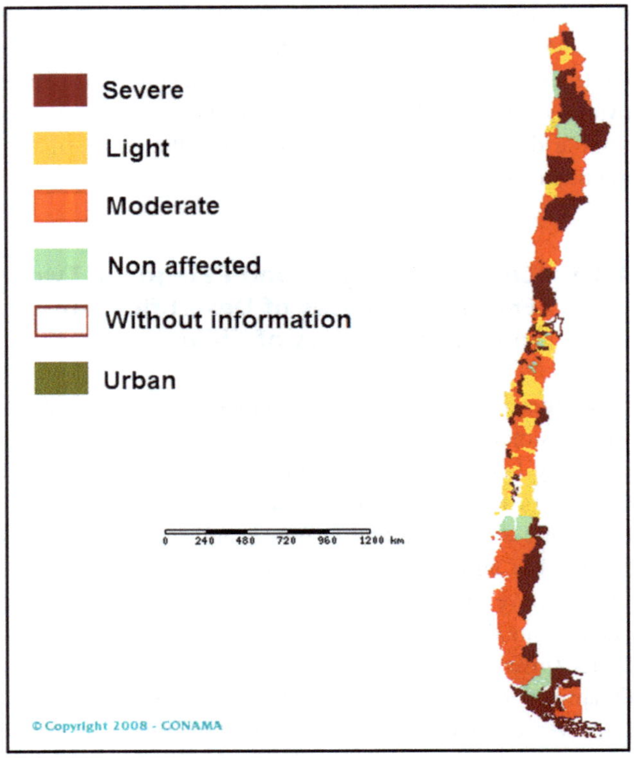

Fig. 4.33 Desertification map of Chile (http://www.mma.gob.cl). Accessed 20 March 2012

the droughts that have affected the region for many decades. Even earlier, natural resources had been severely depleted; at present, further degradation is taking place at rates that could lead to an environmental collapse in the near future. For example, Bahamondes (2003) reported that certain threatening agricultural practices have not diminished notably the large goat population and the gathering of firewood.

Agricultural communities are social organisations of small farmers, joined by family bonds or friendship, living on communitarian ownership, which is basically an undivided and indivisible expanse of land. These communities have their origin in land grants to generally licensed military personnel of the Spanish armies (sixteenth–seventeenth centuries). Many of these lands were subdivided by inheritance and finished up as *minifundia*, while a few remained undivided, either as *haciendas* or as communities. Grazing privileges on the ranges and slopes, which often constitute by far the largest proportion of the community lands, are enjoyed by all members.

Although most of the 180 agricultural communities identified in Chile have holdings between 500 and 10,000 ha and their total area covers almost 1 million ha (25 % of the Region IV), only a small proportion is classified as permanently suitable for agriculture and stock rearing. In fact, only 1,000 ha are irrigated, and a further 9,000 ha are of limited use because of insufficient water. Furthermore, almost 900,000 ha can only be used as rangeland. For the 70,000 inhabitants of the agricultural communities, per capita land with some agricultural potential is extremely low: 0.017 ha of permanent agricultural land, 0.13 ha of cropland with limitations and 0.69 ha of land suitable for dry-farming or grazing. The proportions of these different categories of land vary greatly. All agricultural communities have the same problem of having to subsist on a small area of land of low agricultural value, most of which is covered by unproductive shrub on which

Fig. 4.34 Average annual precipitation (1869–1999) in Vicuña and La Serena (Young et al. 2010) at Region IV of Chile

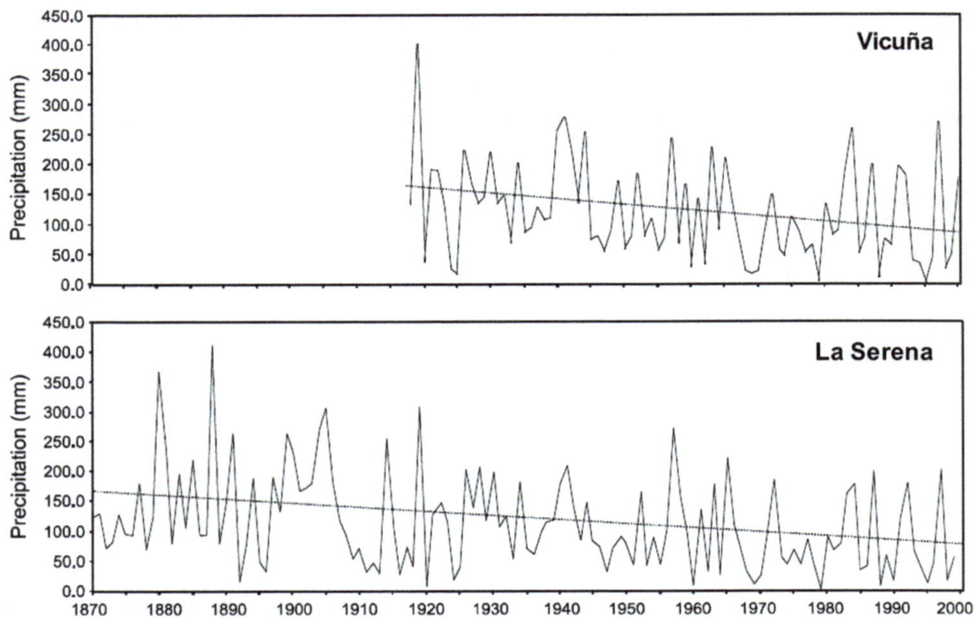

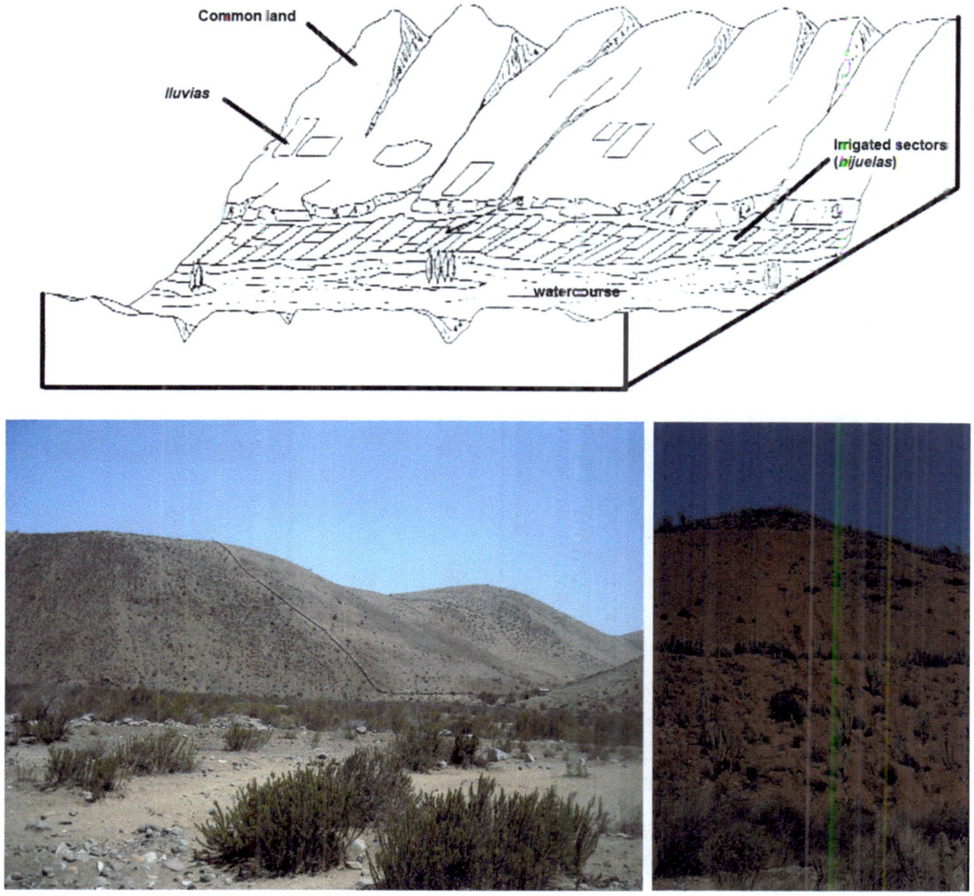

Fig. 4.35 Scheme of land possession types in rainfed zone of Region IV (*above*). Very steeply cropped land (*lluvias*) with erosion features (*below*)

even the goats have difficulties surviving in many years. The widespread failure of crop and animal production, especially since the 1968 drought, has forced a large sector of the rural population to put even greater pressure on the region's natural resources. However, it seems clear that the present rates of extraction of wild plants and firewood cannot be sustained by the ecosystem in the future.

Within the agricultural communities, members generally have the possibility to be the owners of small private plots that are permanently assigned to families, while larger tracts of dryland for crops are often rotated. For several decades, it has been traditional that rainfed soils on steep slopes are widely used in the production of cereals (wheat, barley) and umbelliferae (cumin, anise). This activity net of subsistence has contributed to the degradation of the scarce land available, with an abandonment of agricultural land at certain sites, locally called *lluvias* (Fig. 4.35), where the soil has lost its natural fertility or has reached levels of severe erosion. Cultivation is followed by at least 1 year of fallow (Fig. 4.36) and after harvesting the stubble fields are grazed by farm animals (goat overgrazing).

Land abandonment constitutes a depreciation of environmental capital stock and has many, mostly negative, socioeconomic and environmental consequences. While it can be argued that the *lluvias* are a management system that is not recommended and which should be excluded, it is equally true that farmers are continuing with this practice and will continue in the future, giving validity to the tragedy of the commons hypothesis (Hardin 1968).

In Region IV, plant productivity has decreased over time, caused by a negative water balance (Kalthoff et al. 2006) and intense soil degradation. In the inner rainfed area, plant productivity is low due to lack of adequate water, which induces a short growing season. However, the socio-economic effects of an imminent climate change will be reflected most clearly in the existing marginal system of communities, because farmers with access to supplementary water and financial support will be able to respond better than those poor farmers.

Soil and water conservation in Hyper-arid to Semi-arid zone of Chile is a priority. Integrated management of these cultivated hillsides has been poorly addressed by the national investigations, but abundant documentation relating to the management of runoff in these conditions has been produced (Verbist 2011; Verbist et al. 2009; Sangüesa et al. 2010).

A field study examining an agroforestry and/or rainwater harvesting combination with *A. saligna* was carried out in a

Fig. 4.36 Rills erosion in a *lluvia* (*above*) and in a hillside being grazed by goats (*below*), Region IV

rainfed area of central Chile, between 1996 and 2004. The field experiment, located at Germán Greve Silva Experimental Station of the University of Chile (Santiago), assessed the influence of this synergic combination on some soil fertility parameters at four depths (Salazar 2003; Leiva 2005; Salazar et al. 2006) of a Typic Haploxeroll (see Sect. 4.2.2). Soil physical properties were evaluated later by Bauzá (2009) and Moreno (2012) to obtain improved values. The lack of significant statistical differences observed was attributed to poor contribution of OM, leading to the conclusion that in more arid conditions, addition of organic amendments is essential to benefit plant growth and improve soil quality.

Traditionally, research has tended to propose or suggest techniques that aim to change the use and management of soils (Salazar and Casanova 2011). Even though there are many options being offered, farmers have preferred to maintain their traditional use and management. Therefore, an initiative (Casanova et al. 2011) in Region IV is starting a trial that aims to establish a demonstration area to show different options (*Acacia saligna*, *Atriplex numularia* and *Prosopis chilensis* trees, with and without organic amendments) in soil and water conservation to local farmers (Fig. 4.37). Clearly, although an important biophysical potential for intensification and enhancement of arid zones

Fig. 4.37 Soil and water conservation practices in deeply degraded soils in Region IV, initial (April-2011, *above*) and later (October-2011, *below*) scenarios. Microcatchments, stone-lines and erosion plots

exists, it is necessary to give more attention to socio-economic variables of the system. Many authors (Abdelkdair and Schultz 2005; Montambault and Alavalapati 2005) argue that the integration of agroforestry system with rainwater harvesting will require active participation by farmers in all planned trials.

Finally, it should be emphasised that at least three important institutions are working in this desertified region of Chile:

a. CEZA (Centre of Arid Zones Studies, University of Chile: http://agronomia.uchile.cl/centros/ceza/index.html)
b. CEAZA (Advanced Centre of Arid Zones Studies: http://www.ceaza.cl)
c. CAZALAC (Water Centre for Arid and Semi-Arid Zones in Latin America and the Caribbean: http://www.cazalac.org/eng/index.php).

Desertification in the Patagonia (eastern side of the Andes) is also a significant environmental problem, both due to its severity and to the area it covers. The areas

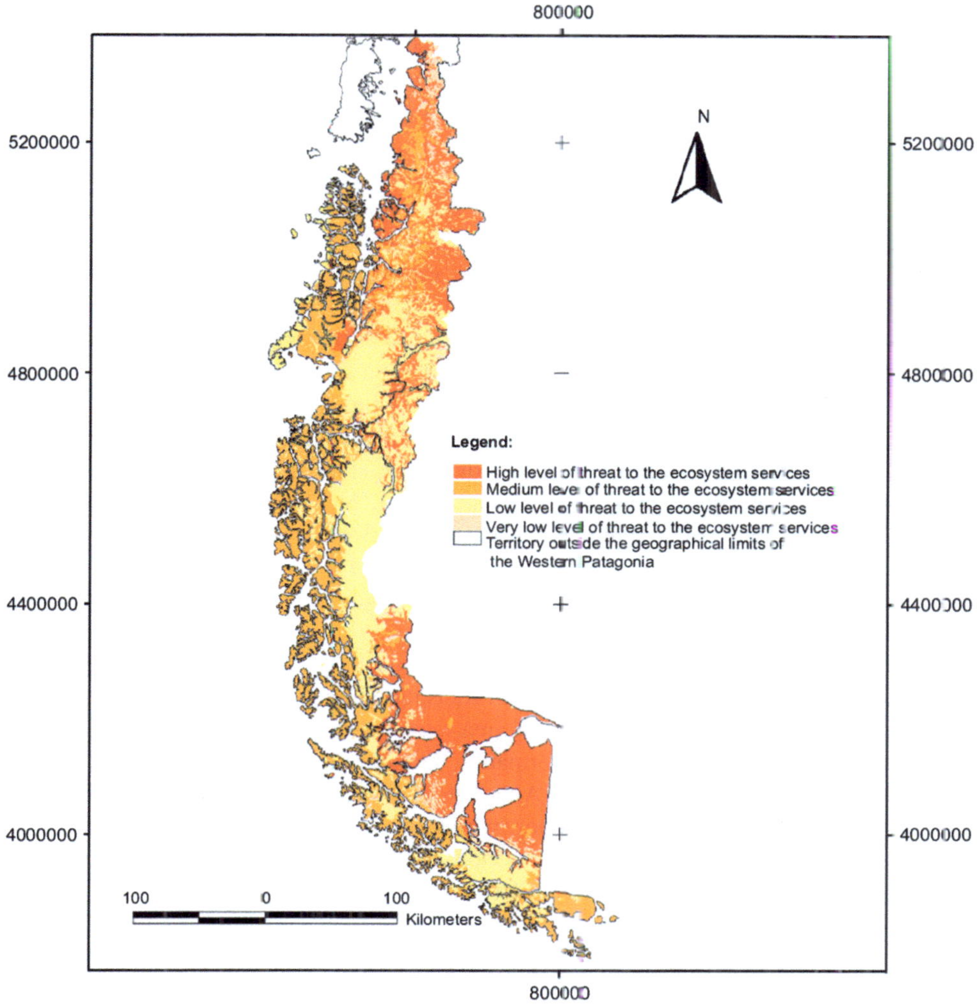

Fig. 4.38 Level of threat to the regional territory and the National System of Protected Areas from Western Patagonia (Martínez-Harms and Gajardo 2008)

classified as irreversible for the development of agricultural and livestock production activities cover 58 % of the whole area, i.e. 73,544,300 ha of Argentinian territory (Mazzonia and Vásquez 2010). Overgrazing by sheep is assigned as the main cause of desertification in the zone, because grazing, by removing perennial grasses and pulverising the surface soil can have a major impact on soil erosion (Chartier and Rostagno 2006). However, hydrocarbon extraction and mining activities have also been incorporated into the system, with the subsequent clearing of extensive areas. For western Patagonia, Holz and Veblen (2011) report that although in both pre-historic and modern times climate variability is the dominant control on years of widespread fires, aboriginal farmers and Euro-Chilean settlers have amplified fire activity (particularly during the twentieth–twenty-first centuries) and shifted the region's fire regimes to new behaviours, which threaten the soil severely (Vött and Endlicher 2001). Moreover, results obtained by Martínez-Harms and Gajardo (2008) show an unbalanced coverage of the National System of Protected Areas in Chilean Patagonia. These authors conclude that despite the high amount of regional area allocated to conservation, key territorial units in the provision of ecosystem services, which are vulnerable to human action, remain without a conservation category (Fig. 4.38).

4.3.2 Easter Island, an Example of Collapse by Soil Degradation

Nowadays, the soils of the Easter Island show the effects of intensive soil degradation processes, particularly soil erosion. In the Island the sheet erosion is evident, with abundant gullies in the northern area. In the 1940s, Díaz (1949) found that soil erosion was an important soil degradation constrain that affected soils with slope >10 %, occupied mainly with pastures covering 50 % of this degraded areas. While in the 1990s, Honorato and Cruz (1999) reported that 20 % of the soils of the island present erosion processes, mainly located in hilly areas (slope >15 %) around the main volcanoes. They noted that other studies in the island estimated severe erosion processes in 57 % of the total area.

However, the causes of these soil degradation processes are not totally clear yet. Louwagie et al. (2006) suggested that physical soil degradation is witnessed by increased erosion on the island since the introduction of grazing sheep (from 1872 to 1985) and cattle (in the 1970s) with ensuing overgrazing, especially on slope positions. Mieth and Bork (2005) added that after the woodland clearance around AD 1,300–1,400 the soils were exposed to the harsh climate conditions, where surface impacts after the destruction of the vegetation realised by agriculture in open land and by the construction of new settlements enabled migrating sheet erosion. Diamond (2007) noted that new information on Easter Island is helping to identify the cause of the massive deforestation that occurred prior to European arrival, but unanswered questions still remain. Even Mieth and Bork (2005) hypothesised that soil erosion may have played a dominant role in the breakdown of Easter Island stone culture.

Mieth and Bork (2005) in the area around Poike volcano, estimated a rate of 8.6 Mg ha^{-1} yr^{-1} of soil transported by erosion. They suggested that gullies in this area were caused by excessive overgrazing, while in its formation the soil rate loss was around 190 Mg ha^{-1} yr^{-1}. Although the same authors highlighted the effects of strong winds over soil surface in the badlands in the Cumming Cape, they noted that water erosion is the main degradation process in this portion of the Easter Island.

4.4 Future of Soil Conservation in Chile

At the present time, combination of climate changes and land cover changes is particularly threatening for soil conservation in Chile. The average net annual deforestation rate (1975–2008) for Central Chile was estimated by Schulz et al. (2010) to be −1.7 %, with a reduction in dryland forest and conversion of shrubland to intensive land uses (farmland) as major trends in this highly dynamic landscape. Moreover, Chile stands to be one of the planet's most vulnerable countries to climate change due to glacial melt and shifts in rainfall patterns. All glaciers in Chile had a net retreat between 1955 and 2007, with mean frontal rates of −22 m yr^{-1} (Le Quesne et al. 2009). According to ENERSIS-Fund (García and Ormazábal 2008), 45 % of the original forest area in continental Chile was lost during the sixteenth century due to factors mainly such as fuel-wood extraction, carbon and wood, fires and land clearing for agriculture production and livestock.

The most direct effect of climate change on erosion by water can be expected to be the effect of changes in rainfall erosivity. Ellies (2000) reported that in central and southern Chile, 5–8 % of the annual precipitation (from 500 to 2,500 mm yr^{-1}) has a high kinetic energy, with a range of erosivity ranging between 27 and 35 MJ ha^{-1} mm^{-1}. Bonilla and Vidal (2011) and Santibañez et al. (2008) described this factor for other important areas of Chile (Fig. 4.39).

Annual precipitation is forecast to change by more than 30 % in some areas of the country by 2040, an amount that illustrates Chile's vulnerability under a future scenario with increased (double) atmospheric CO_2 levels. The central Chile, for example, may see a significant reduction in precipitation, while the Altiplano zone in the far north will experience higher precipitation levels due to tropical cyclone activity. Precipitation will decrease by about 20–25 % between Antofagasta (20 °S) and Puerto Montt (45 °S), but will increase from Chiloé Island to the south. As a consequence of these trends, aridity will increase in north and central Chile down to the Region VIII (CONAMA 2010).

The Chilean government has subsidised soil conservation activities through a major national programme (ISRD). This aimed to improve the productivity of Chilean soils, focusing on the restoration of degraded soils that cannot be used anymore in a sustainable and productive way. The first programme ran from 1999 to 2010, controlled by the Agriculture Ministry, and today a new programme (ISRD-S) is being introduced.

In the ISRD programme, all cash payments were awarded through an invitation to tender, covering about 50–80 % of the total costs of soil conservation practices, including agricultural inputs, labour and technical material. Six different subprogrammes were developed: regeneration of a permanent plant cover, crop rotations rehabilitation and soil conservation by physical structures. After 15 years, almost 31 million dollars were assigned to an annual average of 3,548 users and 102,000 ha.

At the end of ISRD (2010), from a territorial point of view a zone from Region VII to the extreme south concentrated 85 % of invested resources, and Region X alone received almost 32 % of the investment.

However, there is no legislation which regulates land use by land owners in Chile. The government has developed decrees to promote sustainable land use, but its application is not mandatory. In fact, direct legal aspects of soil degradation have in the past been generally neglected at the national level (Cavieres 2000). Today, it appears to be economically beneficial for some land users to exploit soils, transferring the capital loss to society. There is thus an urgent need to enact a Chile Soil Law as soon as possible, including the protection, conservation and recovery of these resources as main objectives. While such legislation is lacking, the State is not fulfilling its constitutional obligation to ensure the protection of one of fundamental natural resources of the nation (Escárate et al. 2005).

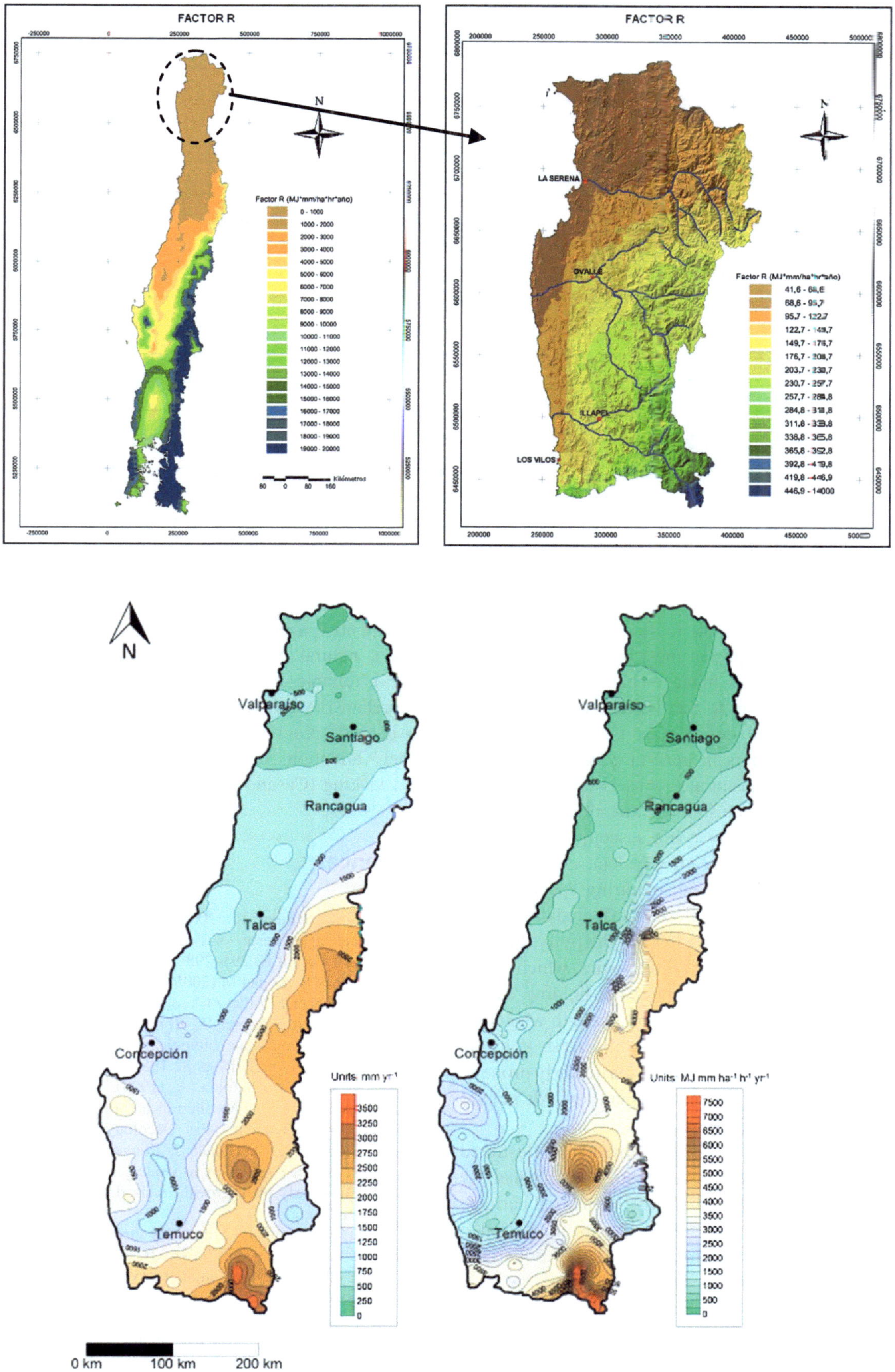

Fig. 4.39 Rainfall erosivity map in Chile (Santibáñez et al. 2008, *above*; Bonilla and Vidal 2011, *below*)

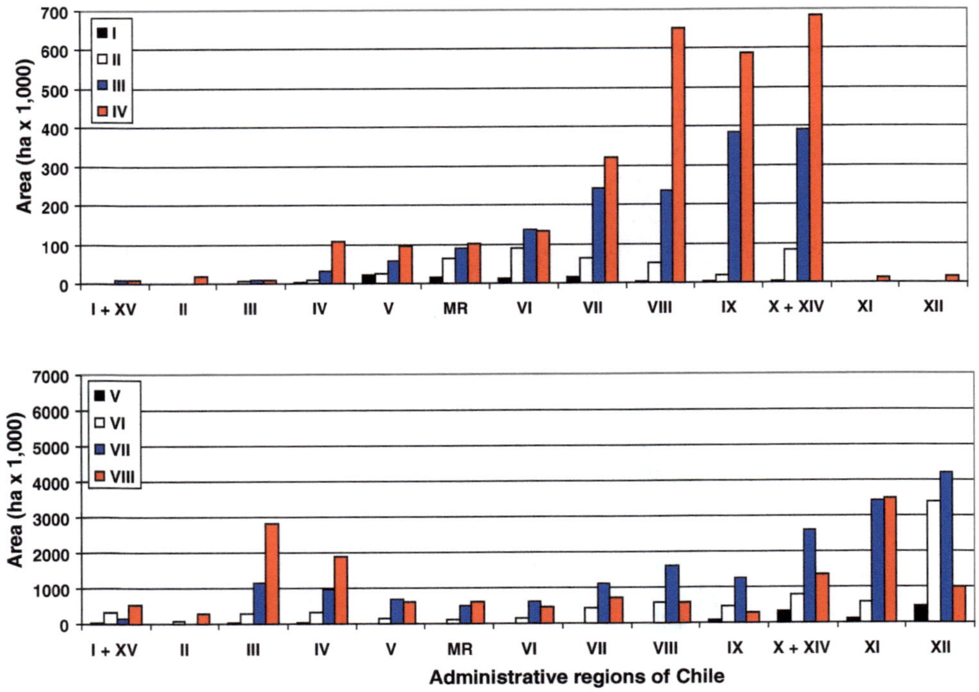

Fig. 4.40 Land capability classes of Chile (http://www.ine.cl, accessed 22 may 2011)

With human-induced soil degradation, soil resilience represents an expression of soil to resist or recover from such perturbation, a concept that must be considered when rejecting or stimulating soil management practices (Casanova 2000). The Hyper-arid to Semi-arid zone is known to be more sensitive to soil degradation than other climate zones of Chile and requires in general considerable management inputs and appropriate conservation practices. From the point of view of the soil scientist, different soils of central Chile have a widely varying susceptibility to degradation due to the vast range of physical, chemical and mineralogical properties involved, requiring specific measures against erosive and/or non-erosive degradation. Alfisols, for example, are subject to low productivity and soil degradation and have several characteristics which make management difficult, including low water-holding capacity, low aggregate stability and high soil strength when dry. In the southern part of the territory, while fresh volcanic deposits with low cohesiveness may be particularly prone to degradation, their progressive development towards mature Andisols generally shows more resilience.

The land capability classes (LCC) along Chile illustrated in Fig. 4.40 represent a trend from north to south of generally better soil quality. As mentioned above, this natural heritage of the country must be protected considering its integration to other natural resources, sustainable productive activities and the socio-economic particularities of stakeholders.

Finally, it is possible to think that it is high time for action, i.e. to pass from rhetoric to practice in soil conservation with sustainable soil use and management programs, including monitoring, impact evaluation, assessment experiments (participative, innovative and adapted to local reality), natural resources inventory (base lines and databases), reliable data generation and above all a solid and comprehensive training of all the actors involved. Although considerable progress in erosion modelling has been made, model validation for major soil zones and ecoregions is often lacking (Casanova et al. 2010b).

References

Abdelkdair A, Schultz R (2005) Water harvesting in a runoff-catchment agroforestry system in the dry lands of Ethiopia. Agroforest Forum 63:291–298

Alister C, Lopez R, Kogan M (2005) Simazine dynamics in a vineyard soil at Casablanca Valley, Chile. Pest Manag Sci 61:1083–1088

Alvear M, Rosas A, Rouanet JL, Borie F (2005) Effects of three soil tillage systems on some biological activities in an Ultisol from southern Chile. Soil Till Res 82:195–202

Aronson J, Rigel A, Shmida A (1993) Reproductive allocation strategies in desert and Mediterranean populations of annual plants grown with and without stress. Oecology 93:336–342

Aronson J, Ovalle C, Avendaño J, Longeri L, Del Pozo A (2002) Agroforestry tree selection in central Chile: biological nitrogen fixation and early plant growth in six dryland species. Agroforest Syst 56:155–166

Arroyo MTK, Pliscoff P, Mihoc M, Arroyo-Kalin M (2005) The Magellanic moorland. In: Fraser LH, Keddy PA (eds) The world's largest wetlands: ecology and conservation. Cambridge University Press, London, UK

Bachmann J, Contreras K, Hartge KH, MacDonald R (2006) Comparison of soil strength data obtained in situ with penetrometer and with vane shear test. Soil Till Res 87:112–118

Badilla-Ohlbaum R, Ginocchio R, Rodríguez P, Céspedes A, González S, Allen H, Lagos G (2001) Relationship between soil copper content and copper content of selected crop plants in central Chile. Environ Toxicol Chem 20:2749–2757

Baginsky C, Seguel O, Contreras A (2010) Impact of the utilization of crop and organic amendments in the soil functionality. University of Chile, Agronomic Science Series N° 17. Santiago, Chile (in Spanish)

Bahamondes M (2003) Poverty-environment patterns in a growing economy: farming communities in Arid Central Chile, 1991–99. World Dev 31(11):1947–1957

Barra R, Popp P, Quiroz R, Bauer C, Cid H, von Tümpling W (2005) Persistent toxic substances in soils and waters along an altitudinal gradient in the Laja River Basin, Central Southern Chile. Chemosphere 58:905–915

Barrow CJ (1991) Soil degradation: development and breakdown of terrestrial environments. Cambridge University Press, Cambridge

Bastías EI, González-Moro MB, González-Murúa C (2004) *Zea mays* L. *amylacea* from the Lluta Valley (Arica-Chile) tolerates salinity stress when high levels of boron are available. Plant Soil 267:73–84

Bauzá F (2009) Physical properties evaluation in a soil with agroforestry and water harvesting in the inner rainfed of the Central Zone of Chile. Dissertation, University of Chile (in Spanish with English summary)

Beekman GB (2007) Climate and national action programs in Latin America. In: Sivakumar MVN, Ndiang'ui N (eds) Climate and land degradation. Springer, Berlin

Benavides C (1992) Soil as a physic system. In: Vera W (ed) Soils, an actualised vision of the resource. University of Chile, Faculty of Agrarian and Forestry Sciences, Santiago, Chile (in Spanish)

Benavides C (2011) Tolerance to salinity in olives, physic conditions of the olive's soils and leaching soils in areas with salinity risk: basic considerations for the Atacama Region. In: Fichet T, Razeto B, Curkovic T (eds) The olive: agronomic study in the Atacama Region. University of Chile, Faculty of Agronomic Sciences, Santiago, Chile (in Spanish)

Bernier R, Alfaro M (2006) Soil acidity and liming effect. Agriculture and Livestock Research Institute (INIA), INIA Bull 151, Osorno, Chile (in Spanish)

Biester H, Kilian R, Franzen F, Woda C, Mangini A, Schöler HF (2002) Elevated mercury accumulation in a peat bog of the Magellanic Moorlands, Chile (53°S) an anthropogenic signal from the Southern Hemisphere. Earth Planet Sci Lett 201:609–620

Boix-Fayos C, Martínez-Mena M, Arnau-Rosalén E, Calvo-Cases A, Castillo V, Albaladejo J (2006) Measuring soil erosion by field plots: understanding the sources of variation. Earth Sci Rev 78:267–285

Boix-Fayos C, Martínez-Mena M, Calvo-Cases A, Arnau-Rosalén E, Albaladejo J, Castillo V (2007) Causes and underlying processes of measurement variability in field erosion plots in Mediterranean conditions. Earth Surf Process Landf 32:85–101

Bonilla C, Vidal KL (2011) Rainfall erosivity in Central Chile. J Hydrol 410:126–133

Bonilla C, Reyes JL, Magri A (2010) Water erosion prediction using the revised universal soil loss equation (RUSLE) in a GIS framework, central Chile. Chil J Agric Res 70:159–169

Botti C (2000) Horticultural trees for arid zones. University of Chile, Faculty of Agronomic Sciences, Department of Agriculture Production, Santiago, Chile (in Spanish)

Botti C, Palzkill D, Muñoz D, Prat L (1998a) Morphological and anatomical characterization of six jojoba clones at saline and non-saline sites. Ind Crop Prod 9:53–62

Botti C, Prat L, Palzkill D, Cánaves L (1998b) Evaluation of jojoba clones grown under water and salinity stresses in Chile. Ind Crop Prod 9:39–45

Brito J, Peña L (1980) Determination of water erosion prediction universal equation R factor in the province of Ñuble. Agric Tecnica (Chile) 40(2):152–155 (in Spanish with English summary)

Campillo R, Rodríguez J (1984) Acidifying effects of urea transformations of two Ancisols of the Lake Regions. Agric Tecnica (Chile) 44(2):131–133

Campillo R, Sadzawka MA (2006) Soil acidification. In: Campillo R (ed) Natural resources management in the incentives system to recovery degraded soils in the Araucanía. Series N°38. Agriculture and Livestock Research Institute, INIA, Regional Research Centre in Carillanca, Temuco, Chile (in Spanish)

Campillo R, Urquiaga S, Undurraga P, Pino I, Boddey RM (2005) Strategies to optimize biological nitrogen fixation in legume/-grass pastures in the southern region of Chile. Plant Soil 273: 57–67

Candan F, Broquen P (2009) Aggregate stability and related properties in NW Patagonian Andisols. Geoderma 154:42–47

Cartes E, Rubilar R, Mardónez O (2009) Comparison of water erosion on plantation and native forests in granitic soils of Bio–Bio's Region, Chile. XIII World Forestry Congress. 18–23 Oct Buenos Aires, Argentina (in Spanish)

Casanova M (2000) Stimulated and rejected managements in soil use. Chil Soc Soil Sci Bull 14:60–76 (in Spanish)

Casanova M, Leiva C, Seguel O, Salazar O, Luzio W (2006) Soil degradation and desertification. In: Luzio W, Casanova M (eds) Advances in knowledge of the Chilean soils. Santiago, Chile (in Spanish)

Casanova M, Haberland J, Seguel O, Luzio W (2007) Effect of gradient and slope aspect in the erosion of rainfed hillsides of central Chile. In: Proceeding XVII Latin-American congress of soil science (León-México) (in Spanish, with English summary)

Casanova M, Seguel O, Luzio W, Nario A, Videla X, Parada AM (2009) Assessment of soil degradation in hillsides. A contribution to a sustainable viticulture. University of Chile and Chilean Commission of Nuclear Energy. Santiago of Chile (in Spanish)

Casanova M, Dercon G, Zapata F (2010a) Soil erosion in Latin America and Caribbean. In: Casanova M, Ruiz G, Zagal E (eds) Proceeding XVI international soil conservation organization congress (ISCO). November, Santiago of Chile (in Spanish)

Casanova M, Dercon G, Zapata F (2010b) Soil erosion assessment: traditional methods. In: Casanova M, Ruiz G, Zagal E (eds) Proceeding XVI international soil conservation organization congress (ISCO). November, Santiago of Chile (in Spanish)

Casanova M, Seguel O, Salazar O, Homer I (2011) Demonstrative unit associated to 'Incentive system for agroenvironmental sustainability of agricultural and livestock soils'. University of Chile-Agriculture Ministery (Agriculture and Livestock Service, SAG)

Castro CP, Soto MV, Fernández R, Märker M, Rodolfi G (2009a) Impact of today geodynamic of Nantoco Valley, basin of Copiapó River, associated to productive reconversion. Rev Geogr Norte Gd 42:81–99 (in Spanish with English summary)

Castro M, Iturrieta R, Fassio C (2009b) Rootstock effect on the tolerance of cv. Hass avocado plants to NaCl stress. Chil J Agric Res 69:316–324

Cavieres A (2000) Criteria for development of a framework law for soil conservation. Chil Soc Soil Sci Bull 14:107–116 (in Spanish)

Cerda P, Jimenez K (2003) Relationship between soil loss, use and management on hillsides commune of La Cruz. Dissertation, University of Santiago, Chile (in Spanish)

Chartier MP, Rostagno CM (2006) Soil erosion thresholds and alternative states in northeastern Patagonian rangelands. Rangeland Ecol Manage 59:616–624

CIREN (2003) Soil survey of the Region X of Chile. Natural Resources Information Centre, Publication N° 123. Santiago, Chile (in Spanish)

CIREN (2007) Soil survey, Region III of Chile (Copiapó and Huasco valleys). Natural Resources Information Centre, Publication N°135. Santiago, Chile (in Spanish)

CONAMA (2010) National Climate Change Action Plan 2008–2012. National Environmental Commission (Chile) Department of Climate Change

Contreras H (1986) Water, life and development. Water use and conservation manual in Rural Zones of Latino America and Caribbean. UNESCO-ROSTLAC. Project DR-PRM. (in Spanish)

Contreras K (2006) Evaluation of the soil structural state using simple parameters: the stress at rest coefficient. Dissertation, Austral University of Chile (in Spanish with English summary)

Cortés A (2011) Physical stabilization of raised beds with avocado trees through organic amendments/cover crops (Region IV, Chile). Dissertation, University of Chile (in Spanish with English summary)

Cuevas J, Ellies A (2001) Pressure transmission and porosity changes generated by traffic in a Hapludand with different soil moisture conditions. R C Suelo Nutr Veg 1(2):48–57 (in Spanish with English summary)

Cuitiño H (1999) Qualitative assessment of water erosion on bare soils of pre-Andean and Central Valley of Region VII. Dissertation, University of Talca, Chile (in Spanish)

Daudin D, Sierra J (2008) Spatial and temporal variation of belowground N transfer from a leguminous tree to an associated grass in an agroforestry system. Agric Ecosyst Environ 126:275–280

De Gregori I, Fuentes E, Rojas M, Pinochet H, Potin-Gautier M (2003) Monitoring of copper, arsenic and antimony levels in agricultural soils impacted and non-impacted by mining activities, from three regions in Chile. J Environ Monit 5:287–295

Delatorre J, Pinto M (2009) Importance of ionic and osmotic components of salt stress on the germination of four quinua (*Chenopodium quinoa* Wild.) selections. Chil J Agric Res 69:477–485

Delatorre J, Arenas J, Campos H (1995) Morphologic comparison of nine quinua ecotypes collected in the Altiplano in the Iquique province. Rev Agric Desierto 1:5–14 (in Spanish)

Diamond J (2007) Easter Island revisited. Science 317:1692–1694

Díaz C (1949) Association of soils of the Easter Island. Agric Tecnica (Chile) 9:115–126 (in Spanish)

Dörner J, Dec D, Peng X, Horn R (2009) Change of shrinkage behavior of an Andisol in southern Chile: effects of land use and wetting/drying cycles. Soil Till Res 106:45–53

Dube F, Thevathasan NV, Zagal E, Gordon AM, Stolpe N, Espinosa M (2011) Carbon sequestration potential of silvopastoral and other land use systems in the Chilean Patagonia. In: Kumar BM, Nair PKR (eds) Carbon sequestration potential of agroforestry systems: opportunities and challenges, advances in agroforestry, p 8

Elizalde R (1970) The survival of Chile. Renewable natural resources conservation, 2nd edn. Chilean Ministry of Agriculture, Chile (in Spanish)

Ellies A (1978) Some aspects of hidrophobicity in soils derived from volcanic ashes. Chil Soc Soil Sci Bull 2(2):344–365 (in Spanish)

Ellies A (1986) Effects of the mechanic consolidation in some Distrandepts and Palehumults of South Chile. Cienc Suelo (Argentina) 2:91–97 (in Spanish)

Ellies A (1988) Mechanical consolidation in volcanic ash soils. In: Drescher J, Horn R, de Boodt M (eds) Impact of water and external forces on soil structure. Catena supplement 11. Cremlingen-Destedt, Germany

Ellies A (1999) Structural changes and stress distribution in soils under machinery traffic. Bosque 20:37–45 (in Spanish)

Ellies A (2000) Soil erosion and its control in Chile—an overview. Acta Geol Hispan 35:279–284

Ellies A (2001) Changes in soil physical properties by drainage. Chil Soc Soil Sci Bull 17:66–72 (in Spanish)

Ellies A, Grez G, Ramírez C (1995a) Changes in wetting properties of soils under different managements. Turrialba 45:42–48 (in Spanish)

Ellies A, Grez R, Ramírez C (1995b) Wettening potential and stability of aggregates in soils with different management. Agric Tecnica (Chile) 55:220–225 (in Spanish with English summary)

Ellies A, Smith R, Horn R (1996) Stress transmission into the profile of some soils. Agro Sur 24:149–158 (in Spanish)

Ellies A, Horn R, Smith R (2000) Effect of management of volcanic ash soil on structural properties. Int Agrophys 14:377–384

Escárate K, Casanova M, Luzio W (2005) Conflict between constitutional guarantees for the soil conservation: a critical review of Chilean legislation. R C Suelo Nutr Veg 5(2):1–8 (in Spanish with English summary)

Espinosa G, Hajek E, Fuentes E (1985) Geographic distribution of landslides associated to disasters in Chile. Amb Des 1(2):81–90 (in Spanish)

Farías E (2009) Changes in soil physical properties on hillsides cropped with vineyard. Dissertation, University of Chile (in Spanish with English summary)

Ferreyra R, Aljaro A, Ruiz R, Rojas L, Oster JD (1997) Behavior of 42 crop species grown in saline soils with high boron concentrations. Agric Water Manage 34:111–124

Ferreyra R, Sellés G, Pinto M, Morales M, Seguel O (2011) Effect of soil air capacity on water relations and vegetative growth of *Thompson Seedless* grafted on different rootstocks: preliminary results. Acta Hortic 889:145–150

Flores C, Morgante V, González M, Navia R, Seeger M (2009) Adsorption studies of the herbicide simazine in agricultural soils of the Aconcagua Valley, Central Chile. Chemosphere 74: 1544–1549

Flores JP, Martínez E, Espinosa M, Muñoz P (2011) Determination of actual and potential erosion Chile territory through geomatics techniques, remote sensing and GIS. Chil Soc Soil Sci Bull 24:21–28 (in Spanish)

Flores-Aqueveque V, Vargas G, Rutllant J, Le Roux J (2009) Seasonality of erosion and eolian particle transport in the coastal Atacama Desert, Chile (23°S). Andean Geol 36(2):288–310

Fuentes I (2010) Mechanical behavior of soil matrix and their individual aggregates depending on soil water tension in three soils of central Chile. Dissertation, University of Chile (in Spanish with English summary)

Fuentes JP, Varnero MT (2011) Carbon quality and biologic properties as soil perturbation indicators in the Mediterranean area of Chile. Chil Soc Soil Sci Bull 24:83–94 (in Spanish)

Fuentes I, Seguel O, Casanova M (2011) Soil physical degradation in the Central Zone versus volcanic soils from South Chile. Chil Soc Soil Sci Bull 24:75–81 (in Spanish)

García N, Ormazábal C (2008) Native trees of Chile. ENERSIS-Huinay Fundation, Chile (in Spanish)

Gayoso J, Alarcon D (1999) Forest soil conservation guide. Austral University of Chile, Valdivia

Ginocchio R (2000) Effects of a copper smelter on a grassland community in the Puchuncaví Valley, Chile. Chemos 41:15–23

Ginocchio R (2011) Consequences of mining activity in soils of Chile. Chil Soc Soil Sci Bull 24:97–105 (in Spanish)

Ginocchio R, Rodríguez P, Badilla-Ohlbaum R, Allen H, Lagos G (2002) Effect of soil copper content and pH on copper uptake of selected vegetables grown under controlled conditions Environ Toxicol Chem 21:1736–1744

Ginocchio R, Carvallo G, Toro I, Bustamante E, Silva Y, Sepúlveda N (2004) Micro-spatial variation of soil metal pollution and plant recruitment near a copper smelter in Central Chile. Environ Pollut 127:343–352

Ginocchio R, Sánchez P, de la Fuente L, Camus I, Bustamante E, Silva Y, Urrestarazu P, Torres J, Rodríguez P (2009) Agricultural soils spiked with copper mine wastes and copper concentrate: implications for copper bioavailability and bioaccumulation. Environ Toxicol Chem 25:712–718

González S (2000) Soil contamination: heavy metals. Chil Soc Soil Sci Bull 14:42–59 (in Spanish)

González S, Bergqvist E (1986) Heavy metal pollution in an area of the coastal dryland of the Region V. Agric Tecnica (Chile) 46(3):299–306 (in Spanish with English summary)

González S, Ite R (1992) Soil metal accumulation in the area under industrial chimney influence (Valparaíso Province, Region V). Agric Tecnica (Chile) 50:214–219 (in Spanish with English summary)

González S, Bergqvist E, Ite R (1984) Heavy metal pollution of an area near a copper smelter in Chile. Catemu, V Region. Agric Tecnica (Chile) 44(1):63–68 (in Spanish with English summary)

González I, Muena V, Cisternas M, Neaman A (2008) Copper accumulation in a plant community affected by mining contamination in Puchuncaví Valley, central Chile. Rev Chil Hist Nat 81:279–291

Goudie A (2008) The history and nature of wind erosion in deserts. Annu Rev Earth Planet Sci 36:97–119

Gualterio H (2006) Trends in land degradation in South America. In: Stefanski R, Pasteris P (eds) Management of natural and environmental resources for sustainable agricultural development. World Meteorological Organization, Workshop Proceedings. Portland, Oregon. (http://www.wmo.int/Templates/wmo.vcf)

Gut B (2008) Trees in Patagonia. Birkhäuser Verlag AG, Basel

Hahm J, Wester D (2004) Effects of surface-applied biosolids on grass seedling emergence in the Chihuahuan desert. J Arid Environ 58:19–42

Hardin G (1968) The tragedy of commons. Science 162:1243–1248

Heredia W, Peirano P, Borie G, Zunino H, Aguilera M (2007) Organic carbon balance in Chilean volcanic soils after human intrusion and under different management practices. Acta Agric Scand B-S P 57:329–334

Higueras P, Oyarzún R, Oyarzún J, Maturana H, Lillo J, Morata D (2004) Environmental assessment of copper-gold-mercury mining in the Andacollo and Punitaqui districts, northern Chile. Appl Geochem 19:1855–1864

Hillel D (2004) Environmental soil physics. Academic Press, San Diego

Holz A, Veblen T (2011) The amplifying effects of humans on fire regimes in temperate rainforests in western Patagonia. Palaeogeogr Palaeoclimatol Palaeoecol 311(1–2):82–92

Homer I, Casanova M (2011) Conservation tillage on hillsides. Chil Soc Soil Sci Bull 24:59–73 (in Spanish)

Honorato R, Cruz G (1999) Erodability and soil erosion of Easter Island. Cien Invest Agric (Chile) 26:27–36 (in Spanish with English summary)

Honorato R, Barrales L, Peña I, Barrera F (2001) USLE erosion model evaluation on six locations between IV and IX Region of Chile. Cien Invest Agric 28:7–14 (in Spanish with English summary)

Hooda PS (2010) Trace elements in soils. Wiley, Chichester

Horn R (2003) Stress-strain effects in structured unsaturated soils on coupled mechanical and hydraulic processes. Geoderma 116:77–88

Hugo G (2008) Trends in land degradation in South America. In: Stefanski R, Pasteris P (eds) Management of natural and environmental resources for sustainable agricultural development. World Meteorological Organization, Workshop Proceedings. Portland, Oregon. (http://www.wmo.int/Templates/wmo.vcf)

Inostroza-Blancheteau C, Soto B, Ulloa P, Aquea F, Reyes-Díaz M (2008) Resistance mechanisms of aluminum (Al^{3+}) phytotoxicity in cereals: physiological, genetic and molecular bases. J Soil Sci Plant Nutr 8:57–71

IREN (1966) Dunes inventory of Chile (29°48′–4°50′S). National Research of Natural Resources Institute and Agriculture Ministry, Department of Soil and Water Conservation, Santiago

IREN-CORFO (1979) Fragility of natural ecosystems of Chile. National Research of Natural Resources Institute- Development Corporation. Santiago

Joel A, Messing I, Seguel O, Casanova M (2002) Measurement of surface water runoff from plots with two different sizes. Hydrol Process 14(7):1467–1478

Kabata-Pendias A (2011) Trace elements in soils and plants, 4th edn. CRC Press, Boca Raton

Kalthoff N, Fiebig-Wittmaack M, Meißner C, Kohlera M, Uriarte, Bischoff-Gau BI, Gonzáles E (2006) The energy balance, evapotranspiration and nocturnal dew deposition of an arid valley in the Andes. J Arid Environ 55:420–443

Keller K (2011) Evaluation of the water dynamics in soils with organic manure applications under drip irrigation. Dissertation, University of Chile (in Spanish with English summary)

Kelm U, Helle S, Matthies R, Morales A (2009) Distribution of trace elements in soils surrounding the El Teniente porphyry copper deposit, Chile: the influence of smelter emissions and a tailings deposit. Environ Geol 57:365–376

Klein D, Fuentes JP, Schmidt A, Schmidt H, Schulte A (2008) Soil organic C as affected by silvicultural and exploitative interventions in *Nothofagus pumilio* forests of the Chilean Patagonia. Forest Ecol Manag 255:3549–3555

Lagos M (2005) Protocol for selection of alternatives for soil conservation. Technical Series N° 18. Agriculture and Livestock Service, SAG. Santiago, Chile (in Spanish)

Lagos E (2006) Characterization of precipitation-runoff process using a rain simulator. Dissertation, University of Talca, Chile (in Spanish with English summary)

Lavín A (1988) Boron toxicity induced by ulexite applications in some drip irrigated fruit crops at Cauquenes, Chile. Agric Tecnica (Chile) 48(2):167–170 (in Spanish with English summary)

Le Quesne C, Acuña C, Boninsegna JA, Rivera A, Barichivich J (2009) Long-term glacier variations in the Central Andes of Argentina and Chile, inferred from historical records and tree-ring reconstructed precipitation. Palaeogeogr Palaeoclimatol Palaeoecol 281:334–344

Leiva C (2005) Soil assessment at seven years of agroforestry and water harvesting management in the inner rainfed of central Chile. Dissertation, University of Chile (in Spanish with English summary)

Lemus M, Navarro G (2003) Manual for development of soil conservation works. Diversion channel—infiltration trench system. National Forestry Corporation, Chile Government (in Spanish)

León O, Carrasco MA (2011) Soil chemical degradation. Chil Soc Soil Sci Bull 24:115–122 (in Spanish)

Louwagie G, Stevenson CM, Langohr R (2006) The impact of moderate to marginal land suitability on prehistoric agricultural production and models of adaptive strategies for Easter Island (Rapa Nui, Chile). J Anthropol Archaeol 25:290–317

Luzio W, Casanova M, Seguel O (2010) Soils of Chile. In: Luzio W (ed) University of Chile. Maval Press, Santiago (in Spanish)

Macaya C, Gallardo M (2007) Rehabilitation manual of soils subjected to extraction of subsoil in agricultural areas. Report, Agriculture and Livestock Service (SAG, Chile) (in Spanish)

Martínez EA, Veas E, Jorquera C, San Martín R, Jara P (2009) Re-introduction of quinoa into Arid Chile: cultivation of two lowland races under extremely low irrigation. J Agron Crop Sci 195:1–10

Martínez-Harms MJ, Gajardo R (2008) Ecosystem value in the Western Patagonia protected areas. J Nat Conserv 16:72–87

Mathieu R, Cervelle B, Remy D, Pouget M (2007) Field-based and spectral indicators for soil erosion mapping in semi-arid mediterranean environments (Coastal Cordillera of central Chile). Earth Surf Proc Land 32:13–31

Mazzonia E, Vasquez M (2010) Desertification in Patagonia. In: Latrubesse EM (ed) Natural hazards and human-exacerbated disasters in Latin America. Developments in Earth Surface Processes, vol 13. Elsevier, UK

McBride MB (1994) Environmental chemistry of soils. Oxford University Press, New York

Mcintyre B, Riha S, Ong C (1997) Competition for water in a hedge-intercrop systems. Field Crop Res 52:151–160

Mieth A, Bork HR (2005) History, origin and extent of soil erosion on Easter Island (Rapa Nui). Catena 63:244–260

MINVU (2010) Soil consume in cities and regions. Estimations from building statistics and satellite images. Observatories of the commission on housing and urban studies. Ministry of Building and Urbanism, Chile (in Spanish)

Molina M, Aburto F, Calderón R, Cazanga M, Escudey M (2009) Trace element composition of selected fertilizers used in Chile: phosphorus fertilisers as a source of long-term soil contamination. Soil Sediment Contam 18:497–511

Montambault J, Alavalapati J (2005) Socioeconomic research in agroforestry: a decade in review. Agroforest Syst 65:151–161

Montenegro G, Fredes C, Mejías E, Bonomelli C, Olivares L (2009) Content of heavy metals in soils near a Chilean copper mining tailing. Agrociencia (Mexico) 43:427–435

Moreno R (2012) Seasonal assessment of soil water content and penetration resistance in an agroforestry system associated to water harvesting in the inner rainfed of central Chile. Dissertation, University of Chile (in Spanish with English summary)

Muena V, González I, Neaman A (2010) Effects of liming and nitrogen fertilization on the development of Oenothera affinis in a soil affected by copper mining. J Soil Sci Plant Nutr 10:102–114 (in Spanish with English summary)

Muñoz C, Ovalle C, Zagal E (2007a) Distribution of soil organic carbon stock in an Alfisol profile in Mediterranean Chilean ecosystems. J Soil Sci Plant Nutr 7:15–27 (in Spanish with English summary)

Muñoz C, Ovalle C, Zagal E (2007b) Influence of trees on soil organic matter in Mediterranean agroforestry systems: an example from the 'Espinal' of central Chile. Eur J Soil Sci 58:728–735

Muñoz C, Monreal CM, Schnitzer M, Zagal E (2008) Influence of *Acacia caven* (Mol) coverage on carbon distribution and its chemical composition in soil organic carbon fractions in a Mediterranean-type climate region. Geoderma 144:352–360

Nario A, Capri E, Balderacchi M, Pino I, Parada AM, Videla X, Luzio W, Casanova M, Seguel O (2009) Pesticide risk management using indicators for vineyards in the central valley of Chile. Integr Environ Assess Manag 5:476–482

Neaman A, Reyes L, Trolard F, Bourrié G, Sauvé S (2009) Copper mobility in contaminated soils of the Puchuncaví Valley, central Chile. Geoderma 150:359–366

Nissen J, Quiroz C, Seguel O, Mac Donald R, Ellies A (2006) Non-saturated water movement in Andisols. J Soil Sci Plant Nutr 6:9–19 (in Spanish with English summary)

Olivares A, Gastó J (1981) *Atriplex repanda*, organization and management of ecosystems with bushes for forage. Agriculture Sciences N°7, University of Chile, Santiago, Chile (in Spanish)

Orellana I (2010) Comparison of two methods to measure the wetting capacity of soils. Dissertation, Austral University of Chile (in Spanish with English summary)

Ormazábal CS (1991) Silvopastoral systems in arid and semi-arid zones of northern Chile. Agroforest Syst 14:207–217

Osorio A, Césped R (2000) Effect of micro-irrigation methods on the salinity of the soil profile in a grape vineyard. I. Electrical conductivity of sodium, chloride and boron at several points of the profile. Chil J Agric Res 60:178–194, 42 (in Spanish with English summary)

Ovalle C, Squella F (1988) Pasture lands with annual prairie in the area with Mediterranean climate influence. In: Ruiz I (ed) Prairies for Chile. Agriculture and Livestock Research Institute, INIA, Santiago, Chile (in Spanish)

Ovalle C, Aronson J, del Pozo A, Avendaño J (1990) The espinal: agroforestry systems of the Mediterranean-type climate region of Chile. Agroforest Syst 10:213–239

Ovalle C, Aronson J, del Pozo A, Avendaño J (1999) Restoration and rehabilitation of mixed espinales in central Chile: 10 year report and appraisal. Arid Soil Res Rehabil 13:369–381

Oyarzún C (1993) Evaluation of the USLE for the prediction of soil erosion in forested areas of the Bio–Bio river basin. Bosque 14(1):45–54

Oyarzún R, Arumí JL, Alvarez P, Rivera D (2008) Water use in the Chilean agriculture: current situation and areas for research development. In: Sorensen ML (ed) Agricultural water management research trends. Nova Publishers, New York

Paruelo JM, Jobbágy EG, Oesterheld M, Golluscio RA, Aguiar RA (2007) The grasslands and steppes of Patagonia and the Río de la Plata Plains. In: Veblen TT, Young KR, Orme AR (eds) The physical geography of South America. Oxford University Press Inc, Oxford

Peña L (1983) Determination of R, K and C factors of the universal equation of estimation erosion for pre Andean Region VII. Agric Tecnica (Chile) 43(2):151–158 (in Spanish with English summary)

Peña L (1985) Water erosion in hilly trumaos. In: Tosso J (ed) Volcanic soils of Chile. Agriculture Ministry, Agriculture and Livestock Research Institute (INIA) Santiago, Chile (in Spanish)

Peña L (1992) Erosion and soil conservation. In: Vera W (ed) Soils, an actualised vision of the resource. University of Chile, Faculty of Agrarian and Forestry Sciences. Santiago, Chile (in Spanish)

Peña-Cortés F, Ailio C, Gutiérrez P, Escalona-Ulloa M, Rebolledo G, Pincheira-Ulbrich J, Rozas D, Hauenstein E (2008) Dunes morphology and dynamic in coastal ridge of Araucania Region in Chile. Antecedents for land conservation and land management. Rev Geogr Norte Gd 41:63–80 (in Spanish with English summary)

Peralta M (1976) Use classification and conservation of soils. Ministry of Agriculture, Chile (in Spanish)

Pérez C, González J (2001) Diagnosis on the state of soil resource degradation in the country. CRI-Quilamapu, Agricultural and Livestock Research Institute (INIA), Chillán–Chile (in Spanish)

Pérez-Quezada JF, Delpiano CA, Snyder KA, Johnson DA, Franck N (2011) Carbon pools in an arid shrubland in Chile under natural and afforested conditions. J Arid Environ 75:29–37

Pfeiffer M, Mascayano C, Aburto F (2010) Soils of Chilean Patagonia in glacial and periglacial environments. Eurasian Soil Sc 43(13):1–9

Pizarro R, Morales C, Vega L, Balocchi F (2009) Design manual for works of recovery and conservation of water and soil. University of Talca, INNOVA-Chile (in Spanish)

Pizarro J, Vergara P, Rodríguez J, Valenzuela A (2010) Heavy metals in northern Chilean rivers: spatial variation and temporal trends. J Hazard Mater 181:747–754

Poblete L (2011) Modification of soil physical properties of a fine textured soil from Region III of Chile cultivated with grape through the use of organic matter on the row-planting. Dissertation, University of Chile (in Spanish with English summary)

Ramírez C, Ellies A, Mac Donald R, Seguel O (2003) Floristic and organic matter changes from native forest to anthropogenic prairies in volcanic soils (Region IX, Chile). J Soil Sci Plant Nutr 3(1):1–12

Rodríguez N (2011) Conditioning for the recultivation of a sandy loam soil disturbed by sand extraction. Dissertation, University of Chile (in Spanish with English summary)

Rodríguez M, Suárez J (1946) The conservation of soil of Chile. General Direction of Agriculture, Chile (in Spanish)

Rodríguez N, Ruz E, Valenzuela A, Belmar C (2000) Effect of tillage on soil erosion losses with a wheat-oat rotation and pasturing in the Andean foothills of the south-central region of Chile. Agric Tecnica (Chile) 60(3):259–269 (in Spanish with English summary)

Romero H, Órdenes F (2004) Emerging Urbanization in the Southern Andes. Environmental impacts of urban sprawls in Santiago of Chile on the Andean piedmont. Mt Res Dev 24(3):197–201

Romo-Kröger CM, Morales JR, Dinator MI, Llona F, Eaton LC (1994) Heavy metals in the atmosphere coming from a copper smelter in Chile. Atmos Environ 28:705–711

Ruiz G (2005) Manual of good soil management practices on hillsides, technical specifications. Technical series N° 20. Chilean Ministry of Agriculture, Agriculture and Livestock Service, SAG. Santiago, Chile (in Spanish)

Ruiz G (2011) System of incentives for agri-environmental sustainability of agricultural soils. Chil Soc Soil Sci Bull 24:31–36 (in Spanish)

Sadzawka MA (2006) Physical-chemical soil properties in the soils I: reaction (pH), acidity and alkalinity. In: Luzio W, Casanova M (eds) Advances in the knowledge of the Chilean soils. University of Chile—Chilean Ministry of Agriculture, Agriculture and Livestock Service (SAG), Maval Press, Santiago, Chile (in Spanish)

Sagardía S, Seguel O, Casanova M (2008) Effect of exudations from *Chaitophorus leucomelas* Koch, parasite of *Populus* sp., on the soil hydraulic properties. In: IV Chilean Congress of Forest Sciences. Talca, Chile, 1–3 Oct, (in Spanish)

Salazar O (2003) Chemical assessment of a soil under agroforestry associated to water harvesting essay in the inner rainfed of central Chile. Dissertation, University of Chile (in Spanish with English summary)

Salazar O, Casanova M (2011) Runoff water harvesting as a strategy for increasing agricultural production on hillslope areas in arid and semi-arid zones. In: Carrey DM (ed) Water recycling and water management. Nova Science Publishers, New York

Salazar O, Casanova M, Benavides C, Luzio W, Vera W (2006) Chemical properties of a soil under agroforestry and water harvesting in the inner rainfed zone of central Chile. Terra 24:471–476 (in Spanish with English summary)

Salazar O, Casanova M, Kätterer T (2011) The impact of agroforestry combined with water harvesting on soil carbon and nitrogen stocks in central Chile evaluated using the ICBM/N model. Agric Ecosyst Environ 140:123–136

Sandoval M, Castillo C, Zagal E, Stolpe N (2007) Hydraulic parameters determined in Andisol under different crop rotations after ten years. R C Suelo Nutr Veg 7(2):32–45 (in Spanish with English summary)

Sangüesa C, Arumí J, Pizarro R, Link O (2010) A rainfall simulator for the in situ study of superficial runoff and soil erosion. Chil J Agric Res 70(1):178–182

Santibáñez F, Santibáñez P, Hernández J, Montaner D (2008) Vulnerability analysis of forestry and agricultural sector, water and soil of Chile against climate change scenarios. AGRIMED-University of Chile

Schulz J, Cayuela L, Echeverria C, Salas J, Rey JM (2010) Monitoring land cover change of the dryland forest landscape of Central Chile. Appl Geogr 30:436–447

Schüller P, Ellies A, Kirchner G (1997) Vertical migration of fallout in agricultural soils from Southern Chile. Sci Total Environ 193:197–205

Schüller P, Sepúlveda A, Trumper R, Castillo A (2000) Application of the ^{137}Cs technique to quantify soil redistribution in Paleohumults from Central-South Chile. Acta Geol Hispan 35(3–4):285–289

Schüller P, Ellies A, Castillo A, Salazar I (2003) Use of the 137Cs technique to estimate tillage- and water-induced soil redistribution rates on agricultural land under different use and management in central-south Chile. Soil Till Res 69:45–51

Schüller P, Bunzl K, Voigt G, Ellies A, Castillo A (2004a) Global fallout ^{137}Cs accumulation and vertical migration in selected soils from South Patagonia. J Environ Radioact 71(43–60):44

Schüller P, Walling D, Sepúlveda A, Trumper R, Rouanet JL, Pino I, Castillo A (2004b) Use of Cs-137 measurements to estimate changes in soil erosion rates associated with changes in soil management practices on cultivated land. Appl Radiat Isot 60:759–766

Schüller P, Iroumé A, Walling D, Mancilla H, Castillo A, Trumper R (2006) Use of ^{7}Be to document soil redistribution following forest harvest operations. J Environ Qual 35:1756–1763

Schüller P, Walling D, Sepúlveda A, Castillo A, Pino I (2007) Changes in soil erosion associated with the shift from conventional tillage to a no-tillage system, documented using Cs-137 measurements. Soil Till Res 94(1):183–192

Schüller P, Iroumé A, Walling D, Castillo A (2010) Use of beryllium-7 to study the effectiveness of woody trash barriers in reducing sediment delivery to streams after forest clearcutting. Soil Till Res 110:143–153

Seguel O, Farías E, Luzio W, Casanova M, Pino I, Parada MA, Videla X, Nario A (2009) Changes in soil physical properties on hillsides vineyard (*Vitis vinifera*). ISTRO 18th Triennial conference. Izmir, Turkey, 15–19 June

Seguel O, Baginsky C, Contreras A, Covarrubias J, González C, Sandoval M (2011) Changes in physical properties of a Typic Haplocambid by annual crop culture. J Soil Sci Plant Nutr 11(1):1–15

Sepúlveda A, Schüller P, Walling D (2008) Use of ^{7}Be to document soil erosion associated with a short period of extreme rainfall. J Environ Radioact 99:35–49

Sierra C, Césped R, Osorio A (2001) Characterization of soil and water salinity in the river in the Copiapó Valley. Agriculture and Livestock Research Institute (INIA), INIA Bull 70, Chile (in Spanish)

Silva M, Lozano U (1986) Description of the main pasture species between arid Mediterranean area and the rainy area. University of Chile. Faculty of Agrarian and Forest Sciences, Department of Animal Production. Educational Publication N° 9. Santiago, Chile (in Spanish)

Silva H, Sagardía S, Seguel O, Torres C, Tapia C, Franck N, Cardemil L (2010) Effect of water availability on growth and water use efficiency for biomass and gel production in Aloe Vera (*Aloe barbadensis* M.). Ind Crop Prod 31:20–27

Sotomayor EM, De la Riva FP, Leiva AJ (1994) Olive adaptation to the edaphoclimatic conditions in the Lluta Valley, Arica-Chile. Acta Hortic 356:189–192

Stocking MA, Murnaghan N (2001) Handbook for the field assessment of land degradation. Earthscan, London

Stolpe N (2005) A comparison of the RUSLE, EPIC and WEPP erosion models as calibrated to climate and soil of south-central Chile. Acta Agric Scand B 55:2–8

Stolpe N, Muñoz C, Zagal E, Ovalle C (2008) Modeling soil carbon storage in the 'Espinal' agroecosystem of central Chile. Arid Land Res Manag 22:148–158

Tchernitchin A, Lapin N, Molina L, Molina G, Tchernitchin N, Acevedo C, Alonso P (2006) Human exposure to lead in Chile. Rev Environ Contam Toxicol 185:93–139

Torres A, Acevedo E (2008) The salinity problem of the water and soil resources in Lluta and Azapa Valleys in northern Chile. Idesia 26:31–44 (in Spanish with English summary)

Traub M (2010) Use of an organic emulsion as an erosion control management in cultivated hillside with vineyards, Region VI of Chile. Dissertation University of Chile (in Spanish with English summary)

Undurraga P, Zagal E, Sepúlveda G, Valderrama N (2009) Dissolved organic carbon and nitrogen in Andisol for six crop rotations with different soil management intensity. Chil J Agric Res 69: 445–454

UNEP (2007) Global environment outlook-4. (Geo-4), United Nations Environment Programme, Nairobi

Valenzuela J, Narvaez C (1983) Some factors associated with boron excess in the Elqui Valley vineyards. Agric Tecnica (Chile) 43(2):145–149 (in Spanish with English summary)

Vargas R, Tokugawa K, Francke S, Makita M (1998) Erosion control manual. Agriculture Ministry of Chile (CONAF) and Japanese International Cooperation Agency (JICA)

Varnero MT, Rojas C, Macaya C, Orellana R (2005) Effects of grass carpet production on agriculture soils. Chil Soc Soil Sci Bull 21:84 (in Spanish)

Verbist K (2011) Climatic and soil physical constraints for efficient rain water harvesting in degraded lands of Chile. Dissertation, University of Ghent, Belgium

Verbist K, CornelisWM, Gabriels D, Alaerts K, Soto G (2009) Using an inverse modelling approach to evaluate the water retention in a simple water harvesting technique. Hydrol Earth Syst Sci 13:1979–1992

Villarroel R (2012) Chemical properties of a soil from Chilean central rainfed under an agroforestry system associated to water harvesting, 12 years old. Dissertation, University of Chile (in Spanish with English summary)

Vött W, Endlicher A (2001) Landscape degradation in the Region XI of Chile within the framework of growing environmental problems in Western Patagonia—a geological and geostatistical approach. Die Erde 132:239–268

Wischmeier WH, Smith DD (1978) Predicting rainfall erosion losses—a guide to conservation planning. Agriculture handbook N° 537. U.S. Department of Agriculture, Washington D.C., USA

Youlton C, Espejo P, Biggs J, Norambuena M, Cisternas M, Neaman A, Salgado E (2010) Quantification and control of runoff and soil erosion on avocado orchards on ridges along steep-hillslopes. Cien Invest Agric 37(3):113–123

Young G, Zavala H, Wandel J, Smit B, Salas S, Jimenez E, Fiebig M, Espinoza R, Diaz H, Cepeda J (2010) Vulnerability and adaptation in a dryland community of the Elqui Valley, Chile. Clim Chang 98:245–276

Zapata F (2003) Handbook for the assessment of soil erosion and sedimentation Using environmental radionuclides. Kluwer Academic Publishers, Dordrecht

Zapata F, Dercon G, Casanova M (2010) Soil erosion assessment: isotopic methods. In: Casanova M, Ruiz G, Zagal E (eds) Proceeding XVI international soil conservation organization congress (ISCO), November, Santiago of Chile

Appendix

This appendix includes physical, chemical and physical–chemical soil characteristics of representative pedons from the four major soil zones (and subdivisions) defined in Table 2.2, Fig. 2.1 and the text of Chap. 2. Some abbreviations used in following tables are:

Soil property/ characteristic	Meaning
Genetic horizon	According to Soil Taxonomy system
2–0.05 (%)	Sand content
0.05–0.002 (%)	Silt content
<0.002 (%)	Clay content
OC (%)	Organic carbon content
SOM (%)	Soil organic matter
pH_{water}	Soil pH measured in water
pH_{KCl}	Soil pH measured in KCl
EC (dS m^{-1})	Soil electrical conductivity at 25 °C
EC_{iw} (dS m^{-1})	Irrigation water electrical conductivity
ESP (%)	Exchangeable sodium percent
SAR (–)	Sodium adsorption ratio
CEC ($cmol_c kg^{-1}$)	Cation exchange capacity
ECEC ($cmol_c kg^{-1}$)	Effective cation exchange capacity
BS (%)	Base saturation, at pH 7
Al_{ox} + ½Fe_{ox} (%)	Acid oxalate extractable aluminium and iron
XRD results:	AT: anorthite, AH: anhydrite, C: calcite, CH: chlorite, G: gypsum, H: halite, K: kaolinite, M: mica, N: nitratite, Q: quartz, S: smectite, listed from highest to lowest amount

2.2.1 Soils of the Hyper-Arid to Semi-Arid Soils

2.2.1.1 Soils of the Altiplano

Skeletal soil in a site between San Pedro de Atacama and El Tatío, Region II of Chile (Luzio et al. 2010)

Depth (cm)	0–20	20–34	34–69	69–105
Genetic horizon	A	C_1	C_2	C_3
Particle size (mm) distribution				
2–0.05 (%)	70.8	54.8	80.0	82.8
0.05–0.002 (%)	15.2	23.2	9.2	9.2
<0.002 (%)	14.0	22.0	10.0	8.0
OC (%)	0.29	0.41	0.05	0.06
pH_{water}	6.50	6.60	6.60	7.20
EC (dS m^{-1})	0.18	10.00	0.14	0.14
Extractable cations ($cmol_c kg^{-1}$)				
Ca	3.6	4.7	2.2	2.7
Mg	1.2	1.8	0.8	0.9
K	0.7	1.1	0.5	1.1
Na	0.4	0.3	0.3	2.2
CEC ($cmol_c kg^{-1}$)	7.0	9.2	4.5	4.9
BS (%)	–	100	–	–

Soil derived from recent volcanic materials, located between Putre and Parinacota, Region XV of Chile

Depth (cm)	0–8	8–15	15–21	21–115
Genetic horizon	A_{11}	A_{12}	A_{13}	C
Particle size (mm) distribution				
2–0.05 (%)	56.0	52.0	47.2	60.0
0.05–0.002 (%)	24.0	20.0	24.8	22.0
<0.002 (%)	20.0	28.0	28.0	18.0
OC (%)	0.28	0.61	0.48	0.08
pH_{water}	7.00	7.20	7.30	7.10
EC (dS m^{-1})	0.32	0.24	0.31	0.31
Extractable cations ($cmol_c kg^{-1}$)				
Ca	3.6	7.5	10.4	8.6

(continued)

M. Casanova et al., *The Soils of Chile*, World Soils Book Series, DOI: 10.1007/978-94-007-5949-7

(continued)

Mg	1.9	3.3	4.4	3.6
K	0.5	1.0	1.2	1.0
Na	0.3	0.4	0.5	0.6
CEC ($cmol_c kg^{-1}$)	8.8	15.6	20.0	13.7

Chirigualla soil (Ustorthent to Haplocambid), Regio XV of Chile (Luzio et al. 2010)

Depth (cm)	0–22	22–43	43–70	70–86
Genetic horizon	A_1	B_w	BC	C
Particle size (mm) distribution				
2–0.05 (%)	60.0	47.0	36.7	42.1
<0.05 (%)	40.0	53.0	63.3	57.9
Water retention −33 kPa (%)	17.7	30.9	31.9	35.1
Water retention −1,500 kPa (%)	10.4	17.5	19.2	21.6
OC (%)	1.3	0.4	0.2	0.1
pH_{water}	6.9	7.2	7.5	7.9
EC ($dS\ m^{-1}$)	0.43	0.47	0.72	0.50
Extractable cations ($cmol_c kg^{-1}$)				
Ca	7.0	15.2	17.7	21.8
Mg	2.8	8.4	8.7	11.7
K	1.0	1.1	1.2	1.1
Na	1.3	2.3	1.8	1.6
CEC ($cmol_c kg^{-1}$)	12.9	24.2	32.9	39.7
ESP (%)	1.00	9.5	5.5	4.0
BS (%)	94	100	90	91
P retention (%)	21.7	25.3	20.4	16.2
Al_{ox} + ½ Fe_{ox} (%)	1.80	0.55	0.43	0.18
Volcanic glass: 0.02–2 mm (%)	44	48	45	39

Pedon at Surire bofedal, Region XV of Chile (Luzio et al. 2002a)

Depth (cm)	0–20	20–29	29–51	51–66	66–90
Genetic horizon	O_{ek1}	O_{ek2}	$2C_{kg1}$	$2C_{kg2}$	$2C_g$
Particle size (mm) distribution					
2–0.05 (%)	14.2	9.4	17.4	24.4	29.8
0.05–0.002 (%)	72.6	67.3	57.0	61.1	48.4
<0.002 (%)	13.2	23.3	25.7	14.5	21.2
OC (%)	8.8	13.8	4.1	2.6	2.0
pH_{water}	8.1	7.9	7.8	8.0	4.9
pH_{KCl}	7.3	7.2	7.1	7.2	4.5
EC ($dS\ m^{-1}$)	16.0	3.9	3.0	2.6	4.0
$CaCO_3$ (%)	38.5	29.6	22.6	45.8	0.0
$CaSO_4$ (%)	–	–	–	–	1.2
Extractable cations ($cmol_c kg^{-1}$)					
Ca	105	65.1	57.4	45.7	9.7
Mg	10.0	9.6	3.8	2.4	8.9
K	262.0	2.7	1.1	0.83	0.04
Na	59.1	18.2	4.1	2.2	2.6
CEC ($cmol_c kg^{-1}$)	37.6	47.2	18.9	8.1	9.8
BS (%)	99	98	–	–	–

Pedon at Caquena bofedal, Region XV of Chile (Luzio et al. 2002a)

Depth (cm)	0–13	13–24	24–36	36–49
Genetic horizon	O_{i1}	O_{i2}	O_{e1}	O_{e2}
Particle size (mm) distribution				
2–0.05 (%)	28.3	27.3	34.3	11.1
0.05–0.002 (%)	43.7	43.6	47.0	47.1
<0.002 (%)	28.0	29.1	18.7	41.7
OC (%)	39.2	41.6	36.3	33.4
pH_{water}	8.5	7.7	7.6	6.7
pH_{KCl}	7.9	7.1	7.1	6.3
EC ($dS\ m^{-1}$)	2.9	2.3	2.4	2.0
$CaSO_4$ (%)	0.7	0.3	–	–
Extractable cations ($cmol_c kg^{-1}$)				
Ca	56.0	52.4	61.3	32.6
Mg	57.2	33.6	35.2	19.8
K	150.0	2.9	2.9	2.5
Na	26.3	14.6	14.3	14.0
CEC ($cmol_c kg^{-1}$)	110.0	96.7	101.0	87.2
BS (%)	99	98	–	–

2.2.1.2 Soils of the Longitudinal Central Valley

Characteristics and soil profile descriptions for three Aridisols from the Atacama Desert (summarised from Ewing et al. 2006)

Pedon[1]	Depth (cm)	Bulk density(Mg m^{-3})	Texture class (clay, %)	Gravel (% by vol.)	XRD <2 mm fraction	pH	Soil structure
A_1	0–4	1.5	lfs (1)	5	Q, AT	7.9	
A_2	4–15	1.9	lfs (1)	5	Q, AT	8.0	Moderate prisms, 10 cm
B_{wk1}	15–28	1.9	lcos (5)	2	AT, Q, C	7.8	Strong plates, 5–10 mm; strong, coarse subangular blocks, (5 %)
B_{wk2}	28–46	1.9	vgrls (2)	50	AT, Q, C	8.2	Massive (loose)
B_{km}	46–107	2.3	vgrls (2)	70	AT, Q, C	8.3	Massive (loose)
BC_{yk}	107–117	2	lvcos	50			Platey
BC_k	117–133	2	lvcos	45 2–4 mm			Massive
BC_{kym}	133–145	2	lvcos	75 2–10 cm			Small (1–2 mm) plates
BC_{ky1}	145–160	2	vcos	60 2–10 cm			Massive
BC_{ky2}	160–174	2	lvcos	50			Massive
BC_{ykm}	174–184	2	lcos	10			Small (1–2 mm) plates
BC_{yk}'	184–195	2	vcos	40			Massive
BC_{kym}	195–204	2	vcos	30			Plates
C_{yk}"	204–	2	vcos	25 all small			Plates
Pedon[2]	Depth (cm)	Bulk density (Mg m^{-3})	Texture class (clay, %)	Gravel (% by vol.)	XRD <2 mm fraction	pH	Soil structure
BC_y	0–3	1.4	lfs (1)	15	Q, AT	7.0	None
B_{yk}	3–13	0.8	lfs (1)	0	G, Q, AT	7.5	Massive and strong plates, 5–10 mm
B_{ykm}	13–34	1.2	lfs (3)	0	G, Q, AT	7.7	Strong prisms, 30 cm wide
B_{tyknm1}	34–81	1.5	scl (21)	0	G, Q, AT	8.1	Massive with some strong prisms, 30 cm wide (10 %)
B_{tyknm2}	81–96	1.6	sl (15)	0			Massive
B_{ykm1}	96–101	1.6	cosl (8)	10 % slightly >2 mm			Moderate plates, 5–10 mm
B_{ykm2}	101–129	1.7	sl (12)	22 % large cobbles			Massive
Pedon[3]	Depth (cm)	Bulk density (Mg m^{-3})	Texture class (clay, %)	Gravel (% by vol.)	XRD <2 mm fraction	pH	Soil structure
BC_{yk}	0–2	1.4	15	17	Q, AT, S, M, CH, K,	7.7	Strong plates, 2 mm thick
B_{yk1}	2–3	0.8		1	G, AH, Q, AT	7.5	consolidated layer
B_{yk2}	3–12	0.6		12	G, Q, AT	7.1	Strong prisms, 5–10 cm; moderate plates in cracks, 5–10 mm
B_{yk3}	12–26	1.2		25	G, Q, AT	6.9	Massive
B_y	26–39	1.5	11	25	G, Q, AT, S, M, CH, K	7.3	Massive
B_{yknzm}	39–71	1.3		10	AH, AT, Q, AH, G, AT	7 3	Strong prisms, ~30 cm thick
B_{ynzk}	71–85	1.2		10		7.4	Massive
B_{yknz}	85–102	1.2		10	Q, AH, H, G, AT, C		Massive
B_{nzyk}	102–122	1.5		0	Q, AH, H, C	7.3	Massive
B_{nzm}	122–146	1.7		0	H, Q, N	7.2	Massive; cracks contain cubic crystals of halite
C_{nzky1}	146–154	1.7	scl (27)	0			Strong plates, 1–2 mm thick
C_{knzy1}	154–180	1.7	scl (27)				Strong plates (1–2 mm) fi massive
C_{knzy2}	180–192	1.6	scl (29)	<5 %–3 mm	M, CH, K, S		moderate subangular blocks, 5–10 mm
C_{knzy3}	192–211	1.7	scl (29)	0			Weak angular blocks, 1–2 cm
C_{knzy4}	211–232	1.7	scl (27)	0			Sedimentary plates weak subangular blocks, 1–2 cm

Pedon[1] : Copiapó pedon, Slope 2 %, S27°01.2790 W70°17.6720, alluvial fan of granitic composition, sandy-skeletal, mixed, thermic, shallow Typic Petrocalcid, 1,215 m a.s.l, MAP 20 mm, MAT 16 °C
Pedon[2] :Altamira pedon, Slope 1 %, 25°45.5870 W70°11.7970; stream terrace at toe of alluvial fan, sandy, gypsic, thermic, shallow Typic Petrogypsid, 1,012 m a.s.l., MAP 10 mm, MAT 16 °C
Pedon[3] :Yungay pedon, Slope 1 %, S24°06.1020 W70°01.0970, Distal alluvial fan of primarily granitic origin, with occasional fine grained mafic gravels, loamy, gypsic, thermic, shallow Petrogypsic Haplosalid, 1,024 m a.s.l., MAP 0 mm, MAT 16 °C
Soil near Pica oasis, Region I of Chile, a typical example of depositional plains soils (Luzio et al. 2010)

Depth (cm)	0–20	20–31	31–105
Genetic horizon/layer	A_1	2B	R
Particle size (mm) distribution			
2–0.05 (%)	70.6	42.6	64.6
0.05–0.002 (%)	12.8	22.8	22.8
<0.002 (%)	16.6	34.6	12.6
OC (%)	0.12	0.16	0.52
pH_{water}	8.4	8.8	8.2
EC (dS m^{-1})	0.6	0.7	4.0
$CaCO_3$ (%)	0.0	0.0	4.51
Extractable cations ($cmol_ckg^{-1}$)			
Ca	6.5	30.0	27.0
Mg	1.6	5.3	1.8
K	1.6	3.9	1.3
Na	0.9	2.4	4.0
CEC ($cmol_ckg^{-1}$)	8.8	3.3	13.7

Pedon at the Pintados salar, Region I of Chile (Luzio et al. 2010)

Depth (cm)	0–30	30–50	50–78	78–85	85–117	117–130
Genetic horizon	A_{yzm}	$2C_1$	$2C_2$	$3C_{km}$	$4C_3$	$5C_4$
Particle size (mm) distribution						
2–0.05 (%)	32.6	38.6	17.4	21.4	67.4	13.4
0.05–0.002 (%)	60.0	27.4	36.0	58.0	22.0	53.2
<0.002 (%)	61.4	43.0	46.6	20.6	10.6	33.4
OC (%)	0.07	0.33	0.28	0.04	0.11	0.06
pH_{water}	8.6	8.8	8.9	9.5	9.1	9.6
EC (dS m^{-1})	54.0	150.0	150.0	64.5	18.8	25.0
$CaCO_3$ (%)	1.86	1.01	0.0	9.5	0.0	0.0
Extractable cations ($cmol_ckg^{-1}$)						
Ca	33.8	8.6	8.3	17.5	1.65	3.9
Mg	0.29	1.0	1.0	0.8	0.4	0.7
K	225.0	16.5	14.0	9.0	6.5	10.0
Na	575.0	63.5	65.0	30.5	12.2	21.5
CEC ($cmol_ckg^{-1}$)	5.0	28.7	20.0	20.0	8.1	22.5

2.2.1.3 Soils of the Valleys

Apacheta soil series (Typic Haplocambids) at Region III of Chile (CIREN 2007)

Depth (cm)	0–11	11–48	48–60	60–80	80–91	91–120
Genetic horizon	A_p	B	$2C_1$	$3C_2$	$4C_3$	$5A_b$
Particle size (mm) distribution						
2–0.05 (%)	23.0	31.0	19.0	33.0	5.0	37.0
0.05–0.002 (%)	51.0	31.0	63.0	43.0	84.0	41.0
<0.002 (%)	26.0	38.0	18.0	24.0	11.0	22.0
Water retention −33 kPa (%)	17.2	22.8	11.5	19.0	6.4	18.7
Water retention − 1,500 kPa (%)	8.8	12.6	5.7	8.3	2.9	7.7
Water saturation (%)	37.4	44.5	24.7	34.9	25.9	33.4
pH_{water}	7.7	8.2	8.5	8.5	8.6	8.6
OC (%)	1.33	1.22	0.17	0.17	0.12	0.29
$CaCO_3$ (%)	4.3	6.0	2.0	3.8	1.5	2.6
$CaSO_4$ (%)	0.12	0.00	0.00	0.00	0.00	0.00
EC (dS m^{-1})	8.1	1.8	1.3	1.3	0.7	1.0
Soluble cations ($mmol_c$ L^{-1})						
Ca	42.9	10.5	6.9	7.7	3.6	5.3
Mg	22.6	3.6	2.3	2.5	1.2	1.9
K	3.1	1.0	0.4	0.4	0.2	0.2
Na	38.0	4.5	3.4	3.7	2.3	2.9
SAR (–)	6.6	1.7	1.6	1.6	1.5	1.5
ESP (%)	8.8	2.4	2.3	2.3	2.1	2.1
Extractable cations ($cmol_ckg^{-1}$)						
Ca	30.80	23.80	14.70	23.50	8.50	23.50
Mg	5.42	4.11	2.17	3.19	1.48	3.39
Na	2.48	0.61	0.30	0.41	0.23	0.43
K	1.35	1.31	0.43	0.52	0.24	0.49
Exchangeable cations ($cmol_ckg^{-1}$)						
K	1.20	1.26	0.42	0.51	0.24	0.48
Na	1.06	0.41	0.22	0.28	0.17	0.33
CEC ($cmol_ckg^{-1}$)	16.8	18.9	9.6	11.8	3.9	13.5
BS (%)	100	100	100	100	100	100
Na saturation (%)	6.3	2.2	2.2	2.4	4.4	2.5

Ramadillas soil series (Sodic Haplocambid), a edon at the lower part of the Copiapó Valley, Region III of Chile (CIREN 2007)

Depth (cm)	0–21	21–39	39–58	58–76	76–87	87–110
Genetic horizon	A_p	B_1	B_2	B_3	$2O_i$	$3B_b$
Particle size (mm) distribution						
2–0.05 (%)	68.0	35.0	59.0	44.0	69.0	55.0
0.05–0.002 (%)	6.0	5.0	19.0	16.0	21.0	15.0
<0.002 (%)	29.0	60.0	22.0	40.0	10.0	30.0

(continued)

(continued)

Water retention −33 kPa (%)	37.6	39.7	38.1	41.9	41.0	38.3
Water retention −1,500 kPa (%)	20.3	23.3	22.6	26.0	22.4	22.1
pH_{water}	7.7	8.1	7.9	7.9	7.8	8.0
OC (%)	2.03	1.28	0.41	0.41	1.04	0.23
$CaCO_3$ (%)	8.5	9.6	1.6	1.9	2.5	15.2
$CaSO_4$ (%)	0.71	0.00	1.09	1.08	1.13	1.09
EC (dS m^{-1})	33.7	9.0	9.0	10.8	9.6	9.0
Soluble cations ($mmol_cL^{-1}$)						
Ca	117.3	26.8	24.6	24.8	22.6	25.4
Mg	123.4	17.7	21.8	34.1	42.6	32.1
K	8.1	1.0	1.0	1.0	0.9	0.8
Na	190.1	62.5	62.9	78.9	63.1	55.2
SAR (–)	17.3	13..2	13.1	14.5	11.1	10.3
ESP (%)	19.6	15.5	15.3	16.8	13.1	12.2
Extractable cations ($cmol_ckg^{-1}$)						
Ca	62.2	48.9	104.0	109.0	129.0	112.0
Mg	19.3	10.7	9.9	17.1	14.8	12.2
Na	16.1	8.78	6.78	9.65	5.91	5.83
K	6.59	1.27	0.95	1.17	0.65	0.81
Exchangeable cations ($cmol_ckg^{-1}$)						
K	5.96	1.19	0.89	1.09	0.59	0.75
Na	1.43	3.73	2.64	3.63	1.44	1.79
CEC ($cmol_ckg^{-1}$)	33.0	31.7	27.3	30.4	27.5	23.6
BS (%)	100	100	100	100	100	100
Na saturation (%)	4.3	11.8	9.7	11.9	5.3	7.6

Paona soil series (Aquic Torriorthents), characteristic soil profile of lower terraces in the Huasco Valley at Region III of Chile (CIREN 2007)

Depth (cm)	0–23	23–45	45–77	77–100
Genetic horizon	A_p	AC_1	AC_3	AC_4
Particle size (mm) distribution				
2–0.05 (%)	50.7	7.5	74.6	65.4
0.05–0.002 (%)	32.4	16.1	18.8	28.8
<0.002 (%)	16.9	6.4	6.6	5.8
Water retention −33 kPa (%)	22.9	14.4	14.2	16.2
Water retention −1,500 kPa (%)	9.9	5.3	5.8	6.6
OC (%)	0.9	–	–	–
pH_{water}	8.2	8.1	8.2	8.2
$CaCO_3$ (%)	4.7	3.9	4.8	5.4
EC (dS m^{-1})	16.6	9.2	8.2	7.4

(continued)

(continued)

Extractable cations ($cmol_ckg^{-1}$)				
Ca	–	–	–	–
Mg	–	–	–	–
Na	0.7	0.1	0.1	0.2
K	2.4	1.0	1.1	1.5
CEC ($cmol_ckg^{-1}$)	13.0	8.9	7.9	8.9
Na saturation (%)	13.5	11.2	13.9	16.8

Chanchoquín soil series (Xerollic Camborthids), characteristic soil profile of intermediate terraces in the Huasco Valley, Region III of Chile (CIREN 2007)

Depth (cm)	0–23	23–62	62–105
Genetic horizon	A_p	B_1	B_3
Particle size (mm) distribution			
2–0.05 (%)	33.7	39.2	29.5
0.05–0.002 (%)	39.3	39.8	52.9
<0.002 (%)	27.1	21.0	17.6
Water retention −33 kPa (%)	25.6	21.9	27.1
Water retention −1,500 kPa (%)	16.5	11.8	12.0
OC (%)	1.2	0.3	0.3
pH_{water}	8.1	8.3	8.3
$CaCO_3$ (%)	14.0	17.1	13.8
EC (dS m^{-1})	0.6	0.5	0.5
Extractable cations ($cmol_ckg^{-1}$)			
Ca	–	–	–
Mg	–	–	–
Na	0.3	0.2	0.2
K	0.5	0.5	0.5
CEC ($cmol_ckg^{-1}$)	23.3	20.7	18.9
BS (%)	2	2	3

Cavancha soil series (Xerollic Haplargids), characteristic soil profile of highest terraces in the Huasco Valley, Region III of Chile (CIREN 2007)

Depth (cm)	0–22	22–33	33–54	54–72
Genetic horizon	A_p	B_1	B_2	B_k
Particle size (mm) distribution				
2–0.05 (%)	51.5	48.7	34.7	33.8
0.05–0.002 (%)	37.8	33.0	32.2	36.1
<0.002 (%)	21.2	18.3	33.1	30.1
Water retention −33 kPa (%)	18.9	17.3	22.7	27.1
Water retention −1,500 kPa (%)	9.1	9.5	15.3	–

(continued)

(continued)

OC (%)	1.0	0.4	0.2	0.2
pH_{water}	8.3	8.2	8.2	8.3
$CaCO_3$ (%)	1.2	0.5	0.4	22.0
EC (dS m^{-1})	1.9	1.1	1.0	1.2
Extractable cations ($cmol_ckg^{-1}$)				
Ca	1.2	1.1	1.0	1.2
Mg	–	–	–	–
Na	–	–	–	–
K	1.3	1.3	1.3	0.5
CEC ($cmol_ckg^{-1}$)	13.8	11.6	17.7	10.3

Puclaro soil series (Typic Torrifluvent), a characteristic poorly developed pedon on the terraces of the Elqui River, Region IV of Chile (CIREN 2005a)

Depth (cm)	0–39	39–100
Genetic horizon	A	C
Particle size (mm) distribution		
2–0.05 (%)	73.1	86.1
0.05–0.002 (%)	18.4	10.8
<0.002 (%)	8.5	3.1
Bulk density (Mg m^{-3})	1.3	–
Water retention −33 kPa (%)	11.4	4.1
Water retention −1,500 kPa (%)	6.9	1.7
OC (%)	0.9	0.2
pH_{water}	7.9	7.9
$CaCO_3$ (%)	1.4	0.0
EC (dS m^{-1})	1.1	0.5
Extractable cations ($cmol_ckg^{-1}$)		
Ca	–	6.3
Mg	–	0.6
K	0.5	0.1
Na	0.2	0.2
CEC ($cmol_ckg^{-1}$)	13.4	7.2
BS (%)	–	100

Altovalsol soil series (Typic Haplocambid), a more developed pedon on the terraces of the Elqui River, Region IV of Chile (CIREN 2005a)

Depth (cm)	0–19	19–36	36–62
Genetic horizon	A_p	B_1	B_2
Particle size (mm) distribution			

(continued)

(continued)

2–0.05 (%)	49.9	48.8	54.3
0.05–0.002 (%)	27.6	29.2	14.3
<0.002 (%)	22.5	22.0	31.4
Bulk density (Mg m^{-3})	1.6	1.6	1.5
Water retention −33 kPa (%)	25.6	25.2	27.3
Water retention −1.500 kPa (%)	15.5	18.3	17.7
OC (%)	1.5	0.7	0.4
pH_{water}	8.1	7.8	8.1
EC (dS m^{-1})	0.9	0.6	0.6
$CaCO_3$ (%)	0.0	0.0	0.6
Extractable cations ($cmol_ckg^{-1}$)			
Ca	14.8	18.3	–
Mg	2.9	4.3	–
K	0.7	0.4	0.4
Na	0.6	0.8	1.2
CEC ($cmol_ckg^{-1}$)	19.1	24.2	20.8
BS (%)	99	98	–

Huatulame soil series (Typic Torriorthent) at the Limarí Valley, Region IV of Chile (CIREN 2005a)

Depth (cm)	0–16	16–59
Genetic horizon	Ap	C1
Particle size (mm) distribution		
2–0.05 (%)	52.2	57.9
0.05–0.002 (%)	39.6	27.8
<0.002 (%)	8.1	14.3
Bulk density (Mg m^{-3})	1.66	–
Water retention −33 kPa (%)	13.2	15.9
Water retention −1,500 kPa (%)	6.2	7.4
OC (%)	1.16	0.35
pH_{water}	7.35	7.62
EC (dS m^{-1})	1.2	0.3
$CaCO_3$ (%)	0.0	0.0
Extractable cations ($cmol_ckg^{-1}$)		
Ca	8.8	2.2
MG	2.0	0.5
K	1.8	1.2
Na	0.4	0.3
CEC ($cmol_ckg^{-1}$)	13.0	4.2
BS (%)	87	84

Huamalata soil series (Typic Haplocambid) at the Limarí Valley, Region IV of Chile (CIREN 2005a)

Depth (cm)	0–13	13–27	27–40	40–63
Genetic horizon	A_p	B_w	BC	C_1
Particle size (mm) distribution				
2–0.05 (%)	55.4	52.8	66.2	67.7
0.05–0.002 (%)	29.4	32.2	25.3	25.5
<0.002 (%)	14.8	15.1	8.6	6.7
Bulk density (Mg m^{-3})	1.78	1.65	1.57	–
Water retention −33 kPa (%)	13.8	13.0	9.5	8.2
Water retention −1,500 kPa (%)	9.2	8.2	5.9	5.2
OC (%)	1.57	1.10	0.64	0.46
pH_{water}	7.20	7.63	7.81	8.03
EC (dS m^{-1})	4.9	1.0	0.6	0.5
Soluble cations ($mmol_cL^{-1}$)				
Ca	42.0	5.4	3.1	2.1
Mg	14.7	1.7	1.0	0.6
K	1.5	0.2	0.1	0.0
Na	10.3	2.9	2.3	2.1
HCO_3 ($mmol_cL^{-1}$)	6.0	3.8	1.2	2.0
Cl ($mmol_cL^{-1}$)	13.2	1.5	0.5	0.4
SO_4 ($mmol_cL^{-1}$)	41.0	4.8	5.0	2.2
SAR	1.9	1.6	1.6	1.8
Extractable cations ($cmol_ckg^{-1}$)				
Ca	15.9	13.5	11.0	9.8
Mg	3.31	2.68	2.32	2.04
K	0.70	0.31	0.14	0.10
Na	0.64	0.35	0.32	0.32
CEC ($cmol_ckg^{-1}$)	19.6	18.7	16.0	14.3
BS (%)	100	90	86	85

Serón soil serie (Typic Haplargid), a representative pedon of developed soils in the Limarí Valley, Region IV of Chile (CIREN 2005a)

Depth (cm)	0–23	23–60	60–90
Genetic horizon	A_p	B_1	B_t
Particle size (mm) distribution (%)			
2–0.05 (%)	60.8	63.3	50.4
0.05–0.002 (%)	26.0	25.9	22.1
<0.002 (%)	13.2	10.8	27.5
Bulk density (Mg m^{-3})	1.66	1.95	1.97
Water retention −33 kPa (%)	16.8	12.5	23.8
Water retention −1,500 kPa (%)	9.6	7.7	15.1
OC (%)	1.22	0.35	0.23
pH_{water}	8.02	8.44	8.27
EC (dS m^{-1})	0.6	0.3	0.2
Soluble cations ($mmol_cL^{-1}$)			

(continued)

(continued)

Ca	5.0	1.5	1.2
Mg	0.9	0.3	0.36
K	0.08	0.03	0.02
Na	1.0	0.7	0.6
HCO_3 ($mmol_cL^{-1}$)	4.8	2.0	1.5
Cl $mmol_cL^{-1}$)	0.4	0.1	0.1
SO_4 ($mmol_cL^{-1}$)	0.9	0.4	0.7
Extractable cations ($cmol_ckg^{-1}$)			
Ca	17.8	16.0	23.2
Mg	2.6	2.33	6.41
K	0.25	0.13	0.11
Na	0.21	0.23	0.35
CEC ($cmol_ckg^{-1}$)	14.0	17.0	21.3
BS (%)	100	100	100

Tuquí soil series (Petrocalcic Calcitorrert) at the Limarí Valley, Region IV of Chile (CIREN 2005a).

Depth (cm)	0–18	18–69	69–35
Genetic horizon	A_p	B_{ss}	BC
Particle size (mm) distribution			
2–0.05 (%)	37.7	24.0	18.8
0.05–0.002 (%)	23.0	20.7	22.4
<0.002 (%)	39.3	55.2	58.7
Bulk density (Mg m^{-3})	1.75	1.93	1.91
Water retention −33 kPa (%)	22.4	29.4	32.8
Water retention −1,500 kPa (%)	14.0	16.4	17.9
OC (%)	1.80	0.35	0.23
pH_{water}	7.74	8.00	8.07
EC (dS m^{-1})	0.9	1.1	1.3
$CaCO_3$ (%)	0.7	11.9	10.7
Soluble cations ($mmol_cL^{-1}$)			
Ca	7.0	4.7	3.7
Mg	1.8	2.3	2.9
K	0.8	0.1	0.1
Na	1.0	5.3	8.0
CO_3 ($mmol_cL^{-1}$)	0.0	0.0	0.0
HCO_3 ($mmol_cL^{-1}$)	5.4	1.8	1.6
Cl ($mmol_cL^{-1}$)	0.7	1.7	1.0
SO_4 ($mmol_cL^{-1}$)	4.1	9.0	11.0
Extractable cations ($cmol_ckg^{-1}$)			
Ca	27.9	30.9	29.9
Mg	4.57	9.35	13.6
K	1.83	0.52	0.65
Na	0.20	1.14	1.83

(continued)

(continued)

CEC ($cmol_ckg^{-1}$)	30.2	26.7	29.7
BS (%)	100	100	100

Tahuinco soil series (Typic Paleargid), a colluvial soil at Choapa Valley, Region IV of Chile (CIREN 2005a)

Depth (cm)	0–12	12–30	30–58
Genetic horizon	A_p	B	B_t
Particle size (mm) distribution			
2–0.05 (%)	58.3	65.4	46.4
0.05–0.002 (%)	23.4	19.7	20.4
<0.002 (%)	18.3	15.0	33.2
Bulk density (Mg m^{-3})	1.50	1.58	1.91
Water retention −33 kPa (%)	17.5	11.5	14.5
Water retention −1,500 kPa (%)	10.8	6.8	9.5
OC (%)	2.5	1.0	0.4
Soluble cations ($mmol_cL^{-1}$)			
Ca	5.5	2.60	2.90
Mg	2.62	0.84	0.99
K	0.23	0.05	0.05
Na	1.84	1.32	2.13
pH_{water}	8.05	8.19	8.58
EC (dS m^{-1})	0.9	0.4	0.5
Extractable cations ($cmol_ckg^{-1}$)			
Ca	16.7	11.1	19.5
Mg	3.8	2.5	5.5
K	0.4	0.2	0.3
Na	0.3	0.2	0.5
CO_3 ($mmol_cL^{-1}$)	0.0	0.0	0.0
HCO_3 ($mmol_cL^{-1}$)	5.6	3.6	3.8
Cl ($mmol_cL^{-1}$)	0.4	0.3	0.5
SO_4 ($mmol_cL^{-1}$)	3.4	0.8	1.1
CEC ($cmol_ckg^{-1}$)	19.9	16.3	19.9
BS (%)	100	86	100

Hualcapo soil series (Fluventic Haploxeroll) on recent terraces of the Aconcagua Valley. Region V of Chile (CIREN 1997a)

Depth (cm)	0–20	20–42	42–72	72–94
Genetic horizon	A_1	B_1	B_2	B_3
Particle size (mm) distribution (%)				
2–0.05 (%)	67.3	67.4	54.7	61.8

(continued)

(continued)

0.05–0.002 (%)	24.3	24.2	34.1	29.2
<0.002 (%)	8.4	8.4	11.2	9.0
Water retention −33 kPa (%)	16.0	15.0	19.0	17.0
Water retention −1,500 kPa (%)	9.0	9.0	10.0	8.0
OC (%)	1.3	0.9	1.0	0.6
pH_{water}	7.6	7.5	7.8	7.8
EC (dS m^{-1})	1.6	1.0	0.6	1.2
Extractable cations ($cmol_ckg^{-1}$)				
Ca	9.2	8.7	11.9	9.8
Mg	1.6	1.5	2.0	1.9
K	0.5	0.4	0.3	0.2
Na	0.2	0.2	0.4	0.4
CEC ($cmol_ckg^{-1}$)	10.6	10.0	13.3	10.9
BS (%)	100	100	100	100

Pocuro soil series (Fluventic Haploxeroll) at remnant terraces of the Aconcagua Valley. Region V of Chile (CIREN 1997a)

Depth (cm)	0–18	18–48	48–82	82–110	110–130
Genetic horizon	A_1	A_2	B_1	B_2	B_3
Particle size (mm) distribution					
2–0.05 (%)	40.8	30.9	28.6	17.0	17.0
0.05–0.002 (%)	41.2	47.8	51.2	59.4	51.7
<0.002 (%)	18.0	21.3	20.2	23.6	31.3
Water retention −33 kPa (%)	23.0	24.0	24.0	29.0	29.0
Water retention −1,500 kPa (%)	13.0	14.0	14.0	15.0	16.0
OC (%)	1.2	1.0	0.6	0.6	0.5
pH_{water}	6.6	6.7	6.8	6.8	6.8
Extractable cations ($cmol_ckg^{-1}$)					
Ca	12.1	13.9	14.2	17.1	18.1
Mg	2.1	2.1	1.9	2.2	2.1
K	0.3	0.1	0.1	0.1	0.2
Na	0.3	0.4	0.4	0.4	0.5
CEC ($cmol_ckg^{-1}$)	15.6	21.3	19.4	21.9	25.4
BS (%)	95	77	86	90	82

Ocoa soil series (Typic Xerochrept), a colluvial soil at piedmont positions in the Aconcagua Valley, Region V of Chile (CIREN 1997a)

Depth (cm)	0–21	21–45	45–63	63–100
Genetic horizon	A_1	B_1	B_2	B_3
Particle size (mm) distribution				
2–0.05 (%)	35.6	41.9	37.2	35.5
0.05–0.002 (%)	42.8	37.7	43.1	44.5
<0.002 (%)	21.6	20.4	19.7	20.0
Water retention −33 kPa (%)	29.0	24.0	21.0	20.0
Water retention −1,500 kPa (%)	22.0	18.0	14.0	13.0
OC (%)	2.6	2.0	1.4	0.5
pH_{water}	7.3	7.5	7.8	7.7
EC (dS m^{-1})	1.4	0.6	0.4	0.5
Extractable cations ($cmol_c kg^{-1}$)				
Ca	17.2	16.3	13.9	11.2
Mg	3.5	3.2	2.6	2.2
K	0.4	0.4	0.4	0.3
Na	0.4	0.3	0.3	0.3
CEC ($cmol_c kg^{-1}$)	22.7	21.5	18.1	15.6
BS (%)	95	94	95	90

Palomar soil series (Typic Medihemist), situated in depressed landscape in the Aconcagua Valley, Region V of Chile (CIREN 1997a)

Depth (cm)	0–18	18–44	44–72
Genetic horizon	A_1	A_2	B
Particle size (mm) distribution			
2–0.05 (%)	7.0	8.2	21.0
0.05–0.002 (%)	51.6	49.2	51.9
<0.002 (%)	41.4	42.6	27.1
Water retention −33 kPa (%)	76.0	70.0	55.0
Water retention −1,500 kPa (%)	40.0	38.0	28.0
OC (%)	20.7	17.1	8.2
pH_{water}	7.8	7.9	7.4
EC (dS m^{-1})	2.8	2.0	2.0
$CaCO_3$ (%)	29.0	35.4	0.5
Extractable cations ($cmol_c kg^{-1}$)			
Ca	–	–	–
Mg	–	–	–
K	0.6	0.1	0.2
Na	1.4	1.5	1.0

2.2.1.4 Soils of the Coastal Range

A pedon near Taltal (25°S, Region II of Chile) in the northern Coastal Range (Luzio et al. 2010)

Depth (cm)	0–5	5–36	36–100
Genetic horizon	A	C_1	C_2
Particle size (mm) distribution (%)			
2–0.05 (%)	63.4	64.4	70.4
0.05–0.002 (%)	22.6	18.6	13.6
<0.002 (%)	14.0	17.0	16.0
OC (%)	5.1	0.9	1.0
pH_{water}	8.8	8.4	8.4
EC (dS m^{-1})	54.0	150.0	150.0
$CaCO_3$ (%)	1.1	0.2	0.2
Exchangeable cations ($cmol_c kg^{-1}$)			
Ca	25.9	4.2	8.1
Mg	8.9	4.6	2.8
K	1.6	1.6	0.6
Na	4.1	3.0	2.1
CEC ($cmol_c kg^{-1}$)	14.9	10.9	7.8

Mincha soil series (Typic Haplocambid) at the Coastal Range, Region IV of Chile (CIREN 2005a)

Depth (cm)	0–8	8–28	28–50	50–70
Genetic horizon	A_1	B_1	B_2	B_3
Particle size (mm) distribution				
2–0.05 (%)	46.6	48.6	22.6	16.5
0.05–0.002 (%)	36.0	34.0	14.0	14.0
<0.002 (%)	17.4	17.4	63.4	69.4
Bulk density (Mg m^{-3})	1.43	1.40	1.51	1.54
Water retention −33 kPa (%)	11.00	11.00	32.50	36.40
Water retention −1,500 kPa (%)	8.25	6.43	22.50	24.70
OC (%)	0.79	0.52	0.42	–
pH_{water}	7.8	7.0	6.5	6.4
EC (dS m^{-1})	0.33	0.31	1.17	1.98
$CaCO_3$ (%)	0.50	0.00	0.00	0.00
Soluble cations ($mmol_c L^{-1}$)				
Ca	1.39	1.31	1.25	1.28
Mg	0.77	0.70	0.75	1.12

(continued)

(continued)

K	0.37	0.24	0.31	0.40
Na	3.63	2.66	10.10	12.80
HCO_3 ($mmol_cL^{-1}$)	3.50	1.75	0.50	0.25
Cl ($mmol_cL^{-1}$)	1.96	2.59	9.45	13.30
SO_4 ($mmol_cL^{-1}$)	1.12	1.06	3.41	10.50
SAR	3.49	2.65	10.10	11.68
Extractable cations ($cmol_ckg^{-1}$)				
Ca	3.70	3.20	7.10	8.70
Mg	1.70	1.60	8.90	9.00
K	0.78	0.66	1.54	1.94
Na	0.25	0.23	3.23	4.48
CEC ($cmol_ckg^{-1}$)	9.24	8.73	24.60	23.90

2.2.1.5 Soils of the Coastal Plains

La Compañía soil serie (Typic Torripsamment). Stabilised dune with certain pedogenic development, near to La Serena city, Region IV of Chile (CIREN 2005a)

Depth (cm)	0–12	12–30	30–50	50–80	80–105	105–120
Genetic horizon	A_p	C_1	C_2	C_3	C_4	C_5
Particle size (mm) distribution						
2–0.05 (%)	85.7	91.5	89.2	92.1	93.8	84.2
0.05–0.002 (%)	10.3	5.4	7.4	5.3	3.6	4.1
<0.002 (%)	4.0	3.1	3.4	2.6	2.6	11.7
Bulk density (Mg m^{-3})	1.63	1.69	1.59	1.75	1.83	1.96
Water retention – 33 kPa (%)	4.0	3.2	2.9	2.9	9.2	9.5
Water retention – 1,500 kPa (%)	2.1	2.0	2.0	2.2	6.8	6.3
OC (%)	0.2	0.2	0.1	0.0	0.0	0.0
pH_{water}	6.3	6.7	6.6	6.9	6.9	7.2
EC (dS m^{-1})	0.8	0.4	0.5	0.5	0.5	0.5
Extractable cations ($cmol_ckg^{-1}$)						
Ca	1.1	0.8	0.8	0.6	1.7	1.6
Mg	0.6	0.8	0.8	0.9	3.7	3.7
K	0.5	0.4	0.4	0.2	0.5	0.5
Na	0.2	0.1	0.1	0.2	0.7	0.6
CEC ($cmol_ckg^{-1}$)	4.4	3.8	3.6	2.6	8.2	8.9
BS (%)	55	55	58	73	80	72

Indurate soils on marine terraces (Calcid Petrocalcid), Region IV of Chile (CIREN 2005a)

Depth (cm)	0–15	15–32	32–55
Genetic horizon	A_p	B_{k1}	B_{k2}
Particle size (mm) distribution			
2–0.05 (%)	50.4	51.4	43.8
0.05–0.002 (%)	30.8	32.2	34.5
<0.002 (%)	18.8	16.4	21.7
Bulk density (Mg m^{-3})	1.39	1.42	–
Water retention −33 kPa (%)	23.0	23.0	25.5
Water retention −1,500 kPa (%)	18.3	14.7	19.5
OC (%)	1.5	0.9	0.5
pH_{water}	8.1	8.1	8.0
EC (dS m^{-1})	0.9	0.9	0.9
$CaCO_3$ (%)	13.0	15.6	32.5
Extractable cations ($cmol_ckg^{-1}$)			
Ca	–	–	–
Mg	–	–	–
K	2.8	1.7	0.6
Na	0.5	0.6	0.7
CEC ($cmol_ckg^{-1}$)	23.2	19.2	16.1

Huentelauquén soil series (Typic Natrargid) on marine terraces at Region IV of Chile (CIREN 2005a)

Depth (cm)	0–7	7–25	25–52	52–81	81–123
Genetic horizon	A_1	A_2	B_{t1}	B_{t2}	B_{t3}
Particle size (mm) distribution					
2–0.05 (%)	26.0	24.0	12.0	24.0	14.0
0.05–0.002 (%)	60.0	60.0	22.0	18.0	22.0
<0.002 (%)	14.0	16.0	66.0	58.0	64.0
Bulk density (Mg m^{-3})	1.20	1.35	1.42	1.48	1.50
Water retention −33 kPa (%)	11.00	10.80	33.20	30.00	29.30
Water retention −1,500 kPa (%)	6.43	6.08	24.40	22.30	20.00
OC (%)	0.75	0.60	0.28	–	–
pH_{water}	7.5	7.5	8.4	8.3	8.4
EC (dS m^{-1})	3.08	3.03	5.29	6.45	5.28
$CaCO_3$ (%)	0.5	0.5	2.4	15.5	20.9
Soluble cations ($mmol_cL^{-1}$)					
Ca	6.35	4.43	2.74	3.06	2.18

(continued)

(continued)

Mg	4.39	3.60	3.95	5.21	3.44
K	1.45	0.45	0.56	0.68	0.54
Na	18.70	20.90	46.50	58.30	46.30
HCO_3 ($mmol_cL^{-1}$)	1.00	1.25	4.00	4.25	3.75
Cl ($mmol_cL^{-1}$)	27.10	28.20	46.30	55.80	47.00
SO_4 ($mmol_cL^{-1}$)	3.04	2.04	6.64	8.82	8.40
Extractable cations ($cmol_ckg^{-1}$)					
Ca	2.30	2.40	5.60	20.00	19.50
Mg	0.92	1.50	10.20	9.90	9.00
K	0.97	0.66	2.41	1.99	1.75
Na	1.19	1.55	9.77	10.00	9.56
SAR	8.07	10.43	25.42	28.67	27.62
CEC ($cmol_ckg^{-1}$)	7.64	7.83	22.9	21.8	20.9
BS (%)	70	78	100	100	100

Tabolango soil series (Typic Palexeralf) on marine terraces at Region V of Chile (CIREN 1997a)

Depth (cm)	0–17	17–44	44–59
Genetic horizon	A	B_{t1}	B_{t2}
Particle size (mm) distribution			
2–0.05 (%)	58.5	24.6	17.7
0.05–0.002 (%)	31.7	19.9	28.3
<0.002 (%)	9.8	55.5	54.0
Water retention −33 kPa (%)	16.0	36.0	43.0
Water retention −1,500 kPa (%)	6.0	29.0	29.0
OC (%)	0.6	0.5	0.2
pH_{water}	5.6	6.3	8.0
EC ($dS\ m^{-1}$)	–	–	4.7
Extractable cations ($cmol_ckg^{-1}$)			
Ca	0.9	12.5	–
Mg	0.7	10.6	–
K	0.4	0.4	0.4
Na	0.2	4.1	8.6
CEC ($cmol_ckg^{-1}$)	6.8	32.2	32.7
Na saturation (%)			6.0
BS (%)	32	86	–

2.2.1.6 Soils of the *serranías*

El Tambo soil series (Typic Haplocambid) at the *serranías*, Region IV of Chile (CIREN 2005a)

Depth (cm)	0–11	11–21	21–32	32–80
Genetic horizon	A_p	B_1	B_2	BC
Particle size (mm) distribution				
2–0.05 (%)	55.8	52.8	66.1	67.7
0.05–0.002 (%)	29.4	32.2	25.33	25.5
<0.002 (%)	14.8	15.1	8.6	6.7
Bulk density ($Mg\ m^{-3}$)	1.26	1.30	1.15	1.20
Water retention −33 kPa (%)	22.7	20.6	11.8	9.1
Water retention −1,500 kPa (%)	12.6	11.4	6.5	5.0
Water saturation (%)	43	41	36	36
OC (%)	1.67	1.10	0.64	0.46
pH_{water}	7.20	7.63	7.81	8.03
EC ($dS\ m^{-1}$)	4.9	1.0	0.6	0.5
Soluble cations ($mmol_cL^{-1}$)				
Ca	42.0	5.4	3.1	2.1
Mg	14.7	1.7	1.0	0.6
K	1.5	0.2	0.1	0.0
Na	10.3	2.9	2.3	2.1
CO_3 ($mmol_cL^{-1}$)	0.0	0.0	0.0	0.0
HCO_3 ($mmol_cL^{-1}$)	6.0	3.8	1.2	2.0
Cl ($mmol_cL^{-1}$)	13.2	1.5	0.5	0.4
SO_4 ($mmol_cL^{-1}$)	21.0	1.8	2.2	1.4
Extractable cations ($cmol_ckg^{-1}$)				
Ca	15.9	13.5	11.0	9.8
Mg	3.31	2.68	2.32	2.04
K	0.70	0.31	0.14	0.10
Na	0.64	0.35	0.32	0.32
CEC ($cmol_ckg^{-1}$)	19.6	18.7	16.0	14.3
BS (%)	100	90	86	85

Combarbalá soil series (Typic Haplargid), inner sector in the Region IV of Chile (CIREN 2005a)

Depth (cm)	0–25	25–45	45–90
Genetic horizon	A_p	B_{t1}	B_{t2}
Particle size (mm) distribution			
2–0.05 (%)	32.6	23.9	18.0
0.05–0.002 (%)	24.0	25.0	25.3
<0.002 (%)	43.4	51.1	56.7

(continued)

(continued)

Bulk density (Mg m^{-3})	1.76	1.95	1.95
Water retention −33 kPa (%)	33.1	37.1	40.2
Water retention −1,500 kPa (%)	17.9	26.8	26.5
Water saturation (%)	68	77	78
OC (%)	1.51	0.75	0.46
pH_{water}	7.51	7.93	8.09
EC (dS m^{-1})	1.1	0.4	0.5
$CaCO_3$ (%)	0.9	1.8	3.3
Soluble cations ($mmol_cL^{-1}$)			
Ca	8.7	2.7	1.7
Mg	1.7	0.6	0.5
K	0.5	0.1	0.1
Na	1.4	1.5	3.3
CO_3 ($mmol_cL^{-1}$)	0.0	0.0	0.0
HCO_3 ($mmol_cL^{-1}$)	5.6	2.0	2.0
Cl ($mmol_cL^{-1}$)	3.0	0.1	0.3
SO_4 ($mmol_cL^{-1}$)	2.3	1.9	2.8
Extractable cations ($cmol_ckg^{-1}$)			
Ca	32.9	42.4	26.0
Mg	5.11	7.71	4.61
K	1.32	0.85	0.31
Na	0.30	0.53	0.74
CEC ($cmol_ckg^{-1}$)	38.7	43.0	47.0
BS (%)	100	100	67

Marquesa soil series (Typic Haplocalcid), soil over calcareous materials in the *serranías*, Region IV of Chile (CIREN 2005a)

Depth (cm)	0–32	32–100
Genetic horizon	A_1	C_k
Particle size (mm) distribution		
2–0.05 (%)	31.6	24.3
0.05–0.002 (%)	54.1	46.4
<0.002 (%)	14.3	29.3
Water retention −33 kPa (%)	26.7	19.7
Water retention −1,500 kPa (%)	15.5	11.5
OC (%)	0.8	0.2
pH_{water}	8.0	8.3
EC (dS m^{-1})	1.1	2.4
$CaCO_3$ (%)	0.9	33.5
Extractable cations ($cmol_ckg^{-1}$)		
Ca	–	–
Mg	–	–
K	0.2	0.1
Na	0.6	1.7
CEC ($cmol_ckg^{-1}$)	15.2	14.8

2.2.2 Soils of the Mediterranean Zone

2.2.2.1 Soils of the Andean Mountains

Los Nevados soil association profile (Acrudoxic Hapludand), Region IX of Chile (CIREN 2002)

Depth (cm)	0–20	20–42	42–80	>80
Genetic horizon	A_1	AC_1	AC_2	C
Particle size (mm) distribution				
2–0.05 (%)	62.1	56.6	71.6	84.3
0.05–0.002 (%)	25.2	36.0	21.0	15.0
<0.002 (%)	12.6	7.3	7.4	0.7
Water retention −33 kPa (%)	34.6	30.6	24.8	18.2
Water retention −1,500 kPa (%)	31.2	21.6	16.3	8.5
OC (%)	9.28	4.19	1.55	0.30
pH_{water}	5.9	6.1	6.1	6.1
Exchangeable cations ($cmol_ckg^{-1}$)				
Ca	0.65	0.29	0.24	0.24
Mg	0.25	0.10	0.06	0.05
Na	0.11	0.03	0.01	0.01
K	0.08	0.04	0.02	0.02
CEC ($cmol_ckg^{-1}$)	26.9	22.1	10.2	6.2
ECEC ($cmol_ckg^{-1}$)	1.51	0.46	0.33	0.32
P retention (%)	97	100	97	81
Al_{ox} + ½Fe_{ox} (%)	2.31	4.11	3.31	2.60
BS (%)	4	2	3	5

Los Riscos soil series (Typic Hydrudand), Region XIV of Chile (CIREN 2003)

Depth (cm)	0–9	9–22	22–40	40–65	65–100
Genetic horizon	A_1	A_2	B_1	B_2	BC
Particle size (mm) distribution					
2–0.05(%)	83.2	79.2	93.2	93.2	89.2
0.05–0.002(%)	7.3	9.3	1.3	1.3	3.3
<0.002(%)	9.5	11.5	5.5	5.5	7.5
Bulk density (Mg m^{-3})	0.53	0.86	0.48	0.41	0.44
Water retention −33 kPa (%)	56.4	32.2	31.9	24.8	35.1
Water retention −1,500 kPa (%)	56.2	26.3	28.8	24.0	30.4
OC (%)	12.6	6.0	4.2	3.7	2.7
pH_{water}	5.6	5.9	6.1	6.2	6.3
Exchangeable cations ($cmol_ckg^{-1}$)					
Ca	2.41	0.39	0.47	0.71	1.00

(continued)

(continued)

Mg	0.90	0.16	0.20	0.25	0.32
Na	0.22	0.09	0.10	0.06	0.05
K	0.26	0.06	0.03	0.03	0.03
Al	0.10	0.03	0.01	0.01	0.00
CEC ($cmol_c kg^{-1}$)	77.0	53.1	50.7	49.1	47.5
ECEC ($cmol_c kg^{-1}$)	3.89	0.73	0.81	1.06	1.40
P retention (%)	99	99	99	99	99
Al_{ox} + ½Fe_{ox} (%)	5.76	7.29	8.86	7.97	9.32
BS (%)	5	1	2	2	3

2.2.2.2 Soils of the Pre-Andean Mountains

Challay soil association (Lithic Haploxeroll), Metropolitan Region of Chile (CIREN 1997a, Valle 2012)

Depth (cm)	0–16	16–42
Genetic horizon	A_1	B
Particle size (mm) distribution		
2–0.05 (%)	43.7	36.3
0.05–0.002 (%)	38.1	37.4
<0.002 (%)	18.2	26.3
Bulk density ($Mg\ m^{-3}$)	1.40	1.44
Water retention −33 kPa (%)	23.6	25.1
Water retention −1,500 kPa (%)	11.1	12.5
OC (%)	1.79	0.77
pH_{water}	5.68	5.76
Exchangeable cations ($cmol_c kg^{-1}$)		
Ca	8.62	17.47
Mg	1.46	3.22
Na	0.15	0.37
K	0.17	0.11
Al	0.46	0.79
CEC ($cmol_c kg^{-1}$)	16.17	22.68
BS (%)	64	93

Bramadero soil series (Humic Haploxerand), Region VIII of Chile (CIREN 1997b, Stolpe et al. 2008)

Depth (cm)	0–17	17–48	48–82	82–120
Genetic horizon	A_1	A_2	AB	B
Particle size (mm) distribution				
2–0.05 (%)	40.8	48.2	41.6	42.3
0.05–0.002 (%)	34.5	29.5	36.9	32.3
<0.002 (%)	24.7	22.3	21.5	25.4
Bulk density ($Mg\ m^{-3}$)	0.95	0.99	1.01	1.08

(continued)

(continued)

Water retention −33 kPa (%)	35.0	35.0	34.0	46.0
Water retention −1.500 kPa (%)	18.0	17.0	20.0	30.0
OC (%)	5.2	3.1	1.9	1.1
pH_{water}	6.0	6.2	6.1	6.0
Exchangeable cations ($cmol_c kg^{-1}$)				
Ca	10.2	7.9	5.4	–
Mg	1.1	1.2	2.0	–
Na	0.1	0.1	0.3	–
K	1.5	1.2	1.0	–
CEC ($cmol_c kg^{-1}$)	31.1	25.8	28.0	–
P retention (%)	90.0	92.0	93.0	85.0
Al_{ox} + ½Fe_{ox} (%)	1.73	2.45	2.65	–
BS (%)	41	40	31	–

Choshuenco soil series (Andic Dystrudept), Region X of Chile (CIREN 2003)

Depth (cm)	0–29	29–54	54–86
Genetic horizon	A_1	B_{w1}	B_{w2}
Particle size (mm) distribution			
2–0.05 (%)	73.0	66.4	62.8
0.05–0.002 (%)	18.8	22.3	21.4
<0.002 (%)	8.2	11.3	15.8
Bulk density ($Mg\ m^{-3}$)	0.93	0.91	0.89
Water retention −33 kPa (%)	23.9	27.4	32.6
Water retention −1,500 kPa (%)	16.1	16.0	17.8
OC (%)	4.38	3.44	[illegible]
pH_{water}	5.6	5.7	5.6
Exchangeable cations ($cmol_c kg^{-1}$)			
Ca	2.39	1.43	1.22
Mg	0.45	0.47	0.39
Na	0.10	0.06	0.06
K	0.10	0.07	0.05
Al	0.16	0.07	0.05
CEC ($cmol_c kg^{-1}$)	25.0	26.9	29.2
ECEC ($cmol_c kg^{-1}$)	3.20	2.10	1.77
P retention (%)	63.0	87.0	92.0
Al_{ox} + ½Fe_{ox} (%)	1.07	1.38	2.41
BS (%)	12	8	6

Cunco soil series (Acrudoxic Hapludand), Region IX of Chile (CIREN 2002)

Depth (cm)	0–19	19–30	30–59	59–86	86–100
Genetic horizon	A_p	$2B_{w1}$	$3B_{w2}$	$3B_{w3}$	3BC
Particle size (mm) distribution					
2–0.05 (%)	52.7	32.0	33.9	40.4	23.1
0.05–0.002 (%)	31.6	43.6	39.9	39.1	44.6
<0.002 (%)	15.7	24.4	26.2	20.5	32.3
Bulk density (Mg m^{-3})	1.05	0.81	0.73	0.78	0.67
Water retention – 33 kPa (%)	26.5	41.0	50.8	47.2	56.3
Water retention – 1,500 kPa (%)	12.4	28.4	28.5	27.1	43.8
OC (%)	3.29	3.26	5.86	5.78	2.67
pH_{water}	6.0	6.3	5.4	5.9	6.0
Exchangeable cations ($cmol_c kg^{-1}$)					
Ca	3.96	1.70	0.90	0.93	0.56
Mg	1.52	1.23	0.25	0.32	0.26
Na	0.09	0.06	0.07	0.08	0.06
K	0.08	0.04	0.04	0.02	0.01
CEC ($cmol_c kg^{-1}$)	21.5	41.8	42.9	37.0	39.7
ECEC ($cmol_c kg^{-1}$)	5.65	3.03	1.31	1.38	0.89
P retention (%)	76.0	99.0	98.0	96.0	99.0
Al_{ox} + ½Fe_{ox} (%)	2.27	4.28	3.82	3.27	3.53
BS (%)	26	7	3	4	2

Antuco soil series (Humic Vitrixerand), Region VIII of Chile (CIREN 1999)

Depth (cm)	0–15	15–35	35–45	45–75
Genetic horizon	A_1	A_2	AC	C_1
Particle size (mm) distribution				
2–0.05 (%)	69.7	91.8	98.0	98.9
0.05–0.002 (%)	27.7	8.0	1.8	1.0
<0.002 (%)	2.6	0.2	0.2	0.1
Bulk density (Mg m^{-3})	1.38	1.59	1.67	1.62
Water retention –33 kPa (%)	11.9	8.8	5.8	5.8
Water retention –1,500 kPa (%)	5.8	4.7	2.3	2.3
OC (%)	2.47	0.88	0.32	0.16
pH_{water}	5.8	6.0	6.3	6.3

(continued)

(continued)

Exchangeable cations ($cmol_c kg^{-1}$)				
Ca	1.28	0.61	0.64	0.49
Mg	0.15	0.11	0.11	0.09
Na	0.03	0.01	0.00	0.01
K	0.13	0.06	0.03	0.03
Al	0.03	0.07	0.00	0.00
CEC ($cmol_c kg^{-1}$)	7.75	3.17	3.45	1.49
ECEC ($cmol_c kg^{-1}$)	1.62	0.86	0.78	0.61
Glass fraction (0.02–2 mm, %)	68.0	40.0	34.0	32.0
P retention (%)	35	43	22	18
Al_{ox} + ½Fe_{ox} (%)	0.80	1.06	0.62	0.56
BS (%)	21	25	23	41

2.2.2.3 Soils of the Longitudinal Central Valley

Maipo soil series (Fluventic Haploxeroll) at Metropolitan Region of Chile (CIREN 1996a)

Depth (cm)	0–16	16–33	33–50	50–74	74–92	92–120
Genetic horizon	A	A_2	B_1	B_2	C_1	C_2
Particle size (mm) distribution						
2–0.5 (%)	42.0	38.0	25.8	33.3	40.3	45.8
0.5–0.002 (%)	31.7	39.7	40.6	34.9	31.9	31.6
<0.002 (%)	27.3	22.3	33.6	31.8	27.8	22.6
Bulk density (Mg m^{-3})	1.5	1.6	1.7	1.9	1.9	1.8
Water retention –33 kPa (%)	21.0	22.0	24.0	23.0	27.0	22.0
Water retention –1,500 kPa (%)	13.0	14.0	16.0	16.0	18.0	13.0
OC (%)	1.5	1.2	1.2	0.8	0.4	0.3
$CaCO_3$ (%)	2.7	2.3	0.0	0.0	0.0	0.0
EC (dS m^{-1})	0.7	0.5	0.4	0.4	0.5	0.8
pH_{water}	8.0	8.0	7.8	7.7	7.7	7.6
Exchangeable cations ($cmol_c kg^{-1}$)						
Ca	–	–	21.3	19.5	19.0	20.0
Mg	–	–	2.3	2.1	2.2	2.1
Na	0.2	0.3	0.5	0.7	0.8	0.9
K	2.2	0.5	0.6	0.6	0.5	0.5
CEC ($cmol_c kg^{-1}$)	17.9	18.1	30.5	28.7	26.4	26.3
BS (%)	–	–	81	80	85	89

Agua del Gato soil series (Petrocalcic Calciaquoll) at Metropolitan Region of Chile (CIREN 1996a)

Depth (cm)	0–19	19–46	46–70	70–85
Genetic horizon	A_p	A_2	AC	$2C_{km}$
Particle size (mm) distribution				
2–0.05 (%)	14.1	11.1	17.1	35.6
0.05–0.002 (%)	55.4	49.9	46.9	41.9
<0.002 (%)	30.5	39.0	36.0	22.5
Bulk density (Mg m^{-3})	1.30	0.77	0.98	1.61
Water retention −33 kPa (%)	35.0	38.0	36.0	26.0
Water retention −1,500 kPa (%)	26.0	30.0	28.0	18.0
OC (%)	3.3	2.6	1.1	0.4
pH_{water}	7.8	7.5	7.5	8.0
EC (dS m^{-1})	0.8	1.3	1.4	1.7
$CaCO_3$ (%)	1.4	0.2	0.3	36.9
Exchangeable cations ($cmol_c kg^{-1}$)				
Ca	–	–	–	–
Mg	–	–	–	–
Na	1.5	2.5	1.6	1.5
K	0.8	0.7	0.7	0.2
CEC ($cmol_c kg^{-1}$)	42.0	50.9	40.9	20.5

Quillayes soil series (Aquic Haploxerept), Region VII of Chile (CIREN 1997b)

Depth (cm)	0–12	12–23	23–29	29–45
Genetic horizon	A_1	A_2	B_1	B_2
Particle size (mm) distribution				
2–0.05 (%)	59.3	50.9	45.1	46.7
0.05–0.002 (%)	28.0	29.1	28.7	25.7
<0.002 (%)	12.7	20.0	26.2	27.6
Water retention −33 kPa (%)	13.0	17.0	21.0	23.0
Water retention −1,500 kPa (%)	5.0	9.0	12.0	14.0
OC (%)	0.6	0.6	0.4	0.4
pH_{water}	6.2	6.3	6.4	6.2
Exchangeable cations ($cmol_c kg^{-1}$)				
Ca	1.7	3.0	4.0	4.2
Mg	0.5	1.1	1.8	2.0
Na	0.1	0.1	0.2	0.2
K	0.2	0.1	0.2	0.3
CEC ($cmol_c kg^{-1}$)	3.9	6.4	9.6	11.0
BS (%)	64	67	65	61

Talca soil series (Ultic Haploxeralf) at Region VII of Chile (CIREN 1997b)

Depth (cm)	0–16	16–70	70–100
Genetic horizon	A_p	B_t	B_2
Particle size (mm) distribution			
2–0.05 (%)	36.7	34.1	55.6
0.05–0.002 (%)	37.4	21.7	23.2
<0.002 (%)	25.9	45.2	21.2
Bulk density (Mg m^{-3})	1.7	1.5	1.6
Water retention −33 kPa (%)	20.0	25.0	22.0
Water retention −1,500 kPa (%)	11.0	20.0	15.0
OC (%)	1.2	0.3	0.2
pH_{water}	5.8	6.4	6.6
Exchangeable cations ($cmol_c kg^{-1}$)			
Ca	4.0	5.2	5.0
Mg	1.1	2.3	2.3
Na	0.1	0.2	0.3
K	0.4	0.2	0.2
CEC ($cmol_c kg^{-1}$)	10.4	16.3	14.4
BS (%)	55	48	53

Metrenco soil series (Typic Paleudult), Region IX of Chile (CIREN 2002)

Depth (cm)	0–13	13–43	43–66	66–85	85–120
Genetic horizon	A_1	B_{t1}	B_{t2}	B_{t3}	B_{t4}
Particle size (mm) distribution					
2–0.05 (%)	11.9	11.5	14.2	13.7	16.3
0.05–0.002 (%)	58.0	35.1	37.8	35.9	38.2
<0.002 (%)	30.1	53.4	48.0	50.3	45.5
Bulk density (Mg m^{-3})	0.98	1.07	0.96	1.11	1.13
Water retention −33 kPa (%)	29.4	33.5	34.7	34.5	32.7
Water retention −1,500 kPa (%)	22.3	26.2	26.6	26.1	25.1
OC (%)	1.94	0.45	0.28	0.35	0.33
pH_{water}	5.3	5.4	5.6	5.4	5.5
Exchangeable cations ($cmol_c kg^{-1}$)					
Ca	3.73	4.62	4.07	4.05	3.89
Mg	1.77	2.48	2.59	2.62	2.50
Na	0.49	0.16	0.05	0.07	0.08
K	0.05	0.09	0.12	0.11	0.09
CEC ($cmol_c kg^{-1}$)	13.3	15.8	17.3	13.1	16.7

(continued)

(continued)

ECEC ($cmol_c kg^{-1}$)	6.30	7.35	6.85	6.91	6.60
P retention (%)	57.0	55.0	55.0	53.0	53.0
Al_{ox} + ½Fe_{ox} (%)	0.79	0.74	0.67	0.59	0.58
BS (%)	33	46	40	52	35

Nueva Imperial soil series (Typic Paleudult), Region IX of Chile (CIREN 2002)

Depth (cm)	0–11	11–25	25–55
Genetic horizon	A_p	B_1	B_2
Particle size (mm) distribution			
2–0.05 (%)	11.1	8.7	14.7
0.05–0.002 (%)	45.3	34.3	31.6
<0.002 (%)	43.6	57.1	53.7
Bulk density (Mg m^{-3})	1.03	1.11	1.02
Water retention −33 kPa (%)	38.0	35.9	39.0
Water retention −1,500 kPa (%)	29.4	29.9	32.0
OC (%)	4.71	2.61	1.79
pH_{water}	5.7	6.4	6.6
Exchangeable cations ($cmol_c kg^{-1}$)			
Ca	8.06	9.95	11.07
Mg	2.12	2.48	3.05
Na	0.22	0.37	0.64
K	0.09	0.04	0.04
CEC ($cmol_c kg^{-1}$)	33.0	34.7	30.0
ECEC ($cmol_c kg^{-1}$)	10.66	12.90	14.81
Al_{ox} + ½Fe_{ox} (%)	1.66	1.41	1.12
BS (%)	32	37	49

Toltén soil series (Acrudoxic Hapludand), Region IX of Chile (CIREN 2002)

Depth (cm)	0–20	20–40	40–70	70–100	100–120
Genetic horizon	A_1	B_1	B_2	B_3	B_4
Particle size (mm) distribution					
2–0.05 (%)	18.7	13.9	9.7	15.0	20.4
0.05–0.002 (%)	50.0	58.4	48.4	56.4	54.5
<0.002 (%)	31.2	27.8	41.9	28.5	25.0
Bulk density (Mg m^{-3})	0.61	0.65	0.93	0.86	0.90
Water retention −33 kPa (%)	57.3	52.4	42.4	40.3	44.4
Water retention −1,500 kPa (%)	36.0	36.6	31.7	33.5	33.1
OC (%)	10.73	3.62	2.31	1.88	1.80
pH_{water}	5.8	6.2	5.9	5.9	5.9
Exchangeable cations ($cmol_c kg^{-1}$)					
Ca	2.68	0.34	0.23	0.28	0.20
Mg	1.10	0.21	0.21	0.25	0.18
Na	0.15	0.10	0.06	0.07	0.10
K	0.14	0.09	0.13	0.11	0.09
CEC ($cmol_c kg^{-1}$)	40.5	28.7	23.7	26.9	24.5
ECEC ($cmol_c kg^{-1}$)	4.29	0.74	0.64	0.73	0.60
P retention (%)	98	100	100	100	100
Al_{ox} + ½Fe_{ox} (%)	2.99	3.97	3.27	3.99	3.92
BS (%)	10	3	3	3	2

(continued)

Corte Alto soil series (Typic Hapludand), Region X of Chile (CIREN 2003)

Depth (cm)	0–18	18–47	47–77	77–122	122–170	170–194
Genetic horizon	A	B_{w1}	B_{w2}	$2C_1$	$2C_2$	$3C_3$
Particle size (mm) distribution						
2–0.05 (%)	20.4	29.1	49.5	80.2	68.9	53.5
0.05–0.002 (%)	33.8	41.5	38.6	18.1	30.6	31.0
<0.002 (%)	45.8	29.4	11.9	1.7	0.5	15.5
Bulk density (Mg m^{-3})	0.86	0.88	0.93	0.85	0.66	0.76
Water retention −33 kPa (%)	41.7	42.7	45.7	55.3	75.6	61.1
Water retention −1,500 kPa (%)	28.7	24.4	24.6	22.3	32.5	26.8
OC (%)	6.76	1.26	0.84	0.88	0.43	0.50
pH_{water}	5.6	5.3	5.8	6.2	6.2	6.3
Exchangeable cations ($cmol_c kg^{-1}$)						
Ca	9.4	3.1	4.8	2.0	1.7	6.0
Mg	2.4	1.0	1.2	0.4	0.2	1.6
Na	0.1	0.2	0.7	0.3	0.2	1.8
K	1.6	1.2	0.1	0.1	0.2	0.2
Al	0.7	1.1	0.2	0.1	0.1	0.2
CEC ($cmol_c kg^{-1}$)	46.8	34.2	32.9	29.5	33.7	30.9
ECEC ($cmol_c kg^{-1}$)	14.2	6.6	7.0	2.9	2.4	9.8
P retention (%)	89.0	96.0	99.0	99.0	99.0	91.0
Al_{ox} + ½Fe_{ox} (%)	1.9	2.2	2.15	5.3	7.0	2.1
BS (%)	34	24	32	14	11	37

Maullín soil series (Hydric Endoaquand), Region X of Chile (CIREN 2003)

Depth (cm)	0–13	13–30	30–52	52–90	90–112
Genetic horizon	A_1	A_2	BA	B_t	B_{qs}
Particle size (mm) distribution					
2–0.05 (%)	78.9	84.8	88.9	73.2	81.2
0.05–0.002 (%)	10.4	4.4	4.4	17.3	9.3
<0.002 (%)	10.8	10.8	6.8	9.5	9.5
Bulk density ($Mg\ m^{-3}$)	0.54	0.60	0.57	0.52	1.06
Water retention – 33 kPa (%)	60.5	34.4	38.8	27.5	26.3
Water retention – 1,500 kPa (%)	47.9	31.1	34.1	19.5	21.0
OC (%)	18.9	13.3	8.0	3.0	5.3
pH_{water}	4.9	5.3	5.4	5.5	5.5
Exchangeable cations ($cmol_c kg^{-1}$)					
Ca	0.40	0.19	0.12	0.10	0.17
Mg	0.26	0.13	0.10	0.03	0.05
Na	0.17	0.10	0.10	0.07	0.10
K	0.26	0.15	0.08	0.03	0.02
CEC ($cmol_c kg^{-1}$)	59.5	53.1	62.9	38.7	41.0
ECEC ($cmol_c kg^{-1}$)	3.30	1.40	0.56	0.24	0.35
P retention (%)	97.0	99.0	99.0	99.0	99.0
Al_{ox} + ½Fe_{ox} (%)	1.54	2.77	6.48	4.44	6.30
BS (%)	2	1	1	1	1

Arenales soil series (Dystric Xeropsamment), Region VIII of Chile (CIREN 1999)

Depth (cm)	0–17	17–36	36–56	56–80	80–100	100–150
Genetic horizon	A_1	C_1	C_2	C_3	C_4	C_5
Particle size (mm) distribution						
2–0.05 (%)	95.6	100.0	99.6	99.4	99.3	98.6
0.05–0.002 (%)	4.4	0.0	0.4	0.6	0.7	1.4
<0.002 (%)	0.0	0.0	0.0	0.0	0.0	0.0
Bulk density ($Mg\ m^{-3}$)	1.53	1.53	1.51	1.52	1.62	1.69
Water retention – 33 kPa (%)	6.8	4.1	3.7	2.9	3.5	3.4
Water retention – 1,500 kPa (%)	2.9	2.1	2.1	2.6	2.6	2.3
OC (%)	0.99	0.11	0.16	0.24	0.11	0.07
pH_{water}	6.2	6.4	6.4	6.3	6.4	6.4
Exchangeable cations ($cmol_c kg^{-1}$)						
Ca	1.84	0.85	0.76	0.93	1.18	1.15
Mg	0.35	0.15	0.26	0.38	0.56	0.87
Na	0.03	0.03	0.02	0.03	0.04	0.04
K	0.13	0.05	0.10	0.14	0.17	0.15
CEC ($cmol_c kg^{-1}$)	3.98	3.3	2.3	1.91	4.56	2.6
ECEC ($cmol_c kg^{-1}$)	2.35	1.08	1.15	1.48	1.95	2.21
P retention (%)	14	13	12	12	18	5
BS (%)	59	33	50	78	43	85

2.2.2.4 Soils of the Coastal Range

Lo Vásquez soil series (Ultic Haploxeralf), Region V of Chile (CIREN 1997a)

Depth (cm)	0–18	18–32	32–43	43–57	57–78
Genetic horizon	A_1	B_{t1}	B_{t2}	B_{t3}	B_4
Particle size (mm) distribution					
2–0.05 (%)	55.8	46.9	48.1	56.8	55.3
0.05–0.002 (%)	29.4	24.7	21.1	14.0	27.5
<0.002 (%)	14.8	28.4	30.8	29.2	17.2
Bulk density ($Mg\ m^{-3}$)	1.6	1.8	1.8	1.8	1.3
Water retention −33 kPa (%)	21.0	24.0	22.0	22.0	20.0
Water retention −1,500 kPa (%)	9.0	9.0	12.0	12.0	11.0
OC (%)	1.1	0.5	0.4	0.2	0.2
pH_{water}	6.6	6.7	6.7	6.7	6.7
Exchangeable cations ($cmol_c kg^{-1}$)					
Ca	8.5	11.9	10.6	13.8	14.8
Mg	2.3	3.4	3.3	4.9	5.3
Na	0.1	0.2	0.3	0.3	0.4
K	0.1	0.2	0.2	0.2	0.1
CEC ($cmol_c kg^{-1}$)	13.8	21.6	22.4	23.9	24.8
BS (%)	30	73	64	80	83

Correltúe soil series (Andic Haplohumult), Region IX of Chile (CIREN 2002)

Depth (cm)	0–7	7–28	28–64	64–98	98–130
Genetic horizon	A_1	A_2	B_{t1}	B_{t2}	B_{t3}
Particle size (mm) distribution					
2–0.05 (%)	18.7	19.5	10.8	9.9	12.2
0.05–0.002 (%)	37.2	42.8	40.7	44.0	40.9
<0.002 (%)	44.1	37.6	48.5	46.1	47.0
Bulk density (Mg m^{-3})	0.67	0.84	0.69	0.71	0.79
Water retention −33 kPa (%)	56.9	62.0	47.9	41.2	37.7
Water retention −1,500 kPa (%)	38.9	39.8	34.4	28.7	27.2
OC (%)	11.99	8.30	1.86	1.32	1.12
pH_{water}	5.6	4.9	5.2	5.4	5.2
Exchangeable cations ($cmol_ckg^{-1}$)					
Ca	1.60	0.14	0.13	0.21	0.33
Mg	0.53	0.05	0.03	0.03	0.05
Na	0.16	0.10	0.05	0.07	0.08
K	0.15	0.07	0.03	0.03	0.02
Al	0.70	0.48	0.01	0.01	0.00
CEC ($cmol_ckg^{-1}$)	36.6	32.2	13.3	18.2	17.2
ECEC ($cmol_ckg^{-1}$)	3.14	0.84	0.25	0.35	0.48
P retention (%)	96.0	97.0	92.0	94.0	89.0
Al_{ox} + ½Fe_{ox} (%)	1.96	2.31	1.40	1.46	1.17
BS (%)	7	1	2	2	3

Los Copihues soil series (Oxiaquic Dystrudept), Region IX of Chile (CIREN 2002)

Depth (cm)	0–19	19–48	48–84	84–120
Genetic horizon	A_p	B_1	B_2	B_3
Particle size (mm) distribution				
2–0.05 (%)	37.5	29.6	26.6	28.3
0.05–0.002 (%)	35.3	27.3	22.8	26.5
<0.002 (%)	27.3	43.1	50.7	45.2
Bulk density (Mg m^{-3})	1.44	1.36	1.31	1.36
Water retention −33 kPa (%)	24.5	21.4	23.5	24.9
Water retention −1,500 kPa (%)	11.3	13.3	15.6	16.0
OC (%)	2.34	0.83	0.59	0.46

(continued)

(continued)

pH_{water}	5.5	5.7	5.9	5.8
Exchangeable cations ($cmol_ckg^{-1}$)				
Ca	2.19	1.53	1.87	2.04
Mg	1.18	0.98	1.21	1.24
Na	0.11	0.10	0.10	0.11
K	0.32	0.19	0.15	0.12
CEC ($cmol_ckg^{-1}$)	15.3	13.9	12.0	17.6
ECEC ($cmol_ckg^{-1}$)	4.06	3.63	3.69	3.66
P retention (%)	47.0	50.0	52.0	50.0
Al_{ox} + ½Fe_{ox} (%)	0.51	0.40	0.36	0.32
BS (%)	25	20	28	20

Hueñi soil series (Andic Dystrudept), Region X of Chile (CIREN 2003)

Depth (cm)	0–14	14–34	34–72	72–120
Genetic horizon	A_1	B	$2C_1$	$2C_2$
Particle size (mm) distribution				
2–0.05 (%)	56.9	62.9	74.9	74.9
0.05–0.002 (%)	23.6	19.6	13.6	13.6
<0.002 (%)	19.4	17.4	11.4	11.4
Bulk density (Mg m^{-3})	0.88	0.96	1.31	1.36
Water retention − 33 kPa (%)	30.5	25.9	12.7	11.1
Water retention − 1,500 kPa (%)	22.5	21.3	8.1	6.4
OC (%)	6.16	3.60	0.93	0.47
pH_{water}	5.27	5.36	5.27	5.23
Exchangeable cations ($cmol_ckg^{-1}$)				
Ca	0.24	0.11	0.12	0.13
Mg	0.23	0.10	0.05	0.05
Na	0.22	0.20	0.11	0.09
K	0.17	0.08	0.04	0.05
CEC ($cmol_ckg^{-1}$)	39.9	40.0	12.5	9.1
ECEC ($cmol_ckg^{-1}$)	2.48	1.74	1.40	1.25
P retention (%)	87.0	87.0	45.0	34.0
Al_{ox} + ½Fe_{ox} (%)	1.37	1.66	0.52	0.36
BS (%)	2	1	3	4

2.2.3 Soils of the Rainy and Patagonian zone

2.2.3.1 Soils of the West of North Patagonia

Matanzas soil series (Oxic Haplustoll), Region VI of Chile (CIREN 1996b)

Depth (cm)	0–20	20–64	64–95
Genetic horizon	A_1	B_1	B_2
Particle size (mm) distribution			
2–0.05 (%)	47.4	39.5	37.5
0.05–0.002 (%)	37.4	43.0	44.4
<0.002 (%)	15.2	17.5	18.1
OC (%)	2.2	1.5	0.4
pH_{water}	5.8	6.4	6.8
Exchangeable cations ($cmol_c kg^{-1}$)			
Ca	7.85	9.06	5.16
Mg	4.30	3.90	4.38
Na	0.40	0.38	0.85
K	1.23	0.69	0.31
CEC ($cmol_c kg^{-1}$)	17.40	16.89	12.58
Na saturation (%)	2.3	2.2	6.8
BS (%)	79	83	85

Pollux soil series (Typic Hapludand), Region XI of Chile (CIREN 2005b)

Depth (cm)	0–23	23–37	37–61	61–77	77–95	95–120
Genetic horizon	A_1	A_2	B	2C	$3A_b$	$3B_b$
Particle size (mm) distribution						
2–0.05 (%)	50.4	55.9	56.3	51.2	48.2	42.1
0.05–0.002 (%)	39.3	36.8	36.5	41.9	43.5	48.4
<0.002 (%)	10.3	7.3	7.2	6.9	8.3	9.5
Bulk density ($Mg\ m^{-3}$)	0.70	0.77	0.78	0.72	0.70	0.69
Water retention −33 kPa (%)	43.1	31.9	31.6	24.7	40.8	46.9
Water retention −1,500 kPa (%)	38.1	22.0	20.0	10.8	20.2	26.6
OC (%)	7.66	4.52	2.96	0.93	1.57	2.26
pH_{water}	6.21	6.42	6.56	6.51	6.56	6.38
Exchangeable cations ($cmol_c kg^{-1}$)						
Ca	17.72	13.37	10.96	4.39	7.05	8.16
Mg	3.48	2.20	2.07	1.11	2.20	2.83
Na	0.07	0.15	0.13	0.07	0.13	0.23
K	0.94	0.43	0.35	0.15	0.23	0.38
CEC ($cmol_c kg^{-1}$)	52.2	43.2	37.3	18.1	36.5	44.3
ECEC ($cmol_c kg^{-1}$)	22.2	16.2	13.5	5.7	9.6	11.6
P retention (%)	81	88	87	59	90	96
Al_{ox} + ½Fe_{ox} (%)	2.01	2.61	3.18	1.54	3.66	4.57
BS (%)	43	37	36	32	26	26

Coyhaique soil series (Andic Distrudept) on alluvial terraces, Region XI of Chile (CIREN 2005b).

Depth (cm)	0–23	23–37	37–61	61–77	77–95	95–120
Genetic horizon	A_1	A_2	B	$2B_b$	$3B_1$	$3B_2$
Particle size (mm) distribution						
2–0.05 (%)	60.5	60.4	54.2	43.0	22.1	32.7
0.05–0.002(%)	27.4	26.6	35.2	40.9	40.7	36.0
<0.002 (%)	12.2	13.0	10.6	16.1	27.2	31.3
Bulk density ($Mg\ m^{-3}$)	0.90	0.90	0.83	0.77	0.85	0.95
Water retention −33 kPa (%)	24.2	23.8	37.7	36.9	34.7	28.0
Water retention −1,500 kPa (%)	12.1	10.8	15.0	15.5	16.6	16.1
OC (%)	2.73	2.38	2.55	1.80	1.33	0.31
pH_{water}	6.51	6.34	6.68	6.87	6.84	7.01
Exchangeable cations ($cmol_c kg^{-1}$)						
Ca	6.26	5.91	9.32	7.91	8.41	8.29
Mg	2.[illegible]9	1.91	2.47	2.89	3.86	3.59
Na	1.25	0.87	0.85	1.38	1.74	1.09
K	0.27	0.08	0.14	0.13	0.24	0.36
CEC ($cmol_c kg^{-1}$)	18.5	18.7	26.0	24.8	26.8	22.0
ECEC ($cmol_c kg^{-1}$)	10.0	8.8	12.8	12.3	14.3	13.4
P retention (%)	37	43	70	66	58	49
Al_{ox} + ½Fe_{ox}(%)	1.00	1.17	2.10	2.06	1.56	1.00
BS (%)	54	47	49	50	53	6[illegible]
Optical density	0.15	0.16	0.27	0.16	0.12	0.08
Melanic index	1.92	2.03	1.64	1.77	1.87	2.02

2.2.3.2 Soils of the East of North Patagonia

Chile Chico soil series (Oxyaquic Haploxeroll), Region XI of Chile (CIREN 2005b)

Depth (cm)	0–13	13–32	32–56	56–70	70–90
Genetic horizon	A_p	B_{w1}	B_{w2}	BC	C
Particle size (mm) distribution					
2–0.05 (%)	38.8	16.3	32.6	38.1	63.0
0.05–0.002 (%)	30.2	34.9	31.7	31.6	22.3
<0.002 (%)	31.0	48.8	35.7	30.3	14.7
Bulk density ($Mg\ m^{-3}$)	0.90	1.00	1.01	1.05	1.03
Water retention −33 kPa (%)	30.0	36.5	26.9	24.0	13.7
Water retention −1,500 kPa (%)	19.4	24.3	14.4	12.3	7.1
OC (%)	3.60	1.97	1.04	0.87	0.35
pH_{water}	6.88	6.99	7.03	7.11	7.08
Exchangeable cations ($cmol_c kg^{-1}$)					
Ca	22.96	29.53	24.69	21.36	12.70
Mg	4.40	6.11	5.54	4.96	3.20
K	1.40	0.80	0.48	0.35	0.29
Na	0.20	0.22	0.26	0.26	0.18
CEC ($cmol_c kg^{-1}$)	31.2	42.4	38.2	33.8	18.9
ECEC ($cmol_c kg^{-1}$)	29.0	36.7	31.0	26.9	16.4
P retention (%)	12	21	18	17	8
Al_{ox} + ½Fe_{ox} (%)	0.26	0.33	0.38	0.30	0.18
BS (%)	93	86	81	80	87
Optical density	0.07	0.08	0.06	0.05	0.02
Melanic index	3.42	4.07	3.30	3.00	4.00

Cochrane soil series (Andic Oxyaquic Dystrudept), Region XI of Chile (CIREN 2005b)

Depth (cm)	0–9	9–40	40–80
Genetic horizon	A	B	2B
Particle size (mm) distribution			
2–0.05 (%)	39.0	39.4	61.1
0.05–0.002 (%)	49.5	41.8	9.6
<0.002 (%)	11.5	18.8	29.3
Bulk density ($Mg\ m^{-3}$)	0.80	0.75	0.95
Water retention −33 kPa (%)	29.9	36.9	34.5
Water retention −1,500 kPa (%)	18.4	13.8	7.6
OC (%)	0.58	2.09	0.58
pH_{water}	6.45	5.25	5.86

(continued)

(continued)

Exchangeable cations ($cmol_c kg^{-1}$)			
Ca	10.55	0.40	0.80
Mg	2.68	0.23	0.35
K	1.31	0.46	0.20
Na	0.04	0.07	0.09
CEC ($cmol_c kg^{-1}$)	27.9	18.9	9.8
ECEC ($cmol_c kg^{-1}$)	14.6	2.6	3.3
P retention (%)	39	60	27
Al_{ox} + ½Fe_{ox} (%)	1.03	1.27	0.43
BS (%)	52	6	15
Optical density	0.25	0.23	0.03
Melanic index	2.07	2.02	1.93

2.2.3.3 Soils of South Patagonia (*Magallanes*)

Las Chinas soil series, Region XII of Chile (CNR 1997)

Depth (cm)	0–20	20–50
Genetic horizon	A	B
pH_{water}	9.25	9.39
EC ($dS\ m^{-1}$)	7.8	10.8
$CaCO_3$ (%)	3.4	2.7
Soluble cations ($cmol_c kg^{-1}$)		
Ca	3.2	3.5
Mg	1.4	1.7
Na	89.1	123.2
K	0.2	0.2
Soluble anions ($cmol_c kg^{-1}$)		
CO_3	0.0	0.0
HCO_3	4.8	4.0
Cl	34.0	51.8
SO_4	53.0	74.0
SAR (–)	59.1	76.8

Podzolic soil near Punta Arenas city, Region XII of Chile (Díaz et al. 1959–1960)

Depth (cm)	0–2	2–6	6–10	10–15	>15
Genetic horizon	A_o	A_2	B_2	C_1	C_2
Particle size (mm) distribution					
2–0.05 (%)	–	44.7	39.3	64.7	62.6
0.05–0.002 (%)	–	47.0	38.5	31.2	30.3
<0.002 (%)	–	8.3	22.2	4.1	7.1

(continued)

(continued)

OC (%)	–	0.73	4.45	0.97	0.44
pH_{water}	5.8	4.9	4.6	5.0	5.2
Exchangeable cations ($cmol_c kg^{-1}$)					
Ca	8.6	1.7	2.1	0.3	0.5
Mg	2.5	0.6	1.1	0.1	0.2
Na	0.3	0.2	0.4	0.2	0.1
K	1.0	0.2	0.3	0.2	0.2
CEC ($cmol_c kg^{-1}$)	33.6	9.3	47.5	21.0	15.9
BS (%)	37	29	8	4	6

2.2.4 Soils of the Insular (Easter-Juan Fernández) and Antarctic Zone

Hotu-Matua soil, Easter Island at Region V of Chile (Díaz 1949)

Depth (cm)	0–50	50–100	100–150
pH_{water}	5.84	6.15	6.34
$N–NO_3$ (%)	0.004	0.004	0.0006
$P–P_2O_5$ (%)	0.014	0.031	0.014
$K–K_2O$ (%)	0.028	0.008	0.015
Ca ($cmol_c kg^{-1}$)	8.13	5.86	6.86
Mg ($cmol_c kg^{-1}$)	4.48	2.78	4.21
CEC ($cmol_c kg^{-1}$)	36.91	30.70	60.66

Orito soil series, Easter Island at Region V of Chile (Alcayaga and Narbona 1969)

Depth (cm)	0–16	16–37	37–62	62–76
Particle size (mm) distribution				
2–0.05 (%)	14.08	9.68	24.97	–
0.05–0.002 (%)	24.85	41.96	38.40	–
<0.002 (%)	60.37	43.39	35.96	–
Water retention −33 kPa (%)	51.9	49.8	68.0	70.2
Water retention −1,500 kPa (%)	34.3	32.0	39.5	37.2
pH_{water}	6.3	6.4	6.4	6.0
OC (%)	4.25	3.08	1.71	1.47
Exchangeable cations ($cmol_c kg^{-1}$)				
Ca	3.5	3.87	3.75	2.01
Mg	5.96	6.72	5.72	4.00
K	0.77	0.91	0.18	0.17
Na	1.08	1.38	2.23	1.98
CEC ($cmol_c kg^{-1}$)	51.15	47.68	48.96	44.62

(continued)

Yunque soil series, Juan Fernández Archipelago at Region V of Chile (IREN 1982)

Depth (cm)	0–7	7–18	18–50
Water retention −33 kPa (%)	63.7	60.9	53.8
Water retention −1,500 kPa (%)	50.5	47.1	44 4
OC (%)	14.62	11.02	9.92
pH_{water}	6.0	6.2	6.8
EC ($dS\ m^{-1}$)	0.26	0.20	0.22
Extractable cations ($cmol_c kg^{-1}$)			
Ca	29.75	31.75	36.00
Mg	17.48	15.62	16.45
K	0.52	0.65	0.73
Na	2.55	2.31	2.76
CEC ($cmol_c kg^{-1}$)	75.62	79.37	71.25
BS (%)	66	63	73

Puerto Inglés soil series, Juan Fernández Archipelago at Region V of Chile (IREN (1982)

Depth (cm)	0–3	3–14	14–58	58–87	87–120
Water retention −33 kPa (%)	51.1	44.4	38.3	35.1	40.1
Water retention −1,500 kPa (%)	41.7	32.6	25.4	22.7	23.7
OC (%)	9.74	1.51	2.36	1.68	0.75
pH_{water}	5.5	5.8	5.75	5.6	5.3
EC ($dS\ m^{-1}$)	0.40	0.14	0.03	0.01	0.08
Extractable cations ($cmol_c kg^{-1}$)					
Ca	16.94	12.95	9.57	10.80	5.00
Mg	18.04	16.34	12.49	12.03	13.16
K	0.86	0.72	0.87	0.97	1.90
Na	1.43	1.27	0.56	0.17	0.12
CEC ($cmol_c kg^{-1}$)	59.37	47.50	43 75	40.62	37.50
BS (%)	63	66	54	59	54

Pedon at Robert Island, Antarctic Chilean territory (Haberland 1992)

Soil depth (cm)	0–3	3–8	8–18
Genetic horizon	A	AC	C
Textural class	Sandy loam	Loamy sand	Sandy loam
Water retention −33 kPa (%)	18.3	16.0	26.2
Water retention −1,500 kPa (%)	12.2	11.7	16.5
OC (%)	1.7	1.1	0.3
pH	5.2	4.8	6.1
Sum of bases ($cmol_c kg^{-1}$)	32.9	24.5	67.3
CEC ($cmol_c kg^{-1}$)	34.0	33.5	51.5

Pedon at Robert Island, Antarctic Chilean territory (Álvarez 1993)

Soil depth (cm)	0–8	8–30	30–47
Genetic horizon /layer	A	C	R
Textural class	Sandy loam	Sandy loam	Loamy sand
Water retention −33 kPa (%)	26.4	25.5	22.2
Water retention −1,500 kPa (%)	15.5	14.6	14.8
OC (%)	0.29	0.13	0.12
pH	7.0	7.55	7.65
Sum of bases ($cmol_c kg^{-1}$)	57.7	60.8	61.7
CEC ($cmol_c kg^{-1}$)	49.3	52.0	56.0

Pedon at Livingstone Island, Antarctic Chilean territory (Henríquez 1994)

Soil depth (cm)	0–8	8–14	14–33	33–70
Genetic horizon	A_1	$2A_2$	$3C_1$	$3C_2$
Textural class	Sandy loam	Sandy loam	Sandy loam	Sandy loam
Water retention −33 kPa (%)	16.2	21.4	18.9	18.7
Water retention −1,500 kPa (%)	11.5	12.6	13.4	13.5
OC (%)	0.39	0.17	0.15	0.15
pH	6.3	7.2	7.4	7.6
Sum of bases ($cmol_c kg^{-1}$)	19.5	25.1	15.3	14.2
CEC ($cmol_c kg^{-1}$)	22.9	21.5	19.4	18.2

Some soil properties of a Lithic Fibristel at Antarctic territory (Michel et al. 2006)

Depth (cm)	0–10	10–20	20–30	30–40	40–50
Particle size (mm) distribution					
2–0.05 (%)	53	57	64	52	51
0.05–0.002 (%)	27	27	24	31	34
<0.002 (%)	20	16	12	17	15
pH_{water}	5.0	4.8	4.6	4.6	4.3
pH_{KCl}	3.9	3.4	3.4	3.4	3.3
N (%)	0.17	0.10	0.15	0.17	0.13
K ($mg\ dm^{-3}$)	144	196	198	182	196
Na ($mg\ dm^{-3}$)	184	162	204	146	174
P ($mg\ dm^{-3}$)	933	562	696	658	950
Al^{3+} ($cmol_c dm^{-3}$)	1.8	3.8	4.6	3.8	6.2
$H^+ + Al^{3+}$ ($cmol_c dm^{-3}$)	18.8	24.4	23.8	29.4	38.9
Ca^{2+} ($cmol_c dm^{-3}$)	3.32	3.70	3.85	3.31	3.69
Mg^{2+} ($cmol_c dm^{-3}$)	1.39	1.34	1.52	1.11	1.21
CEC ($cmol_c dm^{-3}$)	7.68	10.04	11.37	9.32	12.36

Authors' Biographies

Manuel Casanova obtained a degree in Agricultural Engineering (1991) at the University of Chile. He received his post-graduate education at the Department of Soil and Environment, Swedish University of Agricultural Sciences (SLU), Sweden, where he obtained a Master of Science degree majoring in Soil Conservation (1998). Nowadays, he is an Associate Professor and head of the Department of Soil and Engineering, Faculty of Agronomic Sciences, University of Chile, where he teaches and conducts research, mainly on soil and water conservation.

Oscar Seguel obtained a degree in Agricultural Engineering (1997) at the University of Chile. He received his post-graduate education at the Institute of Agrarian Engineering and Soil, Austral University, Chile, where he obtained a PhD degree in the subject of Soil Sciences (2005). Nowadays, he is an Assistant Professor at the Department of Soil and Engineering, University of Chile, Faculty of Agronomic Sciences, where he teaches and conducts research, mainly on soil physics.

Osvaldo Salazar obtained a degree in Agricultural Engineering (2004) at the University of Chile. He received his post-graduate education at the Department of Soil and Environment, Swedish University of Agricultural Sciences (SLU), Sweden, where he obtained a Master of Science degree majoring in Soil Sciences (2003) and a PhD degree in Soil Science specialising in hydrotechnics (2009). He held a post-doc position at the Basic Science and Environment Department, University of Copenhagen, Denmark (2010). Nowadays, he is an Assistant Professor at the Department of Soil and Engineering, Faculty of Agronomic Sciences, University of Chile, where he teaches and conducts research, mainly on soil fertility.

Walter Luzio obtained a degree in Agricultural Engineering (1963) at the University of Chile. He received his post-graduate education at the University of Ghent, Belgium, where he obtained a Master of Science degree majoring in Soil Surveys (1971). He was a Professor at the Department of Soil and Engineering, University of Chile, until his retirement in 2010. He taught and conducted research on pedology, soil classification and soil cartography for the past 40 years.

M. Casanova et al., *The Soils of Chile*, World Soils Book Series, DOI: 10.1007/978-94-007-5949-7,

Index

A

Aeration, 128, 129
Agriculture, 7, 21, 23, 55, 66, 102, 113, 115, 121, 122, 124, 145, 150
Alfisol, 25, 55, 64, 71, 100
Alluvial deposits, 25, 69, 73
Alluvialsoils, 26, 41, 47, 49, 75
Altiplano, 13, 21, 26–28, 100, 150, 159
Aluandic Andosol, 25, 83
Amolana, 45, 49, 110
Andean chain, 5
Andean foreland, 1
Andes, 1, 4–7, 9, 31, 37, 41, 44, 49, 51, 52, 124, 126
Andisol, 25, 29, 35, 59–63, 66, 69, 73, 75, 100, 110, 113, 132, 152
Andosol, 25, 29, 83
Antarctic peninsula, 3, 89
Antofagasta, 1, 7, 52, 53, 150
Apacheta, 45, 48, 162
Archipelago, 5, 9, 34, 67, 80, 83, 84, 179
Argentina, 1, 3, 26, 102, 126
Argillic horizons, 25, 34, 52, 53, 63, 69, 84, 100
Aridisol, 37, 43–45, 100, 111, 113, 132, 133
Aridity, 13, 19, 25, 26, 77, 86, 150
The Austral volcanic zone, 4

B

Bolivia, 1, 13, 17
Brine, 39
Browningia candelaris, 18

C

Calcification, 26
Calcium, 27, 99, 104
Cambic horizon, 26, 47, 49, 52, 53, 56, 65, 69
Cenozoic , 1, 4, 5
Central depression, 4
Chanchoquín, 49, 163
Chilean Antarctic Territory, 1, 4
Chilean territory, 1, 4, 7, 9, 21, 25–27, 55, 179
Choapa basin, 49
Climate zone, 7, 152
Coastal range, 4, 5, 7, 15, 25, 31, 34, 37, 38, 51–53, 56, 63, 68, 100, 107, 122, 125, 134, 167, 175
Colluvial deposits, 60, 70, 72
Colluvial soils, 44, 45, 49
Compaction, 112, 128–133
Concordia line, 1
Crop, 16, 21, 52, 63, 73, 99, 102, 105, 125, 129, 136, 137, 150

D

Degradation, 53, 105, 12[illegible], 125–127, 141, 145, 149, 152
Desert, 7, 17, 26, 40, 113, 126, 160
Desertification, 121, 144, 145, 148
Drainage, 7, 36, 41, 45–49, 59, 64, 69, 108, 125, 135
Dunes, 7, 8, 45, 53, 70, 127, 134, 135

E

Easter Island, 4, 9, 34, 80, 81, 149, 150
El Niño, 11, 122
Elqui River, 47, 164
Entisol, 26, 44, 58, 75, 78, 100, 121
Erosion, 4, 7, 9, 27, 52, 53, 56, 69, 80, 84, 121, 123, 127, 135, 147

F

Fertilisation, 102, 103
Fertilisers, 102, 103, 137, 140, 141
Festuca spp, 18, 58
Fjord, 9, 20
Fluvial deposits, 41
Fluvio-glacial soils, 25, 47, 61, 63, 66, 73, 76
Forearc, 4, 5, 56
Forest, 19, 60, 73, 84, 103, 113, 115, 134, 144, 150

G

Glacial deposits, 25, 51, 68, 71, 73, 76
Glacial soils, 25
Glaciations, 5, 7
Glacier, 9, 16, 66, 71, 150
Gondwana, 1

H

Histosol, 29, 44, 51, 77, 80, 104, 114, 121
Holocene, 7–9, 26, 37, 53, 62, 68, 73, 134
Humidity, 18, 83, 145
Humus, 26, 81, 135
Hyperarid desert, 17

M. Casanova et al., *The Soils of Chile*, World Soils Book Series, DOI: 10.1007/978-94-007-5949-7,

I
Inceptisol, 25, 50, 62, 69, 75, 78, 100, 107, 108
Indurate soils, 34, 53, 168
Iodates, 27

J
Juan Fernández Archipelago, 3, 9, 53, 80, 84
Juan Fernández Ridge, 3, 53, 84, 85

L
Land use, 20, 21, 112, 121, 122, 132, 150
La Niña, 13
Leptosol , 29, 83
Limarí River, 47
Lixisol, 25
Loa River, 44

M
Magellan, 9, 72, 73, 78, 79
Main Andes, 4
Mejillones, 1, 52, 53
Melanisation, 25
Mesozoic, 4
Microbial reaction, 99
Mineral soils, 28, 34, 36, 44, 73, 77, 79, 88
Mining, 41, 105, 121, 137, 139, 140, 149
Miocene, 4, 5, 52, 53, 85
Mollisol, 25, 44, 50, 60, 71, 132
Mudflow, 50

N
Natric horizon, 34, 53, 55
Nitrate, 27, 137
Nitrogen, 102, 104, 115, 116, 137, 143, 144
Nothofagus, 20, 103
Nothofagus alessandrii, 20
Nothofagus antarctica, 20
Nothofagus glauca, 20
Nothofagus pumilio, 20, 103
Nutrient, 89, 99, 102, 105, 113, 115, 122, 142

O
Organic matter (OM), 31, 35, 36, 44, 52, 54, 55, 72, 74, 77, 78, 84, 88, 89, 104, 108, 111, 122
Organic soils, 25, 28, 34, 36, 73, 79, 90
Ornithogenic soils, 34, 88–91

P
Paleozoic, 4, 73
Pampa del Tamarugal, 44
Patagonia, 9, 15, 17, 20, 25, 26, 34, 71–73, 75, 77, 79, 85, 99, 101, 103, 107–109, 113–115, 125, 126, 177, 178
Patagonian Cordillera, 5, 73
Patagonian ice fields, 9, 17
Pedogenesis, 26, 29, 34, 44, 56, 108
Perchlorate, 27
Permeability, 59, 61, 62, 68, 69, 76, 128
Perú, 1, 7, 9, 13
Phaeozem, 25, 83
Phosphorus, 104, 140
Pleistocene, 7, 9, 53, 62, 81
Pollutant, 99, 142
Polylepis rugulosa, 19
Polylepis tarapacana, 19
Porosity , 108, 112, 113, 128, 129, 133, 135
Prosopis chilensis, 19, 148
Prosopis tamarugo, 16, 18, 143
Pudahuel soil series, 25
Pumice, 26, 32, 34, 35, 40, 59, 72, 74

Q
Quaternary, 4, 5, 17, 45, 53, 58, 72

R
Rainfall, 7, 11–13, 15, 18, 37, 58, 59, 102–104, 123, 127, 137, 140, 145, 150
Rainforest, 20, 23
Regosol, 29, 77
Residual soils, 52, 56

S
Salars, 27, 28, 37, 39, 102
Salt, 16, 27, 37, 42, 86, 87, 102, 129, 132, 137, 138
Salt lake, 18, 42
Sand, 7, 37, 38, 47, 54, 59, 61, 66, 68, 77, 109, 126, 135
Sediment, 7, 9, 123, 125, 126
Sedimentary soils, 34, 53
Serranías, 27, 37, 55–58, 169
Silandic Andosol, 25
Skeletal soils, 29, 159
Sodic Haplocambids, 162
Soilmanagement, 103, 112, 115, 132, 152
Soil salinity , 52, 99, 102
Soil sodicity, 55, 57, 102, 132
Soil taxonomy, 25, 82, 83, 88, 89, 93
South America, 1, 3, 11–13, 16, 28, 49, 79, 85, 87, 102
South American central volcanic zone, 1
The South American south volcanic zone, 3
Sphagnum magellanicum, 20, 67, 79
Stratovolcanoe, 4, 80
Sulphate, 27, 39, 105
Sulphur, 39, 86, 105, 137

T
Tatara, 47
Tephras, 26, 34, 38, 39, 58, 59–61, 69, 72, 73, 80, 81, 83
Tierra del Fuego, 5, 20, 72, 73, 78, 85
Tillage, 65, 116, 122, 124, 125, 128, 129, 132, 133, 136, 142
Trichocereus chiloensis, 19
Typic Haplocalcids, 26, 45, 56, 170
Typic Haplocambids , 45, 47, 48, 162
Typic Haplosalids , 45

U
Ultisol, 25, 100, 104, 105, 107, 108, 110, 111, 114, 116, 132, 133, 135, 137
Umbrisol, 25

V
Vegetation, 17–20, 26, 37, 38, 40, 42, 49, 58, 60, 83, 91, 99, 104, 110, 113, 116
Vertisol, 48, 50, 62, 77, 104
Vitric Andosol, 25
Volcanic activity, 58, 72, 78, 81, 127
Volcanic ashes, 66, 107
Volcanoe, 1, 4, 46, 72, 80, 134, 150

W
Weather, 7, 107, 113, 125
World reference base, 25 83

MIX
Papier aus verantwortungsvollen Quellen
Paper from responsible sources
FSC® C105338

If you have any concerns about our products,
you can contact us on
ProductSafety@springernature.com

In case Publisher is established outside the EU,
the EU authorized representative is:
Springer Nature Customer Service Center GmbH
Europaplatz 3, 69115 Heidelberg, Germany

Printed by Libri Plureos GmbH
in Hamburg, Germany